"在实践中成长"丛书

Android 程序设计与开发
（Android Studio 版）

刘国柱 杜军威 编著
QST青软实训

清华大学出版社
北京

内 容 简 介

本书对 Android 技术进行深入剖析和全面讲解，内容涵盖 Android 基本理论、Activity、基础 UI 编程、高级 UI 编程、Intent、BroadcastReceiver、SQLite 数据存储、ContentProvider 数据共享、Service 服务及网络编程等。

书中所有代码基于 Android 5.0 版本，且均在 Android Studio 开发环境下进行调试和运行；内容涉及 Android 5.0、Android 6.0 和 Android 7.0 版本新特性以及 Android Studio 环境常用配置和程序签名。

本书重点突出，强调动手操作能力，以一个项目贯穿所有章节的任务实现，使得读者能够快速理解并掌握各项重点知识，全面提高分析问题、解决问题以及动手编码的能力。

本书适用面广，可作为高校、培训机构的 Android 教材，适合作为计算机科学与技术、软件外包、计算机软件、计算机网络、电子商务等专业的程序设计课程的教材。

本书封面贴有清华大学出版社防伪标签，无标签者不得销售。
版权所有，侵权必究。举报：010-62782989，beiqinquan@tup.tsinghua.edu.cn。

图书在版编目(CIP)数据

Android 程序设计与开发：Android Studio 版/刘国柱等编著.—北京：清华大学出版社，2017
(2021.8重印)
("在实践中成长"丛书)
ISBN 978-7-302-46727-4

Ⅰ.①A… Ⅱ.①刘… Ⅲ.①移动终端－应用程序－程序设计 Ⅳ.①TN929.53

中国版本图书馆 CIP 数据核字(2017)第 048698 号

责任编辑：刘　星　梅奕芳
封面设计：刘　键
责任校对：李建庄
责任印制：沈　露

出版发行：清华大学出版社
网　　址：http://www.tup.com.cn, http://www.wqbook.com
地　　址：北京清华大学学研大厦 A 座　　　　　邮　编：100084
社 总 机：010-62770175　　　　　　　　　　　邮　购：010-83470235
投稿与读者服务：010-62776969，c-service@tup.tsinghua.edu.cn
质量反馈：010-62772015，zhiliang@tup.tsinghua.edu.cn
课件下载：http://www.tup.com.cn，010-83470236

印 装 者：北京国马印刷厂
经　　销：全国新华书店
开　　本：185mm×260mm　　印　张：28.5　　字　数：691 千字
版　　次：2017 年 6 月第 1 版　　　　　　　　印　次：2021 年 8 月第 9 次印刷
印　　数：15501～17500
定　　价：59.00 元

产品编号：072735-01

前言

当今 IT 产业发展迅猛,各种技术日新月异,在发展变化如此之快的年代,学习者已经变得越来越被动。在这种大背景下,如何快速地掌握一门技术并做到学以致用,是很多人关心的问题。一本书、一堂课只是学习的形式,而真正能够达到学以致用的目的,则需要融合在书及课堂上的学习方法,使学习者具备学习技术的能力。

为适应工程教育人才培养课程的改革要求,以能力为导向,培养能够解决复杂工程问题的、高素质的应用型软件人才。青岛科技大学青软国际软件学院与 QST 青软实训积极探索"产教深度融合、校企协同育人"的人才培养模式,实现专业链与产业链、课程内容与职业标准、教学过程与生产过程的对接。通过多年的合作与探索,集高校教师的完备知识体系与企业教师的丰富实践经验,完成本教材。

本书不再是知识点的铺陈,而是致力于将知识点融入实际项目的开发中,达到系统化的学习目的。本书的特色是采用一个"GIFT-EMS 礼记"项目,将所有章节重点技术进行贯穿,每章项目代码会层层迭代不断完善,最终形成一个完整的系统。通过贯穿项目以点连线、多线成面,使得读者能够快速理解并掌握各项重点知识,全面提高分析问题、解决问题以及动手编码的能力。

1. 创新点及优势

1) 面向学习者

以一个完整的项目贯穿技术点,以点连线、多线成面,通过项目驱动学习方法使学习者轻松地将技术学习转化为技术能力。

2) 面向高校教师

为教学提供完整的课程产品组件及服务,满足高校教学各个环节的资源需求。

2. 项目简介

"GIFT-EMS 礼记"项目是一个针对"送礼"的移动端 App,以推荐礼物、购买礼物、送礼攻略等功能为核心,收集时下潮流的礼物和送礼物的方法,为用户呈现热门的礼物攻略,通过"送给 TA"等功能,旨在帮助用户给恋人、家人、朋友、同事制造生日、节日、纪念日的惊喜。

"GIFT-EMS 礼记"系统主要分为 Android 移动端 App 和服务器端两部分,鉴于本书主要讲解 Android 编程,并且服务器端在《Java EE 轻量级框架应用与开发——S2SH》一书中已详细介绍,因此本书中主要介绍 Android 移动端 App 的功能及实现。

在"GIFT-EMS 礼记"移动端 App 的实现过程中,使用了 Application、Activity、Service、Broadcast Receiver、数据存储、网络应用、复杂 UI 等关键技术,以及目前流行的一些实现常见功能的开源类库,例如 JSON 解析库 Gson、日历控件 KCalendar、二维码扫描库

ZBarDecoder 和图片加载库 Universal-Image-Loader 等。

3. 贯穿项目模块

"GIFT-EMS 礼记"移动端 App 贯穿项目的模块实现穿插于本书的所有章节中,每个章节在前一章节的基础上进行任务实现,对项目逐步进行迭代、升级,最终形成一个完整的项目,并将 Android 课程的重点技能点进行强化应用。读者可以按照 Step-By-Step 的方式去学习、研究。

4. 章节任务实现

章	目 标	贯穿任务实现	
第1章 Android 概述	熟悉 Android 开发环境	【任务 1-1】	使用 Android SDK Manager
		【任务 1-2】	使用 Android 模拟器(Intel x86 架构)
		【任务 1-3】	ADB 工具的使用
第2章 Activity 和 Application	项目需求分析及基本架构设计	【任务 2-1】	项目背景介绍及需求分析
		【任务 2-2】	创建项目并编写实体类和 Application 类等基础架构
		【任务 2-3】	编写项目中 Activity、按钮、文本输入框等控件所使用的背景文件
		【任务 2-4】	编写项目的样式文件
第3章 UI 编程基础	主界面及功能 Activity	【任务 3-1】	编写主界面 Activity
		【任务 3-2】	编写各个业务 Activity 的父类 BaseActivity
		【任务 3-3】	编写项目辅助功能对应的 Activity
第4章 UI 进阶	礼品和送礼攻略	【任务 4-1】	礼品和送礼攻略的列表界面
		【任务 4-2】	礼品展示界面
		【任务 4-3】	攻略展示界面
		【任务 4-4】	完成收礼人列表界面
		【任务 4-5】	完成收礼人编辑界面
		【任务 4-6】	完成我的收藏界面
第5章 Intent 与 BroadcastReceiver	用户日程	【任务 5-1】	完成用户日程界面
		【任务 5-2】	完成用户日程编辑界面
		【任务 5-3】	完成用户日程提醒功能
第6章 数据存储	保存用户相关信息数据	【任务 6-1】	完成保存用户登录信息功能
		【任务 6-2】	完成设置信息保存功能
		【任务 6-3】	完成购物袋功能

续表

章	目标	贯穿任务实现
第7章 ContentProvider 数据共享	购买下单	【任务7-1】 完成购买下单功能,可以从通讯录中获取联系人 【任务7-2】 完成订单列表和订单回收站功能
第8章 Service 服务	赠礼留言、二维码扫描机用户日程提醒 Service	【任务8-1】 完成录制赠礼留言功能 【任务8-2】 完成扫描二维码功能 【任务8-3】 完成播放赠礼留言功能 【任务8-4】 完成日程提醒的 Service
第9章 网络编程	移动端 App 与服务器端的交互	【任务9-1】 编写 HttpUtils 类封装采用 HTTP 方式与服务器交互时的 GET、POST 请求调用 【任务9-2】 修改 BaseActivity,完成与服务器交互数据的 Handler 模板 【任务9-3】 修改登录 Activity,改为从服务器验证登录 【任务9-4】 引入 Android-Universal-Image-Loader 库,用于显示网络图片 【任务9-5】 修改礼物类型列表 Activity,改为从服务器查询数据

5. 项目运行截图

首页

登录、注册界面

个人中心

设置及软件更新

礼品中心和礼品攻略

添加日程和日程提醒闹钟界面

6．致谢

本书由青岛科技大学青软国际软件学院与QST青软实训共同编著，刘国柱、杜军威、刘全、李战军、金澄、郭晓丹、江守寰、张瑞全、陶冶、赵克玲、郭全友等多名老师参与本书编写和审核工作，赵克玲负责全书统稿和修订工作。编者均从事计算机教学和项目开发多年，拥有丰富的教学和实践经验，在编写过程中付出了辛勤的汗水。除此之外，青岛科技大学青软国际软件学院的10 000多名学生也参与了本书的试读工作，并从初学者角度对教材提出了许多宝贵的意见，在此一并表示衷心感谢。由于时间有限，书中难免有疏漏和不足之处，恳请广大读者及专家不吝赐教。我们真诚地希望能与读者共同交流、共同成长，待再版时日臻完善，是所至盼。

本书免费提供配套资源：教学PPT、示例源代码，请到book.moocollege.cn下载。

联系方式：

E-mail：QST_book@itshixun.com

400电话：400-658-0166

编　者

2017年4月

目 录

第 1 章 Android 概述 ············ 1

1.1 Android 简史 ············ 1
1.2 Android 系统 ············ 3
  1.2.1 Android 系统架构 ············ 3
  1.2.2 Android 应用程序组件 ············ 4
1.3 Android 开发环境搭建 ············ 5
  1.3.1 下载并安装 JDK ············ 5
  1.3.2 下载并安装 Android Studio ············ 6
1.4 Hello Android 程序 ············ 8
  1.4.1 第一个 Android 项目 ············ 8
  1.4.2 Android 程序结构 ············ 11
1.5 贯穿任务实现 ············ 12
  1.5.1 实现【任务 1-1】············ 13
  1.5.2 实现【任务 1-2】············ 14
  1.5.3 实现【任务 1-3】············ 18
本章总结 ············ 19
  Q&A ············ 20
章节练习 ············ 20
  习题 ············ 20
  上机 ············ 21

第 2 章 Activity 和 Application ············ 22

2.1 Activity ············ 23
  2.1.1 Activity 简介 ············ 23
  2.1.2 创建 Activity ············ 24
  2.1.3 Activity 的生命周期 ············ 25
2.2 资源管理 ············ 31
  2.2.1 资源分类 ············ 31
  2.2.2 资源访问方式 ············ 32
  2.2.3 strings.xml 文本资源文件 ············ 35
  2.2.4 colors.xml 颜色设置资源文件 ············ 38
  2.2.5 dimens.xml 尺寸定义资源文件 ············ 40

2.2.6 styles.xml 主题风格资源文件 …… 43
2.2.7 drawable 图像资源目录 …… 46
2.3 AndroidManifest.xml 清单文件 …… 48
2.4 Android 应用程序生命周期 …… 51
2.5 Application 类 …… 53
   2.5.1 Application 生命周期事件 …… 53
   2.5.2 实现 Application …… 53
2.6 样式和主题 …… 55
2.7 贯穿任务实现 …… 58
   2.7.1 实现【任务 2-1】 …… 58
   2.7.2 实现【任务 2-2】 …… 59
   2.7.3 实现【任务 2-3】 …… 67
   2.7.4 实现【任务 2-4】 …… 70
本章总结 …… 70
   小结 …… 70
   Q&A …… 71
章节练习 …… 71
   习题 …… 71
   上机练习 …… 72

第 3 章 UI 编程基础 …… 73

3.1 Android UI 元素 …… 74
   3.1.1 视图 …… 74
   3.1.2 视图容器 …… 75
   3.1.3 布局管理 …… 78
   3.1.4 Fragment …… 79
3.2 界面布局 …… 79
   3.2.1 线性布局 …… 79
   3.2.2 表格布局 …… 81
   3.2.3 相对布局 …… 84
   3.2.4 绝对布局 …… 88
3.3 事件处理 …… 89
   3.3.1 基于监听的事件处理 …… 89
   3.3.2 基于回调机制的事件处理 …… 94
3.4 Widget 简单组件 …… 102
   3.4.1 Widget 组件通用属性 …… 103
   3.4.2 TextView 文本框 …… 103
   3.4.3 EditText 编辑框 …… 107
   3.4.4 Button 按钮 …… 108

3.4.5 单选按钮和单选按钮组 ··································· 111
3.4.6 CheckBox 复选框 ·· 115
3.4.7 开关控件 ··· 118
3.4.8 图片视图（ImageView）································ 123
3.5 Dialog 对话框 ·· 126
3.5.1 AlertDialog 提示对话框 ······························· 127
3.5.2 ProgressDialog 进度对话框 ·························· 131
3.6 贯穿任务实现 ··· 133
3.6.1 实现【任务 3-1】 ··· 133
3.6.2 实现【任务 3-2】 ··· 141
3.6.3 实现【任务 3-3】 ··· 146
本章总结 ·· 156
    小结 ·· 156
    Q&A ·· 157
章节练习 ·· 157
    习题 ·· 157
    上机 ·· 157

第 4 章　UI 进阶 ·· 159

4.1 Fragment ··· 159
4.1.1 使用 Fragment ·· 160
4.1.2 Fragment 的生命周期 ·································· 168
4.2 Menu 和 Toolbar ··· 176
4.2.1 Menu 菜单 ·· 176
4.2.2 Toolbar 操作栏 ··· 188
4.3 高级组件 ··· 191
4.3.1 AdapterView 与 Adapter ······························ 191
4.3.2 ListView 列表视图 ······································· 193
4.3.3 GridView 网格视图 ······································ 201
4.3.4 TabHost ··· 203
4.3.5 WebView ·· 208
4.4 贯穿任务实现 ··· 211
4.4.1 实现【任务 4-1】 ··· 211
4.4.2 实现【任务 4-2】 ··· 215
4.4.3 实现【任务 4-3】 ··· 225
4.4.4 实现【任务 4-4】 ··· 228
4.4.5 实现【任务 4-5】 ··· 235
4.4.6 实现【任务 4-6】 ··· 240
本章总结 ·· 245

小结 ········· 245
Q&A ········· 245
章节练习 ········· 245
习题 ········· 245
上机 ········· 246

第 5 章　Intent 与 BroadcastReceiver ········· 247

5.1　Intent 意图 ········· 247
　　5.1.1　Intent 原理及分类 ········· 248
　　5.1.2　Intent 属性 ········· 249
　　5.1.3　使用 Intent 启动 Activity ········· 257
　　5.1.4　Intent Filter 过滤器 ········· 267
5.2　BroadcastReceiver ········· 269
5.3　Handler 消息传递机制 ········· 272
　　5.3.1　Handler 简介 ········· 272
　　5.3.2　Handler 的工作机制 ········· 274
5.4　AsyncTask 类 ········· 275
5.5　贯穿任务实现 ········· 277
　　5.5.1　实现【任务 5-1】 ········· 278
　　5.5.2　实现【任务 5-2】 ········· 284
　　5.5.3　实现【任务 5-3】 ········· 293
本章总结 ········· 299
　　小结 ········· 299
　　Q&A ········· 299
章节练习 ········· 300
　　习题 ········· 300
　　上机 ········· 300

第 6 章　数据存储 ········· 302

6.1　数据存储简介 ········· 303
6.2　文件存储 ········· 303
　　6.2.1　I/O 流操作文件 ········· 303
　　6.2.2　读写 SD 卡文件 ········· 307
　　6.2.3　文件浏览器 ········· 310
6.3　使用 SharedPreferences ········· 313
　　6.3.1　SharedPreferences 和 SharedPreferences.Editor 接口 ········· 313
　　6.3.2　SharedPreferences 操作步骤 ········· 314
6.4　SQLite 数据库 ········· 317
　　6.4.1　SQLite 简介 ········· 317

6.4.2　SQLiteDatabase 类 …………………………………………………… 317
　　　6.4.3　SQLite 数据库的创建和删除 ………………………………………… 318
　　　6.4.4　表的创建和删除 ……………………………………………………… 319
　　　6.4.5　记录的插入、修改和删除 …………………………………………… 319
　　　6.4.6　数据查询与 Cursor 接口 …………………………………………… 321
　　　6.4.7　事务处理 ……………………………………………………………… 323
　　　6.4.8　SQLiteOpenHelper 类 ………………………………………………… 323
　　　6.4.9　使用 ListView 滑动分页 ……………………………………………… 327
　本章总结 …………………………………………………………………………… 331
　　小结 ……………………………………………………………………………… 331
　　Q&A ……………………………………………………………………………… 331
　章节练习 …………………………………………………………………………… 331
　　习题 ……………………………………………………………………………… 331
　　上机 ……………………………………………………………………………… 332

第 7 章　ContentProvider 数据共享 …………………………………………… 333

　7.1　ContentProvider 简介 ……………………………………………………… 333
　　　7.1.1　ContentProvider 类 ……………………………………………………… 334
　　　7.1.2　ContentResolver 类 ……………………………………………………… 335
　7.2　开发 ContentProvider 程序 ………………………………………………… 337
　　　7.2.1　编写 ContentProvider 子类 …………………………………………… 337
　　　7.2.2　注册 ContentProvider ………………………………………………… 338
　　　7.2.3　使用 ContentProvider ………………………………………………… 338
　7.3　操作系统的 ContentProvider ……………………………………………… 340
　　　7.3.1　管理联系人 …………………………………………………………… 340
　　　7.3.2　管理多媒体 …………………………………………………………… 346
　本章总结 …………………………………………………………………………… 351
　　小结 ……………………………………………………………………………… 351
　　Q&A ……………………………………………………………………………… 351
　章节练习 …………………………………………………………………………… 351
　　习题 ……………………………………………………………………………… 351
　　上机 ……………………………………………………………………………… 352

第 8 章　Service 服务 ……………………………………………………………… 353

　8.1　Service 简介 ………………………………………………………………… 353
　　　8.1.1　Service 分类 …………………………………………………………… 354
　　　8.1.2　Service 基本示例 ……………………………………………………… 355
　8.2　Service 详解 ………………………………………………………………… 357
　　　8.2.1　Start 方式启动 Service ………………………………………………… 357

8.2.2　Bind 方式启动 Service ······················363
　　8.2.3　混合方式的 Service ························368
　　8.2.4　前台 Service ·····································372
　　8.2.5　Service 中执行耗时任务 ················377
　　8.2.6　远程 Service ···································382
8.3　系统自带 Service ··387
　　8.3.1　NotificationManager ························389
　　8.3.2　DownloadManager ··························390
本章总结 ···391
　　小结 ···391
　　Q&A ··392
章节练习 ···392
　　习题 ···392
　　上机 ···394

第 9 章　网络编程 ··395

9.1　网络编程简介 ··396
9.2　基于 TCP 协议的网络通信 ·······················396
　　9.2.1　Socket ··397
　　9.2.2　ServerSocket ·····································398
9.3　使用 HttpURLConnection ···························403
　　9.3.1　URL 和 URLConnection ··················403
　　9.3.2　HttpURLConnection ·························407
9.4　使用 HttpClient ··411
9.5　使用 WebView 视图浏览网页 ···················413
9.6　Volley 框架 ··416
本章总结 ···419
　　小结 ···419
　　Q&A ··420
章节练习 ···420
　　习题 ···420
　　上机 ···420

附录A　Android 版本新特性 ························422

A.1　Android 5.0 新特性 ····································422
A.2　Android 6.0 新特性 ····································423
A.3　Android 7.0 新特性 ····································424

附录 B　常用的 Android Studio 选项设置 …… 427
B.1　Android Studio 基本配置 …… 427
B.2　Android Studio 快捷键 …… 429
B.3　Android Studio 导入 Eclipse ADT 项目 …… 430
B.3.1　步骤 …… 430
B.3.2　常见问题 …… 431

附录 C　Android 应用程序签名 …… 433
C.1　DOS 命令完成 apk 签名 …… 433
C.2　在 Android Studio 中完成 apk 签名 …… 434

第1章 Android概述

本章任务是熟悉 Android 开发环境及搭建过程：
- 【任务 1-1】 使用 Android SDK Manager。
- 【任务 1-2】 使用 Android 模拟器(Intel x86 架构)。
- 【任务 1-3】 ADB 工具的使用。

知 识 点	Listen(听)	Know(懂)	Do(做)	Revise(复习)	Master(精通)
Android 历史发展	★				
Android 的系统架构	★	★			
Android 的应用程序组件	★	★			
安装 Android Studio 环境	★	★	★	★	
Android 项目的创建及运行	★	★	★	★	★

1.1 Android 简史

目前移动互联网已经深入到人们生活中的方方面面，如社交、购物、旅游、日常工作等，为人们的衣食住行提供了极大的便利，并最终改变了人们的生活方式。传统的 IT 企业都在向移动互联转型,以拓展更广阔的业务空间来获取更大的利润增长点。而移动互联的快速发展离不开 Android(安卓)系统,正是 Android 系统的快速发展奠定了移动互联网的基础。

  Android 是一个以 Linux 为基础的开源操作系统,主要用于智能手机和平板电脑等移动设备,由 Google 领导的开放手持设备联盟(Open Handset Alliance,OHA)持续维护与更新。Android 系统最初由安迪·鲁宾(Andy Rubin)等人设计与开发,开发该系统的最初目的是创建一个先进的数码相机操作系统,但是后来市场规模不够大,加上智能手机市场快速成长,于是 Android 被改造为一款面向智能手机的操作系统。2005 年 8 月 Google 收购了 Android;2007 年 11 月 Google 联合 84 家制造商、开发商及电信营运商共同成立了开放手持设备联盟,来共同研发与改良 Android 系统;随后 Google 以 Apache 免费开放原始码许可证的授权方式公开了 Android 的源码,使得越来越多的手机制造商推出搭载 Android 的智能手机,后来 Android 更逐渐拓展到平板电脑及其他领域中。

  到本书出版时,Android 系统发布过的主要版本如表 1-1 所示。

表 1-1   Android 系统主要版本

版本	代号	日期	描述
Android 1.0	Astro(铁臂阿童木)	2008 年 9 月 23 日	Android 操作系统的第一个正式版本,全球第一台 Android 设备 HTC Dream(G1)就是搭载此操作系统
Android 2.0/2.1	Éclair(闪电泡芙)	2009 年	优化了硬件速度,支持更多的屏幕分辨率,改良的用户界面,支持 HTML 5
Android 2.2	Froyo(冻酸奶)	2010 年 5 月 20 日	基于 Linux 2.6.32 内核,支持将软件安装至扩展内存,加强软件即时编译的速度
Android 2.3	Gingerbread(姜饼)	2010 年 12 月 6 日	基于 Linux 2.6.35 内核,修补 UI,支持更大的屏幕尺寸和更高的分辨率
Android 3.0/3.1/3.2	Honeycomb(蜂巢)	2011 年	基于 Linux 2.6.36 内核,仅供平板电脑使用,支持平板电脑大荧幕、高分辨率
Android 4.0	Ice Cream Sandwich(冰激凌三明治)	2011 年 10 月 19 日	统一了手机和平板电脑使用的系统,应用会自动根据设备选择最佳显示方式
Android 4.1/4.2/4.3	Jelly Bean(果冻豆)	2012 年	基于 Android 4.0 的改善
Android 4.4	KitKat(奇巧巧克力棒)	2013 年 9 月 3 日	修复了以前的漏洞,支持语音打开 Google Now,优化存储器使用
Android 5.0/5.1	Lollipop(棒棒糖)	2014 年 6 月 25 日	采用全新 Material Design 界面,支持 64 位处理器,全面由拼了转用 ART 编译,性能提升四倍
Android 6.0	Marshmallow(棉花糖)	2015 年 5 月 28 日	在软件体验与运行性能上进行了大幅度的优化,使设备续航时间提升 30%
Android 7.0	Nougat(牛轧糖)	2016 年 5 月 18 日	采用 Vulkan 图形处理系统,减少对 CPU 的占用,加入 JIT 编译器,程序安装速度提升 75% 且所占空间减少 50%

第1章 Android概述

> **注意**
>
> 本书基于 Android 5.0(Lollipop 棒棒糖),所有代码都是在该版本基础上进行调试的。Android 5.0 版本经过几年的沉淀,其功能已经十分强大、高效和稳定。

Android 操作系统在全球范围内占据着主导地位。根据权威机构对移动终端市场的统计,截至 2016 年第二季度采用 Android 和 iOS 操作系统的智能手机出货量占全部智能机出货量的 99.1%,其中 Android 全球份额接近 86.2%,占据绝对垄断的地位。

1.2 Android 系统

1.2.1 Android 系统架构

与其他操作系统类似,Android 也采用了分层的架构,Android 软件栈如图 1-1 所示。从架构图看,Android 系统分为 4 层,从高到低分别是应用程序层、应用程序框架层、系统运行库层和 Linux 内核层,各层采用软件栈(Software Stack)的方式进行构建。

图 1-1 Android 软件栈

Android 软件栈是通过一个应用程序框架提供了 Linux 内核和 C/C++ 库的集合,在运行时为应用程序提供相应的服务,并对其进行管理。软件栈的各个层的功能如下所述。

- **Linux 内核层**:核心服务(包括硬件驱动程序、进程和内存管理、安全、网络和电源管理)

都由一个 Linux 内核处理，内核还在硬件和软件栈的其他部分之间提供了一个抽象层。
- **系统运行库层**：在 Linux 内核之上，Android 提供了各种 C/C++ 核心库（例如 Libc 和 SSL）、视频音频相关的媒体库、外观管理器、基于 2D、3D 图形的 SGL 和 OpenGL 图形库、用于本地数据库支持的 SQLite，以及用于集成 Web 浏览器和 Internet 安全的 SSL 和 WebKit。
- **Android 运行时**：可以让一个 Android 手机从本质上与一个移动 Linux 实现区分开来。由于 Android 运行时包含了核心库和 Dalvik 虚拟机，因此 Android 运行时是向应用程序提供动力的引擎，并与之一起形成了应用程序框架的基础。其中 Android 库提供了 Java 核心库和 Android 特定库的大部分功能；Dalvik 虚拟机是一个基于寄存器的 Java 虚拟机，并对其优化从而确保同一设备可以高效地运行多个实例，通过 Linux 内核对线程和底层内存进行管理。
- **应用程序框架层**：提供了用来创建 Android 应用程序的基础类，对硬件访问提供了 API 功能框架，用于管理用户界面和应用程序资源。
- **应用程序层**：所有的应用程序（包括原生和第三方）都在应用层上进行构建；应用层运行在 Android 运行时内，并且使用了应用程序框架中可用的类和服务。

1.2.2 Android 应用程序组件

Android 应用程序是由一些松散耦合的组件构成的，每个应用程序中都会包含一个配置文件 AndroidManifest.xml，其中描述应用程序中所用到的组件及相互关系，还包括一些硬件要求、权限等的声明。当应用程序被安装到系统中时，系统会扫描该配置文件，并将应用程序的组件注册到系统中。

Android 系统的一个典型特点是应用程序的组件允许被其他应用程序调用，从而实现组件间的松散耦合。Android 应用程序中的大部分组件都可以直接运行，例如在原生 Android 系统中，"系统设置"是一个包名为 com.android.settings 的应用程序，其中包含了很多组件，在用户开发应用程序时可以直接打开"系统设置"应用中的 WiFi、蓝牙等设置界面，而无须再编写这些通用性的功能。

Android 应用程序主要包含 4 种组件：Activity、Service、BroadcastReceiver 和 ContentProvider，由若干以上类型的组件集成在一起共同构成一个完整的 Android 应用程序。

- **Activity**（活动）：Activity 是最基本的 Android 应用程序组件，是负责用户交互的最主要组件；一个 Activity 表示一个可视化的用户界面，除非不需要任何用户界面，否则 Android 应用程序应至少包含一个 Activity。
- **Service**（服务）：Service 组件用于提供服务，执行一些持续性的、耗时的且无需用户界面交互的操作。Service 是不可见的，通常用于处理一些无需用户交互但需持续运行的任务，例如从网络上搜索内容、更新 Content Provider、激活 Notification、播放音乐等。
- **BroadcastReceiver**（广播接收器）：BroadcastReceiver 是一种全局监听器，用于接收来自系统和应用程序的广播。在系统中注册 BroadcastReceiver 后，当发生指定的事件时，系统会自动启动应用程序并向 BroadcastReceiver 发出广播，对该事件进行处理，从而实现一种事件驱动的应用程序架构。
- **ContentProvider**（内容提供器）：ContentProvider 组件是一种共享的持久数据存储机制，是在应用程序之间共享数据时的首选方案。应用程序可以通过 ContentProvider 机制向其他应用程序提供数据，也可以访问其他应用程序所提供的 ContentProvider。Android 系统本身提供了大量 ContentProvider 以供应用程序访问系统的数据，例如联系人、媒体库等。

组件与组件之间通过 Intent（意图）关联在一起。Intent 虽不是 Android 应用的组件，但在组件之间传递消息时，Intent 通常作为信息载体。使用 Intent 可以启动、停止 Activity 和 Service，也可以在系统范围内或向指定的 Activity 和 Service 等组件广播消息。Android 系统中大量使用了 Intent，在实际的应用程序开发中也会频繁地使用 Intent 传递信息。

1.3　Android 开发环境搭建

在进行 Android 开发之前需要搭建其开发环境，Android 应用开发需要 JDK（Java SE Developent Kit）、Android SDK 和一个集成开发环境（本书采用 Android Studio）。Android Studio 是 Google 开发的一款面向 Android 开发者的 IDE，基于流行的 Java 语言集成开发环境 IntelliJ 搭建而成，Google 官方将逐步放弃对原来主要的 Eclipse ADT 的支持，并为 Eclipse 用户提供工程迁移的解决方案，因此搭建 Android 开发环境只需要安装 JDK 和 Android Studio 即可。

1.3.1　下载并安装 JDK

开发 Android 应用程序需要 JDK 支持，因此需要下载并安装 JDK。Android 5.0 需要 JDK 7 或更高版本，最新版本的 JDK 须到 Oracle 官方网站进行下载，下载位置：

http://www.oracle.com/technetwork/java/javase/downloads/index.html

需要注意的是：根据开发者所用操作系统及 CPU 架构的不同，需要下载对应版本的 JDK 安装文件。如图 1-2 所示，在安装时，依次选择开发工具（Development Tools）、API 源代码（Source Code）与公用 JRE（Public JRE）。开发工具是必需的，范例程序可供开发者在编写程序时参考，API 源代码可以让开发者了解所使用的 API 实际上是如何实现的，而 JRE 则是执行 Java 程序所必需的，所以推荐将以上 3 部分都选中并进行安装。

单击"下一步"按钮即可开始进行 JDK 的安装。安装 JDK 之后，接着安装 JRE，如图 1-3 所示，单击"下一步"按钮，完成 JRE 的安装。

图 1-2　JDK 安装

图 1-3　安装 JRE

1.3.2 下载并安装 Android Studio

Android Studio 是一款全新的基于 IntelliJ IDEA 的 Android 开发工具。与 Eclipse ADT 插件类似，Android Studio 提供了集成的 Android 开发和调试环境。在 Android Studio 中文社区网站 http://www.android-studio.org 中，可以下载最新版本的 Android Studio 安装文件，如图 1-4 所示矩形区域，选择下载包含 Android SDK 版本的安装文件。

Platform	Android Studio package	Size	SHA-1 checksum
Windows	android-studio-bundle-143.2821654-windows.exe Includes Android SDK (recommended)	1181 MB (1238569296 bytes)	6f7fcdc30800bd8b3fbd5a14c2b9857243144650
	android-studio-ide-143.2821654-windows.exe No Android SDK	270 MB (283805040 bytes)	d8cb3968814b6155f4effe727baf23b18b9f8360
	android-studio-ide-143.2821654-windows.zip No Android SDK, no installer	286 MB (300630577 bytes)	9bec4905e40f0ac16ac7fde63a50f3fbc1eec4d9
Mac OS X	android-studio-ide-143.2821654-mac.dmg	284 MB (298597716 bytes)	4a7ca7532a95c65ee59ed50193c0e976f0272472
Linux	android-studio-ide-143.2821654-linux.zip	284 MB (298125051 bytes)	55d69ad2da0068d818718b26ba43550fbcbeb7e9

图 1-4 下载 Android Studio 安装包

> 本书是基于 Android Studio 2.1.2.0，因此所有代码都是在该版本基础上进行调试。

Android Studio 安装文件下载完成，双击安装文件，等安装文件读取完毕后，出现如图 1-5 所示的安装向导界面。

连续单击 Next 按钮，直到出现如图 1-6 所示的选择安装和存放路径窗口，选择 Android Studio 安装路径和 Android SDK 存放路径。

图 1-5 安装向导界面

图 1-6 选择安装和存放路径

> Android Studio 需要至少 500MB 空间，Android SDK 需要至少 3.2GB 空间，因此在指定安装路径时要确保该路径下的磁盘有足够大的空间。

继续单击 Next 按钮，完成 Android Studio 的安装。最后单击 Finish 按钮后，Android

Studio 就会自行启动,并进入如图 1-7 所示的配置界面,该界面用于导入 Android Studio 的配置文件,如果是第一次安装,请选择第二项(不导入配置文件),然后单击 OK 按钮即可。

图 1-7　配置界面

完成上一步以后,会进入一个欢迎页面,单击 Next 按钮进入设置类型向导页,如图 1-8 所示,该界面有两个选项:Standard(标准)和 Custom(用户自定义),本书建议选择 Standard 选项。

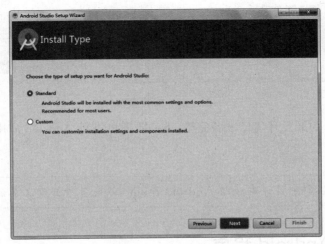

图 1-8　设置类型向导页

单击 Next 按钮,进入配置下载安装界面,如图 1-9 所示,等待下载并安装。

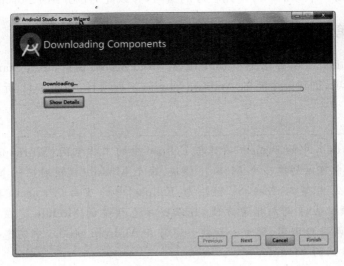

图 1-9　配置下载安装

安装成功后出现如图 1-10 所示的界面，单击 Finish 按钮完成 Android Studio 的安装。

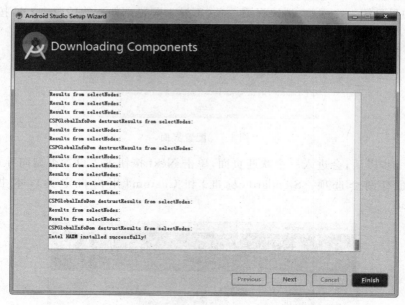

图 1-10　安装成功

至此，Android 的开发环境已准备完毕，下一步即可进行 Android 应用程序开发了。

> 在本章贯穿任务中将详细介绍 Android Studio 以及各种开发工具的使用方式。

1.4　Hello Android 程序

本节将完成第一个 Android 应用程序的编写，并以此为例介绍 Android 项目的结构。

1.4.1　第一个 Android 项目

启动 Android Studio，弹出 Welcome to Android Studio 窗口，单击 Start a new Android Studio project 选项，创建一个新的 Android Studio 工程项目，如图 1-11 所示。

> Android Studio 中的 Project 项目与 Eclipse 中的工作空间（Workspace）类似，在一个 Project 项目中可以创建多个 Module 模块，每个 Module 模块对应一个独立的可执行的应用程序或公共类库，Module 模块与 Eclipse 中的项目（Project）类似。Android Studio 的这种项目管理模式非常方便，可以将多个相关的 Module 建在同一个 Project 中，以便相互之间进行调用、调试和切换。通常在 Android Studio 中创建一个 Project 会同时创建一个默认的 Module。

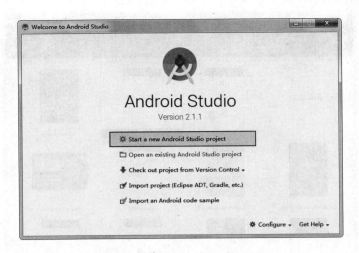

图 1-11　创建 Android Studio project

弹出 Create New Project 窗口,输入应用名(Application name)、公司域(Company Domain)以及指定应用存放目录(Project location),如图 1-12 所示。

图 1-12　开始创建第一个 Android 项目

继续单击 Next 按钮,直至出现 Add an Activity to Mobile 界面,在该界面中选择合适的 Activity 样式模板,如图 1-13 所示。

单击 Next 按钮,将弹出如图 1-14 所示的 Customize the Activity 界面,单击 Finish 按钮,完成 Android Studio 工程项目的创建过程。

在 Android Studio 顶部菜单栏中单击 ▶ 运行按钮,运行 chapter01 项目,如图 1-15 所示。

此时,会出现运行目标的选择对话框,如图 1-16 所示,系统会列出所有已连接的 Android 设备,选择所需要的 Android 设备并单击 OK 按钮,系统会将项目发送到该设备上进行安装并运行。

项目运行结果如图 1-17 所示。

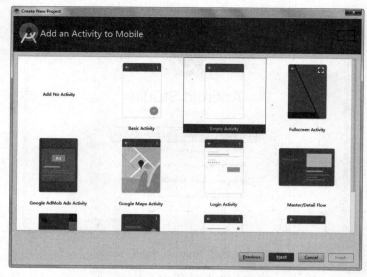

图 1-13　选择 Activity 样式模板

图 1-14　创建 Activity

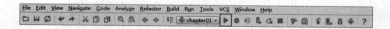

图 1-15　运行第 1 个 Android 项目

图 1-16　选择正在运行的 Android 设备

图 1-17　第 1 个 Android 程序运行结果

> **注意**
>
> 测试应用程序时,除了能够使用真实的 Android 设备外,还可以使用 Android Studio 提供的模拟器。模拟器是一种运行在操作系统上的 Android 环境模拟软件,可以直接运行 Android 应用程序,在本章贯穿任务中将介绍模拟器的使用方式。

1.4.2 Android 程序结构

通过第一个 Android 应用程序,分析一下 Android 应用程序的结构,在 Android Studio 中,提供了多种项目结构类型,如图 1-18 所示。

本书主要介绍两种项目结构类型:Project 项目结构类型;Android 项目结构类型。其中,Project 项目结构类型如图 1-19 所示。

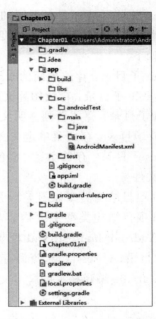

图 1-18　Android Studio 项目结构类型　　　图 1-19　Project 项目结构类型

Project 项目结构类型主要内容如下。

① gradle 目录:gradle 项目产生文件夹(自动编译工具产生的文件);

② idea 目录:IDEA 项目文件夹(开发工具产生的文件);

③ chapter01 目录:模块目录,Android Studio 的 Module 模块;

④ build 目录:编译时产生文件,不需要修改,也不需要纳入项目源代码管理中;

⑤ libs 目录:用于存放项目相关的依赖库;

⑥ java 目录:代码的存放目录;

⑦ res 目录:资源存放目录(包括布局、图像、样式等);

⑧ AndroidManifest.xml 文件:Android 应用程序的声明文件,包含了 Android 系统运行 Android 程序前所必须掌握的重要信息,其中包含应用程序名称、图标、包名称、模块组

成、授权和 SDK 最低版本要求等,而且每个 Android 程序必须在根目录下包含一个 AndroidManifest.xml 文件;

⑨ gradle 目录:项目的 Gradle 编译系统;
⑩ gitgnore 文件:git 版本管理忽略文件,标记出哪些文件不用进入 git 库中;
⑪ build.gradle 文件:gradle 模块自动编译的配置文件;
⑫ chapter01.iml 文件:chapter01 模块的配置文件;
⑬ proguard-rules.pro 文件:代码混淆配置规则文件;
⑭ gradle.properties 文件:gradle 相关的全局属性设置文件;
⑮ HelloWorld.iml 文件:项目的配置文件;
⑯ local.properties 文件:本地属性设置(配置 SDK、NDK、key 等属性)文件;
⑰ settings.gradle 文件:定义项目包含哪些模块的文件;
⑱ External Libraries 目录:项目依赖的库,编译时自动下载。

其中,res 目录下又有多个不同的子目录,在这些子目录下存放着不同类型的文件。res 目录下的主要子目录及其用途如下。

⑲ layout 子目录:存放界面的布局文件;
⑳ drawable 子目录:存放图片文件;
㉑ anim 子目录:存放动画声明文件;
㉒ menu 子目录:存放菜单定义文件;
㉓ values 子目录:存放数组、颜色、尺寸、字符串和样式等资源文件。

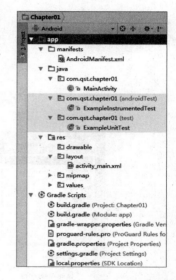

Android 项目结构类型如图 1-20 所示。

Android 项目结构类型主要内容如下:①manifests 目录,存放 AndroidManifest.xml 配置文件;②java 目录,代码存放目录;③res 目录,资源存放目录;④Gradle Scripts 目录,gradle 编译相关的脚本文件。

图 1-20 Android 项目结构类型

注意

　　Project 项目结构类型与 Android 项目结构类型没有本质的区别,可根据个人习惯进行选择,由于 Android 项目结构类型简单明了,本书建议在 Android 项目结构类型下进行代码编写。Project 项目结构类型中,很多文件的名称和功能与 Android 项目结构类型中的文件是相同的,本书不再一一解释。

1.5 贯穿任务实现

　　配置完成 Android 开发环境后,开发者就可以使用其提供的多种开发工具,常用的有 SDK Manager、Android 模拟器和 adb 等工具。

1.5.1 实现【任务 1-1】

Android SDK Manager 用于管理 Android 的 SDK、各种工具以及模拟器的镜像等。由于 Android 版本众多，SDK Manager 提供了一个统一管理的界面。单击 Android Studio 中图标，如图 1-21 所示，打开 SDK Manager。

图 1-21　SDK Manager 图标

打开 SDK Manager 后，界面如图 1-22 所示。

图 1-22　SDK Manager 界面

SDK Manager 启动后会自动检查更新，用户可以按照需求选择是否需要更新，也可以删除已下载的程序版本。SDK Manager 主要管理以下三部分内容。

（1）**SDK Platforms**：Android 各版本的 SDK 和系统镜像文件，如图 1-23 所示。

图 1-23　版本文件

（2）**SDK Tools**：包括开发 Android 应用所需的各种工具（例如模拟器、DDMS 和 adb 等），以及各种扩展工具（例如针对 Google 的某些服务的一些专用库），如图 1-24 所示。

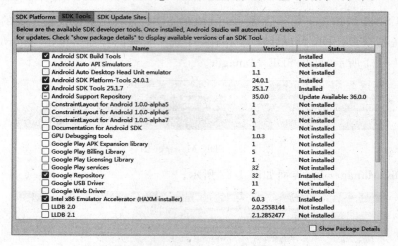

图 1-24 Tools 工具

（3）**SDK Update Sites**：设置 Android SDK 更新网址，如图 1-25 所示。

图 1-25 SDK 网址信息

在本书写作的过程中，Android 7.0 发布。

1.5.2 实现【任务 1-2】

开发 Android 应用程序时，可以使用真实的 Android 设备调试，但是如果需要开发一款能够适配于多个 Android 版本及分辨率的应用，那么使用真机调试就成了一个大问题，因为普通开发者通常无法获取多种不同类型的设备，这时可以使用模拟器来测试。

使用 Android 模拟器，首先单击 Android Studio 菜单栏中的 ▉ 按钮，弹出 Android Virtual Device Manager 对话框，如图 1-26 所示。

单击 Create Virtual Device 按钮，弹出以 Select Hardware 为标题的对话框，如图 1-27 所示。选择一种设备型号，以此型号来创建模拟器。

在设备型号选择界面，可以选择设备的类型、屏幕大小与分辨率等，具体如下：

第 1 章　Android 概述

图 1-26　添加选择模拟器

图 1-27　选择设备型号

Categroy，选择创建设备的类型，包括 TV、Wear、Phone 和 Tablet；Name，手机型号名称；Size，屏幕大小；Resolution，屏幕分辨率；Density，屏幕密度。

> 此处只是选择型号对应的硬件条件，而不会选择该设备在发布时搭载的系统镜像。本书选择 Nexus 5X 型号进行模拟器创建。

单击 Next 按钮，弹出 System Image 对话框，单击上方 x86 Images 选项卡，选择一款需要的系统镜像搭载到模拟器中，如图 1-28 所示。

x86 Images 选项卡中，每列所描述的内容如下：
- Release Name——版本名称；
- API Level——API 级别；
- ABI——模拟的 CPU 类型；
- Target——该服务版本搭载的安卓版本。

· 15 ·

图 1-28 选择合适的系统镜像文件

> Android 模拟器实际上是在 x86 架构上运行的一个 ARM 虚拟机,为了提高模拟器性能,Intel 后来推出了针对 Intel x86 CPU 的镜像,本书默认选择 Android 5.0 版本的 API。

单击 Release Name 后面的 Download 按钮,弹出 SDK Quickfix Installation 对话框,如图 1-29 所示,等待下载安装完成。

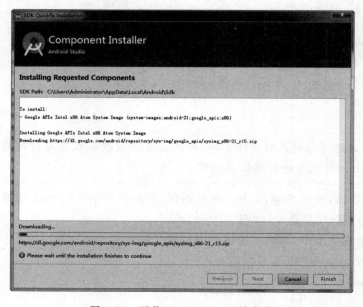

图 1-29 下载 System Image 并安装

下载完毕后，单击 Finish 按钮，然后选中之前下载完成的 API，单击 Next 按钮，弹出 Android Virtual Device(AVD)对话框，如图 1-30 所示。

图 1-30　完成创建

单击 Finish 按钮，完成 AVD 的创建并弹出 Your Virtual Devices 对话框，如图 1-31 所示。单击 ▶ 按钮，启动模拟器。

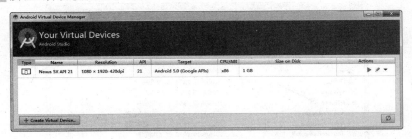

图 1-31　模拟器选择界面

模拟器启动中和启动后的界面如图 1-32 所示。

(a) 启动中　　　　(b) 启动后

图 1-32　AVD 启动中和启动后的界面

模拟器启动后，即可直接运行编写好的 Android 应用程序。按照 1.4.1 节步骤运行应用程序，出现如图 1-33 所示的设备选择界面，可以看到已启动的 Nexus 5X API 21 模拟器，选择并单击 OK 按钮运行该应用程序。

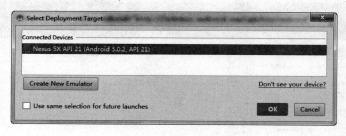

图 1-33　选择设备

1.5.3　实现【任务 1-3】

ADB(Android Debug Bridge)是 Android 应用程序开发中非常重要的一个工具，通过 ADB 可以连接到设备、安装或卸载应用程序，还可以直接执行 Linux Shell，从而使开发人员能够方便的操作设备。

adb.exe 位于 Android SDK\platform-tools\ 目录下，需要使用命令窗口运行，如图 1-34 所示。

图 1-34　platform-tools 目录

打开 Windows 的 cmd 窗口，切换到 adb.exe 所在路径，执行 adb 命令，结果如下所示：

```
D:\tool\adt\sdk\platform-tools>adb
Android Debug Bridge version 1.0.32
…//省略
```

输出结果说明 ADB 启动成功。接下来介绍 ADB 的常用操作。

（1）查看设备列表。adb devices 命令可以查看当前连接的设备列表，例如：

第 1 章　Android概述

```
D:\tool\adt\sdk\platform-tools>adb devices
List of devices attached
emulator-5554    device
```

（2）安装、卸载应用程序。adb install 命令可以安装 apk，例如通过 adb install f:/giftems.apk 命令将 f 盘上的 giftems.apk 安装到模拟器上：

```
D:\tool\adt\sdk\platform-tools>adb install f:/giftems.apk
152 KB/s (3147164 bytes in 20.148s)
        pkg: /data/local/tmp/giftems.apk
Success
```

adb uninstall 命令可以卸载指定包名的应用程序，例如将包名为 com.qst.giftems 的应用程序从模拟器上卸载：

```
D:\tool\adt\sdk\platform-tools>adb uninstall com.qst.giftems
Success
```

（3）在计算机与设备之间传输文件。adb push 命令可以将计算机中的文件复制到设备中，例如将电脑 f:/a.txt 复制到设备的 /dev 目录下，并改名为 b.txt：

```
adb push f:/a.txt /dev/b.txt
```

adb pull 命令可以将设备中的文件复制到计算机中，例如将设备的 /dev/b.txt 复制到计算机的 f:/ 目录，并改名为 a.txt：

```
adb pull /dev/b.txt f:/a.txt
```

（4）直接进入 Linux Shell。adb shell 命令可以直接进入 Linux Shell，然后按照 Linux 的 Shell 语法编写 Shell 指令，例如：

```
D:\tool\adt\sdk\platform-tools>adb shell
root@generic_x86_64:/ # cd /system/app
root@generic_x86_64:/system/app # ls
BasicDreams
Browser
Calculator
...
```

上述操作中，首先通过 adb shell 命令进入设备 Shell 环境；然后调用 cd 命令切换到 /system/app 目录，即 Android 存放预装应用的目录；再使用 ls 命令列出所有的预装应用。

本章总结

- Android 是一个以 Linux 为基础的开源操作系统，用于智能手机和平板电脑等移动设备。
- Android 系统分为 4 层，从高层到低层分别是应用程序层、应用程序框架层、系统运行库层和 Linux 内核层。
- Android 应用程序主要包含 4 种组件：Activity、Service、BroadcastReceiver 和 Content Provider。
- Activity 是最基本的 Android 应用程序组件，一个 Activity 表示一个可视化的用户界面。
- Service 组件用于提供服务，专门用于执行一些持续性的、耗时的并且无需用户界面交互的操作。
- BroadcastReceiver 用于使应用程序监听到匹配指定标准的广播信息。
- ContentProvider 组件是一种共享的持久数据存储机制，是在应用程序之间共享数据的首选方案。
- Android Studio 是 Google 开发的一款面向 Android 开发者的 IDE。
- Android 程序在 AVD 虚拟机上运行。

Q&A

问题：简述 Android 应用程序主要组件。

回答：Android 应用程序主要包含 4 种组件：Activity、Service、BroadcastReceiver 和 ContentProvider。Activity 是最基本的 Android 应用程序组件，一个 Activity 表示一个可视化的用户界面；Service 组件表示一种服务，专门用于执行一些持续性的、耗时的且无需用户界面交互的操作；BroadcastReceiver 用于使应用程序监听到匹配指定标准的广播信息；ContentProvider 组件是一种共享的持久数据存储机制，是在应用程序之间共享数据的首选方法。

章节练习

习题

1. Android 是一个以_____为内核基础的开源操作系统。
   A. Linux　　　　　B. Windows　　　　　C. iOS　　　　　D. Java
2. Android 采用了分层的架构，下面_____不属于 Android 的分层。
   A. Linux 内核层　　B. 应用程序层　　C. 系统运行库层　　D. POJO 层
3. 下面关于 Android 的系统运行库说法错误的是_____。
   A. Android 的系统运行库是 Android 的核心服务，在硬件和软件栈的其他部分之间提供了一个抽象层
   B. Android 的系统运行库中包含了各种 C/C++核心库，例如 Libc 和 SSL 等
   C. Android 的系统运行库中包含用于 2D 和 3D 图形的 SGL 和 OpenGL 的图形库
   D. Android 的系统运行库中包含用于本地数据库支持的 SQLite
4. Android 应用程序主要包含 4 种组件，其中_____通常就是一个单独的屏幕。
   A. Activity　　　　　　　　　　　　　B. Intent

C. BroadcastReceiver　　　　　　　D. SQLite

5. 下面_____不属于 Android 应用程序的组件。

　　A. Activity　　　　　　　　　　　B. ContentProvider

　　C. BroadcastReceiver　　　　　　　D. Intent

6. Android 应用程序主要包含_____、_____、_____和_____4 种组件，组件之间通过_____进行信息传递。

7. 简述 Android 系统的优势。

上机

训练目标：Android 开发环境搭建。

培养能力	熟练搭建 Android 开发环境		
掌握程度	★★★★★	难度	容易
代码行数	30	实施方式	编码强化
结束条件	搭建好环境并测试		

参考训练内容

（1）下载并安装 JDK、Android Studio，配置并启动 AVD；

（2）编写一个 Android 应用程序，在屏幕中央显示"Hello Android, this is QST!"

第 2 章

Activity和Application

本章任务是完成"GIFT-EMS 礼记"需求分析，创建项目并完成基本架构设计：
- 【任务 2-1】 项目背景介绍及需求分析。
- 【任务 2-2】 创建项目并编写实体类和 Application 类等基础架构。
- 【任务 2-3】 编写项目中 Activity、按钮、文本输入框等控件所使用的背景文件。
- 【任务 2-4】 编写项目的样式文件。

第2章 Activity和Application

知 识 点	Listen（听）	Know（懂）	Do（做）	Revise（复习）	Master（精通）
Activity 的创建及生命周期方法	★	★	★	★	★
Android 的资源分类及访问	★	★	★	★	
AndroidManifest.xml 清单文件	★	★	★	★	
Android 应用程序生命周期	★	★			
Application 类及生命周期事件	★	★	★	★	
样式和主题	★	★	★		

第 1 章从整体上对 Android 应用的结构进行了分析，至此，读者对 Android 应用已经有了一个大致的轮廓。从本章开始，逐步解剖 Android 应用所涉及的各种资源与组件。

2.1 Activity

Activity 用于提供可视化用户界面的组件，可以与用户进行交互来完成某项任务，例如拨号、拍照或发送 E-mail 等。Activity 是 Android 应用程序中最基本的组成单位，每一个 Activity 被赋予一个窗口，用于绘制用户界面。一个 Activity 对象代表一个单独的窗口，该窗口通常充满屏幕，但也可以小于屏幕而浮于其他窗口之上。

2.1.1 Activity 简介

通常一个 Android 应用程序由多个松散耦合的 Activity 组成，而一个应用程序中会有一个 Activity 被指定为主界面（Main Activity），即启动应用程序时第一个呈现给用户的界面。Activity 是 Android 应用程序中使用频率最高的组件，在具体实现时，每个 Activity 都被定义为一个独立的类，并继承 android.app.Activity 类或其子类。

Activity 基类及其子类的继承层次如图 2-1 所示。

图 2-1 Activity 类继承层次

一般情况下,用户建立自己的 Activity 继承 Activity 基类即可,但在不同的应用场景下,有时需要继承 Activity 的子类。例如,界面需要显示一个列表,则可以让应用程序继承 ListActivity 类;而界面需要实现带标题的功能时,则可以继承 AppCompatActivity 类。

Activity 类中常用的方法如表 2-1 所示。

表 2-1　Activity 常用方法

方　　法	功　能　描　述
setContentView(int layoutResID)	设置 Activity 界面布局
onCreate(Bundle savedInstanceState)	Activity 生命周期的方法,用于第一次创建 Activity
onStart()	Activity 生命周期的方法,用于启动 Activity
onPause()	Activity 生命周期的方法,用于暂停 Activity
onStop()	Activity 生命周期的方法,用于停止 Activity
OnDestroy()	Activity 生命周期的方法,用于销毁 Activity
onResume()	Activity 生命周期的方法,用于将 Activity 由暂停状态恢复使用
onRestart()	Activity 生命周期的方法,用于将 Activity 由停止状态恢复使用
onKeyDown(int keyCode, KeyEvent event)	键盘按键按下时的动作事件处理方法
onKeyUp(int keyCode, KeyEvent event)	键盘按键抬起时的动作事件处理方法
onTouchEvent(MotionEvent event)	监听屏幕的触摸事件处理方法
openContextMenu(View view)	开启上下文菜单
setResult(int resultCode)	返回数据给上一个 Activity
startActivityForResult(Intent intent, int requestCode)	携带数据并跳转 Activity
finish()	结束当前 Activity

2.1.2　创建 Activity

Activity 是 Android 应用中最重要、最常见的应用组件,Android 开发的一个重要组成部分就是如何开发 Activity。创建一个自定义 Activity 时,需要继承 Activity 基类。

本书主要介绍两种方式实现 Activity 类:

① 通过继承 android.app.Activity 基类的方式实现 Activity;

② 通过继承 android.support.v7.app.AppCompatActivity 类的方式实现 Activity。

1. 继承 Activity 基类

下述代码通过继承 Activity 基类的方式实现自定义的 BaseActivity 类。

【代码 2-1】 BaseActivity.java

```java
import android.app.Activity;
import android.os.Bundle;
public class BaseActivity extends Activity {
    @Override
    public void onCreate(Bundle savedInstanceState) {
        super.onCreate(savedInstanceState);
        setContentView(R.layout.activity_main);}
}
```

在使用 Eclipse 工具开发 Android 应用时，BaseActivity 自动继承的是 android.app.Activity 类；当使用 Android Studio 开发工具时，BaseActivity 自动继承的是 android.support.v7.app.AppCompatActivity 类，因 AppCompatActivity 是 Activity 类的子类，所以只需将其改成 Activity 即可。运行结果如图 2-2 所示。

2. 继承 AppCompatActivity 类

Android Studio 在 API 22 之后，当创建 Android 应用时，MainActivity 会自动继承 android.support.v7.app.AppCompatActivity 类，该类是 Activity 的子类。AppCompatActivity 类用来替代已过时的 ActionBarActivity 类。AppCompatActivity 与 ActionbarActivity 功能类似，都能添加标题栏，而且 AppCompatActivity 类继承 FragmentActivity 类，并能够兼容低版本。

下述代码通过继承 AppCompatActivity 类的方式实现 Activity。

【代码 2-2】 MainActivity.java

```java
import android.support.v7.app.AppCompatActivity;
import android.os.Bundle;
public class MainActivity extends AppCompatActivity {
    @Override
    public void onCreate(Bundle savedInstanceState) {
        super.onCreate(savedInstanceState);
        setContentView(R.layout.activity_main);}
}
```

运行结果如图 2-3 所示。

图 2-2　继承 Activity 类　　　　图 2-3　继承 AppCompatActivity 类

图 2-2 与图 2-3 相比较可以看出，当 MainActivity 继承 Activity 类时，运行的程序界面缺少标题栏。

> 在实际开发过程中，Activity 与 AppCompatActivity 在方法应用上并无太大区别，可根据实际需要选择合适的 Activity 的基类或者子类进行开发。

2.1.3　Activity 的生命周期

在 Android 系统中，Activity 是由 Activity 栈进行管理的。当一个新的 Activity 启动时，将被放置到栈顶，成为运行中的 Activity，前一个 Activity 保留在栈中，不再放到前台，直到新的 Activity 退出为止。

一个Activity可以启动另一个Activity,以完成不同的动作,当新的Activity启动时,前一个Activity就会停止,并由系统将该Activity保留在一个Back Stack栈上。新启动的Activity被推送到栈顶,并获得用户的焦点。Back Stack栈符合简单的"后进先出"原则,当用户完成当前Activity并单击Back按钮时,该Activity会被弹出栈,并被销毁,然后恢复之前的Activity。

当一个Activity因新的Activity启动而停止时,将调用Activity生命周期中的回调方法并改变其状态。一个Activity可能会收到多个回调方法,这源于Activity自身的状态变化。无论系统创建、停止、恢复还是销毁Activity,每个回调方法提供完成适合当前状态的指定行为。当Activity停止时,应该释放所占用的资源,如网络数据库连接等;当Activity恢复时,可以重新获得必要的资源和恢复被中断的动作。这些状态都属于Activity的生命周期。

Activity有4种本质区别的状态。

- **运行状态(Active/Running)**:在屏幕的前台(Activity栈顶),称为活动状态或者运行状态。
- **暂停状态(Paused)**:如果一个Activity失去焦点,但是依然可见(一个新的非全屏的Activity或者一个透明的Activity被放置在栈顶),此时为暂停状态。处于暂停状态的Activity依然保持活力(保持所有的状态、成员信息和窗口管理器保持连接),当系统内存极低时将被杀掉。
- **停止状态(Stopped)**:如果一个Activity被另外的Activity完全覆盖掉,则此时为停止状态。停止状态虽然保持所有状态和成员信息,但是窗口被隐藏,不再可见;当系统内存需要被其他应用使用时,处于Stopped状态的Activity将被杀掉。
- **销毁状态(Killed)**:如果一个Activity是Paused或者Stopped状态,则系统可以将其从内存中删除。Android系统有两种删除方式,要么要求该Activity结束,要么直接杀掉其进程。当该Activity再次显示给用户时,必须重新开始和重置前面的状态。

Activity的状态转换过程如图2-4所示,矩形框表明Activity在状态转换之间的回调接口,开发人员可以重写实现该方法以便实现相应的功能,而椭圆形则表明Activity所处的某个状态。

从图2-4可以看出,Activity生命周期中有三个关键的循环。

- **整个生命周期**:从onCreate()开始到onDestroy()结束。Activity在onCreate()中设置所需的"全局"状态,在onDestroy()中释放所有的资源。例如,某个Activity有一个在后台运行的线程用于从网络下载数据,则该Activity可以在onCreate()中创建线程,在onDestroy()中停止线程。
- **可见生命周期**:从onStart()开始到onStop()结束。在这段时间内,Activity在屏幕中可见,但有可能不在前台,不能和用户交互。在这两个接口之间,需要保持显示给用户的UI数据和资源等,例如,在onStart()中注册一个IntentReceiver来监听数据变化导致UI的变动,当不再需要显示时可以在onStop()中将其注销。onStart()和onStop()可以被多次调用,从而实现Activity在可见和隐藏之间切换。
- **前台生命周期**:从onResume()开始到onPause()结束。在这段时间内,该Activity处于所有Activity的最前面,用户可以与之进行交互。Activity可以经常性地在运

第2章 Activity和Application

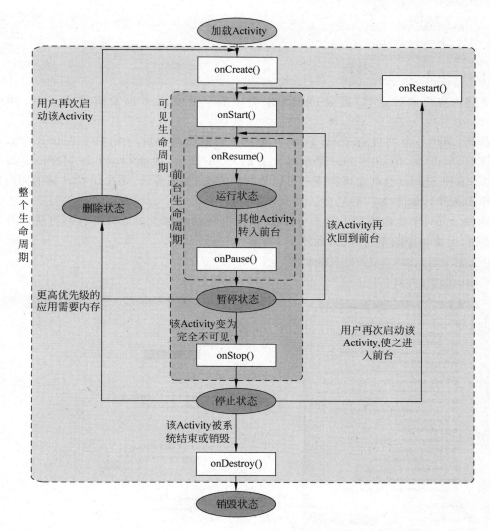

图 2-4　Activity 的状态转换

行状态和暂停状态之间切换，例如，当设备准备休眠时、当 Activity 处理结果被分发时或新的 Intent 被分发时。因此在 onResume() 和 onPause() 方法中的代码应该属于非常轻量级的处理操作。

Activity 整个生命周期的任一方法都可以被重写。所有 Activity 都需要通过 onCreate() 方法进行初始化，大部分 Activity 需要 onPause() 方法来提交更改过的数据，onFreeze() 方法用来恢复在 onCreate() 方法中所设置的状态。

【示例】 Activity 类的定义

```
public class Activity extends ContextThemeWrapper {
    protected void onCreate(Bundle icicle){...}
    protected void onStart(){...}
    protected void onRestart(){...}
    protected void onResume(){...}
    protected void onFreeze(Bundle outIcicle) {...}
```

```
    protected void onPause(){...}
    protected void onStop(){...}
    protected void onDestroy(){...}
}
```

下述内容将通过重写(覆盖)Activity 类中的 7 个状态方法来演示 Activity 的生命周期。

首先,新建一个项目 ActivityTest,在创建项目时自动生成一个 MainActivity 类,双击打开该类,单击 Android Studio 菜单栏上方的 Code 按钮,选择 Override Methods 菜单选项。Android Studio 会列出该类所有可以重写的方法(如需多选,则按住 Ctrl 键的同时鼠标左键单击选择),如图 2-5 所示。

选择 Activity 生命周期中的 7 个方法(系统已经自动加上了 onCreate()方法),并单击 OK 按钮,生成相应的重写方法,如图 2-6 所示。

图 2-5　创建 Activity　　　　　图 2-6　创建完成 Activity

在每一个生命周期的方法中添加日志输出代码,代码如下所示。

【代码 2-3】 ActivityTest.java

```
public class ActivityTest extends AppCompatActivity {
    private static final String TAG = "ActivityTest";
    @Override
    public void onCreate(Bundle savedInstanceState) {
        super.onCreate(savedInstanceState);
        setContentView(R.layout.main);
        Log.d(TAG, "执行了 onCreate()方法");   }
    @Override
    protected void onStart() {
        super.onStart();
        Log.d(TAG, "执行了 onStart()方法");}
    @Override
```

```
    protected void onResume() {
        super.onResume();
        Log.d(TAG, "执行了 onResume()方法");}
    @Override
    protected void onStop() {
        super.onStop();
        Log.d(TAG, "执行了 onStop()方法");}
    @Override
    protected void onPause() {
        super.onPause();
        Log.d(TAG, "执行了 onPause()方法");}
    @Override
    protected void onRestart() {
        super.onRestart();
        Log.d(TAG, "执行了 onRestart()方法");}
    @Override
    protected void onDestroy() {
        super.onDestroy();
        Log.d(TAG, "执行了 onDestroy()方法");}
}
```

上述代码中使用 Log.d()方法记录日志信息,Log 日志类能够记录程序运行过程中的相关信息,其常用方法如表 2-2 所示。

表 2-2 Log 类的常用方法

方　　法	功能描述	方　　法	功能描述
Log.e()	记录错误信息	Log.d()	记录调试信息
Log.w()	记录警告信息	Log.v()	记录详细的信息
Log.i()	记录一般提示性信息		

单击 Android Studio IDE 窗口底部的"6：Android Monitor"按钮,弹出 Android Monitor 窗口,通过 LogCat 窗口可以查看数据的传递过程,帮助完成调试过程。

启动模拟器并运行 ActivityTest 应用,可以在模拟器上看到执行结果。默认显示一个 Hello World 的界面,单击"返回"键离开程序,此时程序已经出现在模拟器的安装程序列表中。下面在模拟器上运行该应用程序,观察 LogCat 的显示结果。

1. 测试一

进入模拟器的应用程序列表界面,单击 ActivityTest 图标打开此应用,可在 LogCat 窗口中看到此 Activity 启动时的状态转换过程,如图 2-7 所示。

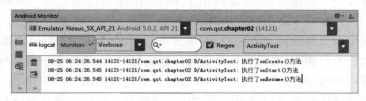

图 2-7 LogCat 窗口显示的 Activity 启动时的生命周期过程

单击"返回"键退出应用，Activity 的返回时的状态转换过程如图 2-8 所示。

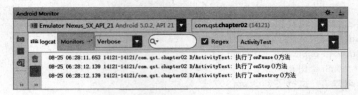

图 2-8　LogCat 窗口显示的 Activity 返回后的生命周期过程

在 LogCat 窗口中，Tag 列的值为 ActivityTest 的行即是需要关注的信息。系统按顺序依次调用 onCreate()、onStart() 和 onResume() 三个方法来创建 Activity，单击"返回"键时依次调用 onPause()、onStop() 和 onDestroy() 方法来结束 Activity。

2．测试二

在模拟器上启动 ActivityTest，然后单击 Home 键；再单击模拟器下方的"拨号"键打电话，接着单击"返回"键离开拨号器应用；最后单击模拟器上首页界面下方中间图标调出模拟器所有应用的列表，单击 ActivityTest 图标打开该应用程序。LogCat 的输出结果如图 2-9 所示。

图 2-9　测试二 LogCat 显示的 Activity 生命周期状态

当运行通话应用时，系统调用 onStop() 方法将 ActivityTest 应用切换至暂停状态，当 ActivityTest 再度呈现到屏幕上时，依次运行 onRestart()、onStart() 和 onResume() 三种方法。

在本程序中需要使用类方法 Log.d() 输出日志信息，通过 LogCat 窗口来查看输出的信息。通过上述的测试可见，当用户第一次打开 Android 应用时，应用展现在用户的手机桌面，并获取用户的输入焦点。在启动过程中，Android 系统调用了 Activity 一系列的生命周期方法，并建立应用组件和用户之间的联系。当用户启动了应用中的另外一个 Activity，或者直接切换到另外一个应用时，系统也调用了 Activity 生命周期中的一系列方法使应用可以在后台运行。

在 Activity 生命周期的回调方法中，开发者可以定义 Activity 在用户第一次进入和重新进入应用时的行为。例如，当设计一个流媒体播放器时，在用户切换到另外一个应用时暂停视频并停止网络连接，当用户切换回来的时候，重新连接网络并从用户之前暂停的点继续播放。深入理解 Activity 生命周期中各个方法的功能对于编写一些复杂的程序是非常有帮助的。

2.2 资源管理

在 Android 应用程序中,资源扮演着非常重要的角色。所谓资源,就是在代码中使用的外部文件,包括图片、音频、动画和字符串等。在传统的程序开发过程中,需要用到很多常量、字符串等资源,如果在程序中直接使用这些资源,会给阅读和维护源码带来很多麻烦。因此,Android 架构对这些字符串、数值等资源的定义做了进一步的改进:Android 允许将应用中所用到的各种资源集中在 res 目录中定义,并为每个资源自动生成一个编号,在应用程序中可以直接通过编号来访问这些资源。

在 Android 应用程序中,除了 res 目录外,assets 目录也用于存放资源,这两个目录的区别如下。

(1) 通常在 assets 目录中存放的是应用程序无法直接访问的原生资源,应用程序需要通过 AssetManager 类以二进制流的形式来读取资源。

(2) 而对于 res 目录中的资源,Android SDK 会在编译时自动在 R.java 文件中为这些资源创建索引,程序可以通过资源清单类 R.java 对资源进行访问。

2.2.1 资源分类

在 Android 开发中常用的资源包括文本字符串(strings)、颜色(colors)、数组(arrays)、动画(anim)、布局(layout)、图像和图标(drawable)、音频视频(media)以及其他应用程序所使用的组件等。资源文件是使用频率最高的一类文件,无论是 string、drawable 还是 layout,这些资源都是经常使用到的,而且为开发提供了很大的方便。

Android 的资源可分为两大类:

(1) **原生资源**:是指无法通过由 R 类进行索引的原生资源(如 MP3 文件等),该类资源保存在 assets 目录下,且 Android 程序不能直接访问,必须通过 android.content.res.AssetManager 类以二进制流的形式进行读取和使用。

(2) **索引资源**:是指可以通过 R 类进行自动索引的资源(如字符串),该类资源保存在 res 目录下,在应用程序编译时索引资源通常被编译到应用程序中。

本书经常提到的 Android 应用资源是指位于 res 下的应用资源,Android SDK 会在编译该应用时在 R 类中为其创建对应的 ID 索引项。Android 应用资源的类型及存放目录如表 2-3 所示。

表 2-3　Android 应用资源的类型及存放目录

目　　录	资　源　描　述
/res/animator/	存放定义属性动画的 XML 文件
/res/anim/	存放定义了补间动画(tweened animation)或逐帧动画(frame by frame animation)的 XML 文件;该目录下也可以存放定义 property animations 的 XML 文件,推荐分类存放
/res/color/	存放定义不同状态下颜色列表的 XML 文件

续表

目 录	资 源 描 述
/res/drawable/	存放能转换为绘制资源(Drawable Resource)的位图文件(后缀为.png、.9.png、.jpg、.gif的图像文件)或者定义了绘制资源的XML文件；res/drawable-xxxx是存放图片的目录,每个图片需要准备4种分辨率版本：drawable-hdpi(存放高分辨率版本)、drawable-xdpi(存放超高分辨率版本)、drawable-ldpi(存放低分辨率版本)、drawable-mdpi(存放中等分辨率版本)
/res/layout/	存放各种界面布局文件,每个Activity对应一个XML文件
/res/menu/	存放为应用程序定义的各种菜单资源,包括选项菜单、子菜单、上下文菜单
/res/raw/	存放直接复制到设备中的任意文件。该资源无须编译,添加到应用程序编译产生的压缩文件中即可；当使用这些资源时,可以调用Resources.openRawResource()方法来获取,该方法的参数是资源的ID,即R.raw.somefilename
/res/values/	存放定义多种类型资源的XML文件。资源的类型包括字符串、数据、颜色、尺寸和样式等类型。这些XML文件的根标签都是< resources ></resources >标签对；在该标签中添加不同的子元素则代表不同的资源,例如：string/integer/bool子标记,代表添加一个字符串值、整数值或布尔值；color子标记,代表添加一个颜色值；array子标记或string-array、int-array子元素,代表添加一个数组；style子标记,代表添加一个样式；dimen子标记,代表添加一个尺寸。 各种常数都可以定义在/res/values/目录下的资源文件中,为了方便添加和修改等操作,Android通常使用不同的文件存放不同类型的值,如：arrays.xml,定义数组资源；colors.xml,定义颜色资源；dimen.xml,定义尺寸资源；strings.xml,定义字符串资源；styles.xml,定义样式资源
/res/xml/	存放任意的原生XML文件,这些文件可以使用Resources.getXML()方法来访问
/assets/	存放各种资源文件,包括音频文件、视频文件等。使用AssetManager类访问这些资源

需要注意,res文件夹下的子文件夹必须按规范来命名,否则会报类似"invalidresource directory name **"的错误提示信息,除了表2-3中提供的默认文件夹外,一般可以用"默认文件夹名＋短横线＋配置相关的限定符"构成需要的资源文件夹,用于区别不同屏幕分辨率、不同机型特点(是否带键盘等)以及不同的本地化资源等。

一旦将应用程序的各种资源存放在Android应用的res目录下,就可以在Java程序代码中访问这些资源,也可以在其他XML资源中访问这些资源。

2.2.2 资源访问方式

2.2.1小节对Android资源类型进行分析,本节从简单资源入手,详细介绍各类资源的访问。在没有明确说明资源不能在XML资源文件中使用时,该资源都是既可以在XML资源文件中使用,又可以在Java代码中使用。

在Android应用中,资源访问的方式有两种：

（1）一种是在Java源代码中访问资源,既可以访问res资源,也可以访问assets原生资源；

（2）另一种是在XML文件中进行访问资源。

下面分别介绍在Java代码中如何访问res和assets资源,以及如何在XML文件中使用资源。

1. Java代码访问res资源

每个res资源都会在项目的R类中自动生成一个代表资源编号的静态常量,在Java代码中通过R类可以访问这些res资源,其语法如下所示。

第2章 Activity和Application

【语法】

> [packageName.]R.resourceType.resourceName

其中：

- packageName 是包名，除了应用程序自动生成的 R 类外，Android 系统中还有一个可访问的 R 类，即 android.R；通过限定 R 的包名可以指定使用哪一个 R 类，例如 android.R.drawable.ic_delete 表示 Android 系统中的 ic_delete 图片资源，而 qst.chapter02.R.ic_delete 则表示当前项目（chapter02 项目的包名是 qst.chapter02）中的 ic_delete 图片资源；
- resourceType 是资源类型，代表 R 类中的资源类型，例如 R.layout 表示布局文件资源，R.drawable 表示图片资源等。
- resourceName 是资源名称，可以是资源文件的名称，也可以是定义资源的 XML 文件中的资源标签的 name 属性值。
- android.content.res.Resources 类是 Android 资源访问控制类，该类提供了大量方法来获取实际资源。通过 android.content.Context 类的 getResources() 方法可以获取 Resources 对象；由于 Activity 继承了 Context 类，所以在 Activity 中可以直接调用 getResources() 方法来获取 Resources 对象。Resources 类中提供的访问资源的方法如表 2-4 所示。

表 2-4 Resources 类的资源访问方法

方法	功能描述
int getColor(int id)	对应 res/values/colors.xml
Drawable getDrawable(int id)	对应 res/drawable/
XmlResourceParser getLayout(int id)	对应 res/layout/
String getString(int id)	对应 res/values/strings.xml
CharSequence getText(int id)	对应 res/values/strings.xml
InputStream openRawResource(int id)	对应 res/raw/
void parseBundleExtra(String tagName, AttributeSet attrs, Bundle outBundle)	对应 res/xml/
String[] getStringArray(int id)	对应 res/values/arrays.xml
float getDimension(int id)	对应 res/values/dimens.xml

通过调用 Resources 类中的方法，可访问到对应的资源。

2. Java 代码访问 assets 原生资源

通过 Resources 类的 getAssets() 方法可获得 android.content.res.AssetManager 对象，该对象的 open() 方法可以打开指定路径的 assets 资源的输入流，从而读取到对应的原生资源。在 Android Studio 新建项目是不会自动创建 assets 文件夹的，需要开发者手动创建该文件夹，创建和使用 assets 的步骤如下所示。

(1) 首先，右击相应的项目，如图 2-10 所示，选择 New→Folder→Assets Folder，再单击 Finish 按钮完成创建。

(2) 将一张图片复制到 assets 文件夹中，然后打开 res 文件目录，右击 layout 文件夹，选择 New→XML→Layout XML File 菜单项，创建一个新的 XML 布局文件，如图 2-11 所示。

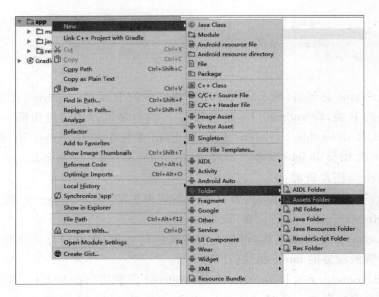

图 2-10　创建 assets 文件夹

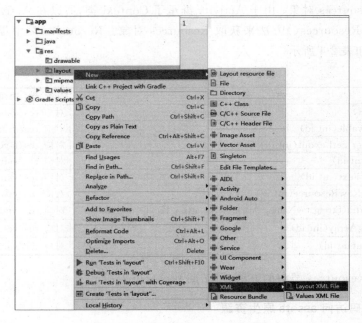

图 2-11　创建新的 XML 布局文件

（3）创建一个名为 assets_layout 的 XML 文件,代码如下所示。

【代码 2-4】　assets_layout.xml

```
<?xml version = "1.0" encoding = "utf-8"?>
< LinearLayout xmlns:android = "http://schemas.android.com/apk/res/android"
    android:layout_width = "match_parent"
    android:layout_height = "match_parent">
    < ImageView
        android:layout_width = "wrap_content"
```

```
            android:layout_height = "wrap_content"
            android:id = "@ + id/im1" />
</LinearLayout>
```

> **注意**
>
> ImageView 为安卓开发的常用组件之一，用于显示图片资源，本章仅限于演示如何显示 assets 中的图片资源，该控件的详细介绍参见第 3 章。

（4）创建一个新的 Activity 类，代码如下所示。

【代码 2-5】 Assets_ActivityDemo.java

```java
public class Assets_ActivityDemo extends AppCompatActivity {
    ImageView iv;
    @Override
    public void onCreate(Bundle savedInstanceState) {
        super.onCreate(savedInstanceState);
        setContentView(R.layout.assets_layout);
        iv = (ImageView) findViewById(R.id.im1);
        try {
            InputStream is = getResources().getAssets().open("android.jpg");
            Bitmap bitmap = BitmapFactory.decodeStream(is);
            iv.setImageBitmap(bitmap);
        } catch (IOException e) {
            e.printStackTrace();}}
}
```

上述代码中 getResources().getAssets().open("android.jpg")为 Java 代码来读取 asstes 文件中名为 android.jpg 的图片文件，并将该图片以 ImageView 的形式显示出来。代码运行结果如图 2-12 所示。

3. XML 文件中使用资源

在 XML 文件中引用其他资源的语法如下所示。

图 2-12 读取 assets 中的图片文件

【语法】

```
@[packageName:]resourceType/resourceName
```

其中，packageName、resourceType 和 resourceName 的含义与 Java 代码中访问资源时相同，需要注意前面必须有一个@符号。

2.2.3　strings.xml 文本资源文件

res/values 目录用于存放常用的一些简单的 XML 资源文件。Android 常用的定义资源的 XML 文件有以下几个：strings.xml 用于定义文本内容的资源文件；colors.xml 用于定义颜色设置的资源文件；dimens.xml 用于定义尺寸的资源文件；styles.xml 用于定义主题风格的资源文件。其中，/res/values/strings.xml 是最重要的文本资源文件，其格式比较简单,示例代码如下所示。

【示例】 strings.xml 文本资源文件

```xml
<!--文本资源文件 res\values\strings.xml -->
<?xml version = "1.0" encoding = "utf-8"?>
<resources>
    <string name = "title">Resources</string>
    <string name = "message">Hello World!</string>
    <string name = "error">Wrong resource!</string>
</resources>
```

在定义程序所使用的文本资源文件后,可以将 strings.xml 中所定义的字符串变量在 Java 代码中使用,或者在其他 XML 文件的资源文件中使用。

在 Java 代码中访问字符串资源的语法格式如下。

【语法】

```
R.string.字符串名
```

【示例】 Java 代码中访问字符串

```java
CharSequence app_name = getString(R.string.title);
CharSequence display = getString(R.string.message);
```

在 XML 文件中访问字符串资源的语法格式如下。

【语法】

```
@string/字符串名
```

其中:@是前置符号;string 是字符串的标记名称;字符串名则是<string>标签的 name 属性值,且需要使用斜杠(/)与前面的 string 标记进行间隔。

【示例】 XML 文件中访问字符串

```
android:app_name = "@string/title"
android:display = "@string/message"
```

当需要应用程序能够支持国际化时,可以创建不同的 values 目录,例如 values-en 表示英语、values-es 表示西班牙语等,然后将不同语言的 strings.xml 文件分别放在所属的 values 目录中即可。

打开 res 目录下的 values 文件夹,双击打开 strings.xml 文件进行编辑,示例代码如下所示。

【代码 2-6】 strings.xml

```xml
<resources>
    <string name = "app_name">My Application</string>
    <string name = "app_class">Java 代码访问 strings 文字资源</string>
    <string name = "app_xml">XML 文件访问 strings 文字资源</string>
</resources>
```

下述代码演示如何在 XML 文件中访问字符串。创建一个名为 strings_layout 的 XML

文件,代码如下所示。

【代码 2-7】 strings_layout.xml

```xml
<?xml version = "1.0" encoding = "utf-8"?>
<LinearLayout xmlns:android = "http://schemas.android.com/apk/res/android"
    android:layout_width = "match_parent"
    android:layout_height = "match_parent"
    android:orientation = "horizontal"
    >
    <TextView
        android:layout_width = "wrap_content"
        android:layout_height = "wrap_content"
        android:textAppearance = "?android:attr/textAppearanceLarge"
        android:text = "@string/app_xml"
        android:textColor = "#000"
        android:id = "@ + id/tv1" />
    <TextView
        android:layout_width = "wrap_content"
        android:layout_height = "wrap_content"
        android:textAppearance = "?android:attr/textAppearanceLarge"
        android:text = "Large Text"
        android:textColor = "#000"
        android:id = "@ + id/tv2" />
</LinearLayout>
```

上述代码是一个 Activity 对应的界面布局文件,其中 LinearLayout 是常用的线性布局,TextView 是文本组件。代码中@string/app_xml 为 XML 文件读取 strings.xml 文件中名为 app_xml 的字符串,并将该字符串以 TextView 的形式显示出来。

本章所涉及的 XML 布局文件都是基于能够掌握各资源的实际应用为目标,因此采用最简单的线性布局和文本组件进行演示,并没有过多涉及布局和 UI 组件的具体内容。有关 UI 界面布局和组件的详细内容参见第 3 章。

下述代码演示如何在 Java 代码中访问字符串。创建一个对应的 Activity 类,代码如下所示。

【代码 2-8】 Strings_ActivityDemo.java

```java
public class Strings_ActivityDemo extends AppCompatActivity {
    TextView tv2;
    public void onCreate(Bundle savedInstanceState) {
        super.onCreate(savedInstanceState);
        setContentView(R.layout.strings_layout);
        tv2 = (TextView) findViewById(R.id.tv2);
        tv2.setText(R.string.app_class);}
}
```

上述代码中 R.string.app_class 为 Java 代码来读取 strings.xml 文件中名为 app_class 的字符串,并将该字符串以 TextView 的形式显示出来。运行结果如图 2-13 所示。

图 2-13　XML 文件和 Java 代码分别访问 strings.xml 资源

setText()方法设置了显示的文字为字符串资源 app_class 的内容,更多详细内容参见第 3 章。

2.2.4　colors.xml 颜色设置资源文件

Android 中的颜色是通过红(Red)、绿(Green)、蓝(Blue)三原色,以及一个透明度(Alpha)来表示的。颜色值总是以"♯"号开头,接下来是 Alpha-Red-Green-Blue 形式。其中 Alpha 部分可以省略,即完全不透明。

颜色值的声明有以下几种方式:
- ♯RGB:分别指定红、绿、蓝三原色的值(只支持 0~f 这 16 级颜色)表示颜色;
- ♯ARGB:分别指定透明度、红、绿、蓝三原色的值(只支持 0~f 这 16 级颜色和 16 级透明度)表示颜色;
- ♯RRGGBB:分别指定红、绿、蓝三原色的值(只支持 0~ff 这 16 级颜色)表示颜色;
- ♯AARRGGBB:分别指定透明度、红、绿、蓝三原色的值(只支持 0~ff 这 16 级颜色和 256 级透明度)表示颜色。

颜色资源存储在/res/values/colors.xml 文件中,使用<color>标记进行定义,其语法格式如下:

【语法】

```
<color name = color_name>♯color_value</color>
```

下述代码演示如何进行颜色的定义,代码如下所示。

【示例】　使用<color>标记定义颜色

```
<?xml version = "1.0" encoding = "utf-8"?>
<resources>
    <color name = "text_color">♯F00</color>
    <color name = "translucent_blue">♯800000ff</color>
</resources>
```

将已定义好的颜色值保存在 colors.xml 资源文件中,然后在 Java 代码和 XML 文件中可以对这些颜色值进行访问。

在 Java 代码中访问颜色的语法格式如下。

【语法】

R.color.颜色名

【示例】 Java 代码中访问颜色

```
int color1 = getResources().getColor(R.color.blue);
int color2 = getResources().getColor(R.color.translucent_blue);
```

注意

　　getColor(int id)在 API 23(Android 6.0)以上版本中已过时,可以使用 ContextCompat.getColor(Context context,int id)方法进行替代,该方法能够同时兼容高低版本。

在 XML 文件中访问颜色的语法格式如下。

【语法】

@color/颜色名

其中:@是前置符号;color 是颜色标记名称;颜色名则是<color>标签的 name 属性值,且需要使用斜杠(/)与前面的 color 标记进行间隔。

【示例】 在 XML 文件中访问颜色

```
android:titleColor = "@color/blue"
android:textColor = "@color/translucent_blue"
```

打开 res 目录下的 values 文件夹,双击打开 colors.xml 文件进行编辑,代码如下所示。

【代码 2-9】 colors.xml

```xml
<?xml version = "1.0" encoding = "utf-8"?>
<resources>
    <color name = "colorPrimary">#3F51B5</color>
    <color name = "colorPrimaryDark">#303F9F</color>
    <color name = "colorAccent">#FF4081</color>
    <color name = "color_xml">#ff0000</color>
    <color name = "color_java">#0000ff</color>
</resources>
```

下述代码演示如何在 XML 文件中访问颜色。创建一个新的 XML 布局文件,代码如下所示。

【代码 2-10】 color_layout.xml

```xml
<?xml version = "1.0" encoding = "utf-8"?>
<LinearLayout xmlns:android = http://schemas.android.com/apk/res/android
    android:layout_width = "match_parent"
    android:layout_height = "match_parent"
    android:orientation = "vertical">
```

```xml
<TextView
    android:layout_width = "wrap_content"
    android:layout_height = "wrap_content"
    android:textAppearance = "?android:attr/textAppearanceLarge"
    android:text = "XML 文件访问 colors 资源(红色)"
    android:id = "@ + id/tv3"
    android:textColor = "@color/color_xml" />
<TextView
    android:layout_width = "wrap_content"
    android:layout_height = "wrap_content"
    android:textAppearance = "?android:attr/textAppearanceLarge"
    android:text = "Java 文件访问 colors 资源(蓝色)"
    android:id = "@ + id/tv4" />
</LinearLayout>
```

上述代码中@color/color_xml 为 XML 文件读取 colors.xml 文件中名为 color_xml 的 RGB 颜色代码,并将该颜色以 TextView 的形式显示出来。

下述代码演示如何在 Java 代码中访问颜色。创建一个对应的 Activity 类,代码如下所示。

【代码 2-11】 Color_ActivityDemo.java

```java
public class Color_ActivityDemo extends AppCompatActivity {
    TextView tv4;
    @Override
    public void onCreate(Bundle savedInstanceState) {
        super.onCreate(savedInstanceState);
        setContentView(R.layout.color_layout);
        tv4 = (TextView) findViewById(R.id.tv4);
        tv4.setTextColor(getResources().getColor(R.color.color_java));
    }
}
```

上述代码中 getResources().getColor(R.color.color_java) 为 Java 代码读取 colors.xml 文件中名为 color_java 的 RGB 颜色代码,并将该颜色以 TextView 的形式显示出来。运行结果如图 2-14 所示。

图 2-14　XML 文件和 Java 代码分别访问 colors.xml 资源

2.2.5　dimens.xml 尺寸定义资源文件

在 XML 布局或 Java 代码中都可以使用像素、英寸和磅值等尺寸单位,且无须更改源码,直接使用这些尺寸资源来本地化 Android UI 和设置其样式即可。Android 可以采用以下单位,如表 2-5 所示。

第2章 Activity和Application

表 2-5 各种计量单位表

测量单位	描 述	资源标记	示例
像素	实际的屏幕像素	px	10px
英寸	物理测量单位	in	6in
毫米	物理测量单位	mm	4mm
点	普通字体测量单位	pt	12pt
密度独立像素（density-independent pixels）	相对于 160dpi 屏幕的像素	dp	3dp
比例独立像素（scale-independent pixels）	对于字体显示的测量	sp	10sp

Android 官方推荐使用 dp 和 sp 进行表示，在具体开发中根据实际情况而定，也可使用其他的单位。在/res/values/dimens.xml 中使用< dimen >标签来定义元素的尺寸，示例代码如下所示。

【示例】 dimens.xml

```
<?xml version = "1.0" encoding = "utf-8"?>
<resource>
    <dimen name = "one_pixel">1px</dimen>
    <dimen name = "two_inches">2in</dimen>
    <dimen name = "double_density">2dp</dimen>
    <dimen name = "fourteen_sp">14sp</dimen>
</resource>
```

在 Java 代码中访问尺寸资源的语法格式如下。

【语法】

```
R.dimen.尺寸名
```

【示例】 Java 代码中访问尺寸资源

```
float dimen = getResources.getDimension(R.dimen.one_pixel);
float dimen = getResources.getDimension(R.dimen.fourteen_sp);
```

在 XML 文件中访问尺寸资源的语法格式如下。

【语法】

```
@dimen/尺寸名
```

其中：@是前置符号；dimen 是尺寸标记名称；颜色名则是< dimen >标签的 name 属性值，且需要使用斜杠（/）与前面的 dimen 标记进行间隔。

【示例】 XML 中访问尺寸资源

```
android:textSize = "@dimen/fourteen_sp"
android:textSize = "@dimen/double_density"
```

打开 res 目录下的 values 文件夹，双击打开 dimens.xml 文件进行编辑，代码如下所示。

【代码 2-12】 dimen_layout.xml

```
<resources>
    <dimen name = "activity_horizontal_margin">16dp</dimen>
```

```xml
<dimen name="activity_vertical_margin">16dp</dimen>
<dimen name="dimen_java">30sp</dimen>
<dimen name="dimen_xml">20sp</dimen>
</resources>
```

下述代码演示如何在 XML 文件中访问尺寸。创建一个新的 XML 布局文件，代码如下所示。

【代码 2-13】 dimen_layout.xml

```xml
<?xml version="1.0" encoding="utf-8"?>
<LinearLayout xmlns:android=http://schemas.android.com/apk/res/android
    android:layout_width="match_parent"
    android:layout_height="match_parent"
    android:orientation="vertical">
    <TextView
        android:layout_width="wrap_content"
        android:layout_height="wrap_content"
        android:textAppearance="?android:attr/textAppearanceLarge"
        android:text="XML 文件访问 dimens 资源"
        android:textSize="@dimen/dimen_xml"
        android:id="@+id/tv5" />
    <TextView
        android:layout_width="wrap_content"
        android:layout_height="wrap_content"
        android:textAppearance="?android:attr/textAppearanceLarge"
        android:text="Java 代码访问 dimens 资源"
        android:id="@+id/tv6" />
</LinearLayout>
```

上述代码中 @dimen/dimen_xml 为 XML 文件读取 dimens.xml 文件中名为 dimen_xml 的文字大小，并将该格式以 TextView 的形式显示出来。

下述代码演示如何在 Java 代码中访问尺寸。创建一个对应的 Activity 类，代码如下所示。

【代码 2-14】 Dimen_ActivityDemo.java

```java
public class Dimen_ActivityDemo extends AppCompatActivity{
    TextView tv6;
    @Override
    public void onCreate(Bundle savedInstanceState) {
        super.onCreate(savedInstanceState);
        setContentView(R.layout.dimen_layout);
        tv6 = (TextView) findViewById(R.id.tv6);
        tv6.setTextSize(getResources().getDimension(R.dimen.dimen_java));    }
}
```

上述代码中 getResources().getDimension(R.dimen.dimen_java) 为 Java 代码读取 dimens.xml 文件中名为 dimen_java 的文字大小，并将该格式以 TextView 的形式显示出来。运行结果如图 2-15 所示。

图 2-15　XML 文件和 Java 代码分别访问 dimens.xml 资源

2.2.6　styles.xml 主题风格资源文件

res/values/styles.xml 是样式资源文件，是一种更高级的 XML 资源文件，其中可以声明多个＜style＞样式元素，并在每个＜style＞元素中使用＜item＞元素来引用其他资源，并定义每一个细节风格，其语法格式如下。

【语法】

```
<style name = style_name>
    <item name = item_name>Hex value|string value|reference</item>
</style>
```

下述代码演示主题风格的定义方法。

【代码 2-15】　styles.xml

```
<?xml version = "1.0" encoding = "utf-8"?>
<!-- 一个完整的 res\values\styles.xml 主题风格资源文件 -->
<resources>
    <style name = "ThemeNew">
        <item name = "window">@drawable/screen_frame</item>
        <item name = "color">@drawable/background_color</item>
        <item name = "foreground">#FF000000</item>
        <item name = "background">#FFFFFFFF</item>
        <item name = "textColor">#FFFF0000</item>
        <item name = "textSize">14sp</item>
        <item name = "menuTextColor">?TextColor</item>
        <item name = "menuTextSize">?TextSize</item>
    </style>
</resources>
```

上述代码中定义了一个 name 为 ThemeNew 的样式，用于定义完整的窗口格式，例如将 window 定义为 drawable/screen_frame 格式，并指定 foreground 前景颜色和 background 背景颜色，而 menuTextColor 与 menuTextSize 分别用于定义菜单文字颜色和字体大小。

需要特殊说明的是，通过"?"和@两个符号都可以引用 XML 文件中的资源，其区别如下:使用"?"来引用在同一 XML 文件中定义的资源;使用@符号则引用跨文件的资源。

在 Java 代码中访问样式资源的语法格式如下。

【语法】

```
R.style.样式名
```

【示例】 Java 代码中访问样式资源

```java
setTheme(R.style.ThemeNew);
setTheme(R.style.myStyle);
```

在 XML 文件中访问样式资源的语法格式如下。

【语法】

```
@style/样式名
```

其中:@是前置符号;style 是样式标记名称;样式名则是<style>标签的 name 属性值,且需要使用斜杠(/)与前面的 style 标记进行间隔。

【示例】 XML 中访问样式资源

```
android:app_name = "@style/ThemeNew"
android:display = "@style/myStyle"
```

打开 res 目录下的 values 文件夹,双击打开 styles.xml 文件进行编辑,代码如下所示。

【代码 2-16】 styles.xml

```xml
<resources>
    <style name = "AppTheme" parent = "Theme.AppCompat.Light.DarkActionBar">
        <item name = "colorPrimary">@color/colorPrimary</item>
        <item name = "colorPrimaryDark">@color/colorPrimaryDark</item>
        <item name = "colorAccent">@color/colorAccent</item>
    </style>
    <style name = "blue_textview">
        <item name = "android:layout_width">wrap_content</item>
        <item name = "android:layout_height">wrap_content</item>
        <item name = "android:textSize">25sp</item>
        <item name = "android:textColor">#0000FF</item>
        <item name = "android:textStyle">bold</item>
    </style>
    <style name = "red_textview">
        <item name = "android:layout_width">wrap_content</item>
        <item name = "android:layout_height">wrap_content</item>
        <item name = "android:textSize">40sp</item>
        <item name = "android:textColor">#FF0000</item>
        <item name = "android:textStyle">italic</item>
    </style>
</resources>
```

上述代码中,colorPrimary 用于设定 ActionBar 的颜色;colorPrimaryDark 用于设定状态栏的颜色;colorAccent 用于设定 EditText、RadioButton 和 CheckBox 等控件被选中时的颜色,如图 2-16 所示。

图 2-16　通过 styles.xml 文件设置 Activity 样式

下述代码演示如何在 XML 文件中访问样式。创建一个新的 XML 布局文件，代码如下所示。

【代码 2-17】　style_layout.xml

```xml
<?xml version = "1.0" encoding = "utf-8"?>
<LinearLayout xmlns:android = http://schemas.android.com/apk/res/android
    android:layout_width = "match_parent"
    android:layout_height = "match_parent"
    android:orientation = "vertical">
    <TextView
    android:layout_width = "wrap_content"
    android:layout_height = "wrap_content"
    android:textAppearance = "?android:attr/textAppearanceLarge"
    android:text = "XML 文件访问 styles 资源(加粗蓝色)"
    android:id = "@ + id/tv7"
    style = "@style/blue_textview"/>
    <TextView
    android:layout_width = "wrap_content"
    android:layout_height = "wrap_content"
    android:textAppearance = "?android:attr/textAppearanceLarge"
    android:text = "Java 代码访问 styles 资源(斜体红色)"
    android:id = "@ + id/tv8" />
</LinearLayout>
```

上述代码中 @style/blue_textview 为 XML 文件读取 styles.xml 文件中名为 blue_textview 的文字样式，以该样式作为 TextView 中的字体样式显示出来。

下述代码演示如何在 Java 代码中访问样式。创建一个对应的 Activity 类,代码如下所示。

【代码 2-18】 Style_ActivityDemo.java

```java
public class Style_ActivityDemo extends AppCompatActivity{
    TextView tv8;
    @Override
    public void onCreate(Bundle savedInstanceState) {
        super.onCreate(savedInstanceState);
        setContentView(R.layout.style_layout);
        tv8 = (TextView) findViewById(R.id.tv8);
        tv8.setTextAppearance(this,R.style.red_textview);   }
}
```

在上述代码中,R.style.red_textview 为 Java 代码读取 styles.xml 文件中名为 red_textview 的文字样式,以该样式作为 TextView 中的字体样式显示出来。运行结果如图 2-17 所示。

图 2-17　XML 文件和 Java 代码分别访问 styles.xml 资源

> setTextAppearance()方法用于将 Style_ActivityDemo 中所显示的文本设置为样式资源中的 red_textview 样式,更多详细内容参见第 3 章。

2.2.7　drawable 图像资源目录

Android 应用程序中所使用的小图标、图像或背景图像都要放在资源目录 res/drawable 下。目前 Android 所支持的图像格式有.png、.jpg 和.gif 等格式,只要将所使用的图像放到 res/drawable 目录下,就可以在 Java 源代码或 XML 资源文件中进行引用。

Java 代码中访问图像资源的语法格式如下。

【语法】

```
R.drawable.图像文件名
```

【示例】 Java 代码中访问图像资源

```
getDrawable(R.drawable.icon);
setBackgroundDrawable(getResources().getDrawable(R.drawable.background));
```

在 XML 中访问图像资源的语法如下所示。

【语法】

```
@drawable/图像文件名
```

其中：@是前置符号；drawable 是图像标记名称；图像文件名不带后缀，且需要使用斜杠(/)与前面的 drawable 标记进行间隔。

【示例】 XML 文件中访问图像资源

```
android:icon = "@drawable/app_icon"
android:background = "@drawable/background"
```

在 res 目录中找到 drawable 文件夹，将两张图片复制粘贴到该文件夹中，如图 2-18 所示。

图 2-18　把图片资源放进 drawable 目录下

下述代码演示如何在 XML 文件中访问图片资源。创建一个新的 XML 布局文件，代码如下所示。

【代码 2-19】 drawable_layout.xml

```xml
<?xml version = "1.0" encoding = "utf-8"?>
<LinearLayout xmlns:android = "http://schemas.android.com/apk/res/android"
    android:layout_width = "match_parent"
    android:layout_height = "match_parent"
    android:orientation = "vertical">
    <TextView
        android:layout_width = "match_parent"
        android:layout_height = "match_parent"
        android:layout_weight = "1"
        android:background = "@drawable/bgimg">
    </TextView>
    <TextView
        android:layout_width = "match_parent"
        android:layout_height = "match_parent"
        android:layout_weight = "1"
        android:id = "@+id/tv10">
    </TextView>
</LinearLayout>
```

上述代码中@drawable/bgimg 为 XML 文件读取 drawable 文件夹中名为 bgimg 的图片，并将该图片作为 TextView 的背景显示出来。

下述代码演示如何在 Java 代码中访问图片资源。创建一个对应的 Activity 类，代码如下所示。

【代码 2-20】 Drawable_ActivityDemo.java

```java
public class Drawable_ActivityDemo extends AppCompatActivity {
    TextView tv10;
    @Override
    public void onCreate(Bundle savedInstanceState) {
        super.onCreate(savedInstanceState);
        setContentView(R.layout.drawable_layout);
        tv10 = (FrameLayout) findViewById(R.id.tv10);tv10.setBackgroundDrawable
            (getResources().getDrawable(R.drawable.bgimg1));}
}
```

上述代码中 getResources().getDrawable(R.drawable.bgimg1)为 Java 代码读取 drawable 文件夹中名为 bgimg1 的图片,并将该图片作为 TextView 的背景显示出来。运行结果如图 2-19 所示。

图 2-19 XML 文件和 Java 代码分别访问 strings.xml 资源

有关动画、Nine-Patch 可延伸图像(*.9.png)、菜单文件、XML 文件以及原始文件的访问请查阅相关资料,界面布局资源文件将在下一章详细介绍。

2.3 AndroidManifest.xml 清单文件

AndroidManifest.xml 清单文件是整个 Android 应用程序的全局描述配置文件,也是每一个 Android 应用程序必须有的且放在根目录下的文件。AndroidManifest.xml 清单文件对该应用的名称、所使用的图标以及所包含的组件等信息进行描述和说明。

AndroidManifest.xml文件通常包含以下几项信息。

（1）声明应用程序的包名，包名是用来识别应用程序的唯一标志。

（2）描述应用程序组件，包括组成应用程序的Activity、Service、Broadcast Receiver和Content Provider等，以及每个组件的实现类和其细节属性。

（3）确定宿主应用组件进程。

（4）声明应用程序拥有的权限，是其可以使用API保护的内容与其他应用程序所需的权限，同时声明了与其他应用程序组件交互所需的权限。

（5）定义应用程序所支持API的最低等级。

（6）列举应用程序必须链接的库。

伴随着应用程序的开发过程，程序开发者可能需要随时修改AndroidManifest.xml清单文件的内容。Android SDK文档对AndroidManifest.xml清单文件的结构、元素及元素的属性进行详细说明。而在使用这些元素及元素的属性前，需要先了解一下元素在命名和结构等方面的规则，具体如下。

- 元素：在所有的元素中只有<manifest>和<application>是必需的，且只能出现一次。如果一个元素包含其他子元素时，则必须通过子元素的属性进行赋值；处于同一层次的元素，元素之间没有先后顺序。
- 属性：元素的属性大部分是可选的，只有少数属性是必须设置的。可选的属性，即使不存在，也有默认的数值项说明。除了根元素<manifest>，其他所有元素属性的名字都是以"android:"作为前缀。
- 定义类名：所有的元素名都对应其在SDK中的类名；当开发者自己定义类名时，必须包含类的包名，当类与application处于同一个包中时，包名可以简写为"."。
- 多数值项：当某个元素有超过一个数值时，必须通过重复的方式来说明该元素的某个属性具有多个数值项，且不能将多个数值项一次性说明在一个属性中。
- 资源项说明：当需要引用某个资源时，可以采用@[package:]type:name格式进行引用，例如，<activity android:icon="@drawable/icon" ...>。
- 字符串值：类似于其他语言，如果字符中包含有字符\，则必须使用转义字符\\。

AndroidManifest.xml清单文件的示例代码如下所示。

【示例】 **AndroidManifest.xml**

```xml
<?xml version="1.0" encoding="utf-8"?>
<manifest xmlns:android="http://schemas.android.com/apk/res/android"
    package="com.qst.chapter02">
    <application
        android:allowBackup="true"
        android:icon="@mipmap/ic_launcher"
        android:label="@string/app_name"
        android:supportsRtl="true"
        android:theme="@style/AppTheme">
        <activity android:name=".MainActivity">
            <intent-filter>
                <action android:name="android.intent.action.MAIN" />
```

```xml
            <category android:name = "android.intent.category.LAUNCHER" />
        </intent-filter>
      </activity>
   </application>
</manifest>
```

上述 AndroidManifest.xml 清单文件中,除了头部的 XML 信息说明外,各节点的说明如下。

(1) <manifest>节点是根节点,其属性包括 schemas URL 地址、包名(com.qst.chapter02),以及程序的版本说明。

(2) <application>节点是<manifest>的子节点,一个 AndroidManifest.xml 中必须含有一个<application>标签,该标签声明了应用程序的组件及其属性。<application>标签的属性中包括程序图标和程序名称,其中@表示引用资源,例如@drawable/icon 表示引用 drawable 资源中的 icon,可以在项目工程的 res/drawable 中找到。

(3) <activity>节点是<application>的子节点,其属性包括 activity 的名称和标签名,应用程序中用到的每一个 Activity 都需要在此处声明为一个<activity>元素。

(4) <intent-filter>节点是<activity>子节点,用于声明 Activity 的 Intent 过滤规则,例如上述文件中指定了 name = "android.intent.action.MAIN" 的<action>和 name = "android.intent.category.LAUNCHER" 的<category>,说明当前程序启动时使用该 Activity 作为程序入口。

【示例】 应用程序权限申请

```xml
<?xml version = "1.0" encoding = "utf-8"?>
<manifest xmlns:android = "http://schemas.android.com/apk/res/android"
    package = "com.qst.chapter02">
    <application
        android:allowBackup = "true"
        android:icon = "@mipmap/ic_launcher"
        android:label = "@string/app_name"
        android:supportsRtl = "true"
        android:theme = "@style/AppTheme">
        <activity android:name = ".MainActivity">
            <intent-filter>
                <action android:name = "android.intent.action.MAIN" />
                <category android:name = "android.intent.category.LAUNCHER" />
            </intent-filter>
        </activity>
    </application>
    <uses-permission android:name = "**android.permission.SEND_SMS**"></uses-permission>
</manifest>
```

上述代码中,加粗黑体部分用于说明该软件需要发送短信的功能权限。

Android 定义了百余种 permission 权限,可供开发人员使用,详见 Android 开发官网 http://developers.androidcn.com/reference/android/Manifest.permission.html。

除了使用 Android 预定义的权限外,Android 系统还允许应用程序声明自定义的权限。如果一个应用程序声明了自定义的权限,当其他应用程序使用所声明的这个权限组件时,则

第2章 Activity和Application

必须使用< uses-permission >设置此权限。自定义权限使用< permission >元素声明,其语法格式如下。

【语法】

```
< permission
    android:label = "自定义权限"
    android:description = "@string/test"
    android:name = "com.example.project.TEST"
    android:protectionLevel = "normal"
    android:icon = "@drawable/ic_launcher">
</permission >
```

android:label 是权限标题,用于在安装应用程序时向用户提示。

android:description 是权限的详细描述,description 属性只能通过资源声明,而不能直接赋予 string 值,例如此处使用@string/test。

android:name 是权限名称,当其他应用程序引用该权限时需要使用此名称。

android:protectionLevel 是权限级别,拥有 normal、dangerous、signature 和 signatureOrSystem 四种级别。

Android 的四种不同权限级别的区分如下。

normal:低风险权限,在安装应用程序时系统会自动授予权限给应用程序,而不会向用户提示。

dangerous:高风险权限,系统不会自动授权,安装应用程序时会给用户提示信息,用户可根据提示信息确定是否安装此应用程序。

signature:签名权限,在其他应用程序引用声明该权限的组件时,系统会检查两个应用程序的签名是否一致,如果不一致则不允许调用。

signatureOrSystem:签名或系统权限,此权限主要针对 Android 系统的预装应用程序,要求引用该权限的应用程序具有和系统同样的签名。此权限主要用于设备生产商提供的应用程序,因为普通应用程序通常无法获取系统签名。

2.4 Android 应用程序生命周期

所谓应用程序的生命周期,是指应用程序进程从创建到消亡的整个过程。在 Android 中,多数情况下每个程序都是在各自独立的 Linux 进程中运行的。当一个程序或其某些部分被请求时,进程就"出生";当该程序没有必要继续运行且系统需要回收此进程所专用的内存时,该进程就"死亡"。因此,Android 程序的生命周期是由系统控制而非程序自身直接控制,这与桌面应用程序有一定的区别,桌面应用程序的进程也是在其他进程或用户请求时被创建,但经常在程序结束时执行一个特定的动作(如从 main 方法 return)而导致进程结束。

简而言之,Android 应用程序的生命周期是指在 Android 系统中进程从启动到终止的所有阶段,即 Android 程序启动到停止的全过程,程序的生命周期是由 Android 系统进行调度和控制的。但由于手机的内存是有限的,随着打开的应用程序数量的增多,可能造成应用程序响应时间过长或者系统假死的情况，因此在系统内存不足的情况下，Android 系统便会舍车保帅，选择性地来终止一些重要性较低的应用程序，以便回收内存供更重要的应用程序使用。

Android 根据应用程序的组件及组件当前运行状态将所有的进程按重要性程度从高到低划分为 5 个优先级：前台进程、可见进程、服务进程、后台进程和空进程。图 2-20 展示了 Android 系统进程从高到低的优先级。

图 2-20　Android 系统进程优先级

下面按照进程的优先级由高到低的顺序介绍 Android 系统中的进程。

1．前台进程

前台进程是指显示在屏幕最前端并与用户正在交互的进程，是 Android 系统中最重要的进程。前台进程包括以下 4 种情况：进程中的 Activity 正在与用户进行交互；进程服务被 Activity 调用，而且该 Activity 正在与用户进行交互；进程服务正在执行生命周期中的回调方法，如 onCreate()、onStart() 或 onResume() 方法；进程的 BroadcastReceiver 正在执行 onReceive() 方法。

Android 系统在多个前台进程同时运行时，可能会出现资源不足的情况，此时可清除部分前台进程，以保证主要的用户界面能够及时响应。

2．可见进程

可见进程是指部分程序界面能够被用户看见，却不在前台与用户交互，不能响应界面事件（其 onPause() 方法已被调用）的进程。如果一个进程包含服务，且该服务正在被用户可见的 Activity 调用，则此进程同样被视为可见进程。

Android 系统一般存在少量的可见进程，只有在特殊的情况下，Android 系统才会为保证前台进程的资源而清除可见进程。

3．服务进程

服务进程是指由 startService() 方法启动服务的进程。服务进程具有以下特性：没有用户界面；在后台长期运行。例如，后台 MP3 播放器或后台上传下载数据的网络服务，都是服务进程。

除非 Android 系统不能保证前台进程或可见进程所必要的资源，否则不会强行清除服务进程。

4．后台进程

后台进程是指不包含任何已经启动的服务，且没有任何用户可见的 Activity 的进程。后台进程不直接影响用户的体验。Android 系统中一般存在数量较多的后台进程，因此这些进程会被保存在一个列表中，以保证在系统资源紧张时，系统会优先清除用户较长时间没有用到的后台进程。

5. 空进程

空进程是指不包含任何活跃组件的进程。通常保留这些空进程，是为了将其作为一个缓存，在其所属的应用组件下一次需要时，以缩短启动的时间。

在系统资源紧张时，Android 系统首先会清除空进程，但为了提高 Android 系统应用程序的启动速度，Android 系统会将空进程保存在系统内存中，当用户重新启动该程序时，空进程会被重新使用。

2.5 Application 类

android.app.Application 类代表当前运行的应用程序。应用程序启动时，系统会自动创建对应 Application 类的实例，并一直伴随应用程序的生命周期，而且始终维持一个实例。对于同一个应用程序，由于系统保证只会存在一个 Application 实例，即在所有组件中获取的是同一个 Application 对象，因此 Application 特别适合保存应用程序中多个组件都需要访问的对象。

通过扩展 Application 类，可以完成以下 3 项工作：①对 android 运行时广播的应用程序级事件（如低内存）做出响应；②在应用程序组件之间传递对象；③管理和维护多个应用程序组件所使用的资源。

2.5.1 Application 生命周期事件

Application 类为应用程序的创建和终止、低可用内存和配置的改变提供了事件处理程序，通过重写以下方法，可以实现上述几种情况的应用程序行为。

（1）**onCreate**()：在创建应用程序时调用该方法。通过重写 onCreate()方法来实例化应用程序单态，也可以创建和实例化任何应用程序状态变量或共享资源。

（2）**onLowMemory**()：一般只会在后台进程已经终止但前台应用程序仍然缺少内存资源时会被调用，通过重写 onLowMemory()方法来清空缓存或者释放不必要的资源。

（3）**onTrimMemory**()：作为 onLowMemory 的一个特定应用程序的替代选择，在 Android 4.0（API level 13）中引入。当运行时决定当前应用程序是否应该尝试减少其内存开销（通常在其进入后台时）。该方法包含一个 level 参数，用于提供请求的上下文。

（4）**onConfigurationChanged**()：与 Activity 不同，当配置发生改变时，应用程序对象不会被终止和重启；如果应用程序使用的值依赖于特定的配置，则重写该方法来重新加载这个值，或者在应用程序级别处理配置的改变。

注意

在重写这些方法时必须调用父类的事件处理程序。

2.5.2 实现 Application

如果需要实现自定义的 Application，则需要继承 Application 类，编写示例项目步骤如下所述。

1. 创建一个类继承 Application 类

在 Android Studio 中新建一个类，如图 2-21 所示。单击 OK 按钮后，创建了一个内部代码为空的 Application，如图 2-22 所示。

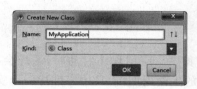

图 2-21 创建 Application 子类窗口

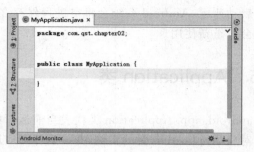

图 2-22 生成 Application 子类的初始窗口

在代码窗口中，添加全局中所使用的成员变量 name，以及对应的 getter 和 setter 方法。在 onCreate() 方法中做一些初始化工作，完善后的代码如下所示。

【代码 2-21】 MyApplication.java

```
public class MyApplication extends Application {
    private String name;
    @Override
    public void onCreate() {
        super.onCreate();
        setName("北京");    //初始化全局变量
    }
    public String getName() {
        return name;}
    public void setName(String name) {
        this.name = name;}
}
```

2. 在 Activity 中使用 Application 类

在应用程序的 MainActivity 中添加代码，并使用上一步所创建的 MyApplication 类，具体代码如下所示。

【代码 2-22】 MainActivity.java

```
public class MainActivity extends Activity {
    private MyApplication app;
    @Override
    public void onCreate(Bundle savedInstanceState) {
        super.onCreate(savedInstanceState);
        setContentView(R.layout.activity_main);
        app = (MyApplication) getApplication();              //获得 MyApplication 对象
        Log.d("应用程序 Name 原来的值为：", app.getName()); //获取进程中 Name 全局变量的值
```

```
        app.setName("青岛");                                        //修改 Name 的值
        Log.d("应用程序 Name 修改后的值为:", app.getName());}   //Name 的值改变
}
```

AndroidManifest.xml 清单文件的代码如下所示。

【代码 2-23】 AndroidManifest.xml

```
<?xml version="1.0" encoding="utf-8"?>
<manifest xmlns:android="http://schemas.android.com/apk/res/android"
    package="com.qst.myapplication">
    <application
        android:allowBackup="true"
        android:icon="@mipmap/ic_launcher"
        android:label="@string/app_name"
        android:supportsRtl="true"
        android:theme="@style/AppTheme"
        android:name="com.qst.chapter02.MyApplication">
        <activity android:name="com.qst.chaper02.MainActivity">
            <intent-filter>
                <action android:name="android.intent.action.MAIN" />
                <category android:name="android.intent.category.LAUNCHER" />
            </intent-filter>
        </activity>
    </application>
</manifest>
```

3. 运行并查看结果

启动 Application 时,系统会创建一个进程 ID,该应用的所有 Activity 都在此进程上运行。因此,在创建 Application 时对全局变量进行初始化,且同一个应用中的所有 Activity 都可以访问某一全局变量,即在同一个应用中,当某个 Activity 对某个全局变量进行修改时,则在其他 Activity 中获取该全局变量的值也会发生改变。

在 LogCat 视图中查看代码中 Log.e() 方法所输出的信息,如图 2-23 所示。

图 2-23　数据传递显示窗口

多窗口或其他组件(例如 Service 等)可以使用同样的方法完成数据传递。至于 Application 类的其他功能请参考 Android 官方文档。

2.6　样式和主题

在实际应用中,可以用样式和主题来统一格式化各种屏幕和 UI 元素。

样式是一个包含一种或者多种格式的集合,一般作为一个单位用于 XML 布局的某个

元素中。例如,通过样式来定义文本的字号大小和颜色,然后将其应用于 View 元素的某个特定的实例上。

主题是一个包含一种或者多种格式的集合,通常将其作为一个单位应用到 Activity 中。例如,在定义主题时,为 Window Frame 和 Panel 的前景和背景定义一组颜色,并定义菜单中的文字大小和颜色属性,然后将所定义的主题应用到程序中所有的 Activity 中。

样式和主题都可视为资源,Android 系统中提供了一些默认的样式和主题资源,用户可以根据需求来自定义所需的主题和样式资源。

创建自定义的样式和主题的步骤如下。

(1) 在 res/values 目录下新建一个名为 style.xml 的文件,并增加< resources >元素作为根元素。

(2) 对于样式或主题,需要为< style >元素增加一个全局唯一的 name 属性,也可以为其增加一个 parent 属性。在编码过程中,可以通过 name 属性来应用该样式,而 parent 属性用于标识当前样式继承于哪个样式。

(3) 在< style >元素内部声明一个或者多个< item >元素,每一个< item >元素用于定义一个 name 属性,并在元素的内部进行赋值。

(4) 引用在其他 XML 中已经定义的资源。

下边是一个声明样式的实例,代码如下所示。

【示例】 样式的声明

```
<?xml version = "1.0" encoding = "utf - 8"?>
< resources >
    < style name = "SpecialText" parent = "@style/Text">
        < item name = "android:textSize">18sp</item >
        < item name = "android:textColor">#008</item >
    </style>
</resources >
```

上述代码中,通过< item >元素来为< style >样式定义一组格式的值。< item >中的 name 属性值可以是一个字符串,而< item >标签内容可以为具体的值或是其他资源的引用。

< style >元素的 parent 属性用于说明当前样式可以从指定的资源中继承这些样式。除此之外,元素还能从任何包含该样式的资源中继承样式。在开发过程中,程序中的资源能够直接或者间接地继承 Android 的标准样式资源。对于程序员而言,只需定义特定的样式即可。

下面演示了如何引用 XML 布局文件中所定义的样式,代码如下所示。

【示例】 样式的引用

```
< EditText id = "@ + id/text1"
    style = "@style/SpecialText"
    android:layout_width = "fill_parent"
    android:layout_height = "wrap_content"
    android:text = "Hello, World!" />
```

上述代码中,EditText 组件所呈现的样式即为前面在 XML 文件中所定义的样式。与样式相似,主题也可以使用< style >元素进行声明。引用方式也非常相似,区别在于主题一般需要在 AndroidManifest.xml 文件中的< application >和< activity >元素中进行引用,即在整个程序或者某个 Activity 使用某个主题,但有时也会在 View 中使用主题。下述代码用于声明一个主题。

【示例】 主题的声明

```xml
<?xml version = "1.0" encoding = "utf - 8"?>
< resources >
    < style name = "CustomTheme">
        < item name = "android:windowNoTitle"> true </item >
        < item name = "windowFrame">@drawable/screen_frame </item >
        < item name = "windowBackground">@drawable/screen_background_white </item >
        < item name = "panelForegroundColor"> #FF000000 </item >
        < item name = "panelBackgroundColor"> #FFFFFFFF </item >
        < item name = "panelTextColor">?panelForegroundColor </item >
        < item name = "panelTextSize"> 14 </item >
        < item name = "menuItemTextColor">?panelTextColor </item >
        < item name = "menuItemTextSize">?panelTextSize </item >
    </style >
</resources >
```

上述代码中,使用了@符号和"?"符号来引用资源。其中,@符号所引用的资源是已经定义的资源(Android 框架内置的资源或当前项目中已经定义的资源);而"?"符号所引用的资源是在当前主题中定义的资源。在 AndroidManifest.xml 或 Activity 中通过< item >元素的 name 属性来引用该标签所定义的资源。

1. 在 AndroidManifest.xml 中设置主题

当需要所有 Activity 都使用同一主题时,可以在 AndroidManifest.xml 文件中通过< application >标签的 android:theme 属性来指定所需的主题,代码如下所示。

```xml
< application android:theme = "@style/CustomTheme">
```

如果只想让应用程序中的某个 Activity 使用某一主题,则可以在指定的< activity >标签中引用所需的主题,代码如下所示。

```xml
< activity android:theme = "@style/CustomTheme ">
```

Android 中提供了多种内置的资源,开发人员可以根据需要进行切换。例如使用对话框主题来让应用中的 Activity 看起来像一个对话框,代码如下所示。

```xml
< activity android:theme = "@android:style/Theme.Dialog">
```

虽然系统提供的主题相对比较美观,但在实际应用中仍需要进行调整时,可以将该主题当作父主题,然后进一步扩展用户自己的主题。例如修改 Theme.Dialog 主题,可以通过继承 Theme.Dialog 来生成一个新的主题,代码如下所示。

```xml
< style name = "CustomDialogTheme" parent = "@android:style/Theme.Dialog">
```

上述代码中,用户自定义主题继承了 Theme.Dialog 后,再按照特定的要求来自定义主题。开发人员可以修改从 Theme.Dialog 中继承的任意 item 元素,在 AndroidManifest.xml 文件中引用时使用 CustomDialogTheme 而不是 Theme.Dialog 主题。

2. 在程序当中设置主题

在实际开发中,除了可以在 AndroidManifest.xml 中设置主题外,还可以在程序中设置主题。在 Activity 中通过 setTheme()方法来动态地加载一个主题。需要注意的是,应该在初始化任何 View 之前设置主题,例如在调用 setContentView(View) 和 inflate(int, ViewGroup)方法前设置主题。这样能够保证将当前主题应用到该程序的所有 UI 界面中,示例代码如下所示。

【示例】 在 Activity 程序中设置主题

```
protected void onCreate(Bundle savedInstanceState) {
    super.onCreate(savedInstanceState);
    ...setTheme(android.R.style.Theme_Light);
    setContentView(R.layout.linear_layout_3);
}
```

> **注意**
>
> 在 Activity 中加载主题时,主题中不能包括任何系统启动该 Activity 所使用的动画,这些动画将在程序启动前显示。

2.7 贯穿任务实现

本章任务是完成"GIFT-EMS 礼记"需求分析,创建项目并完成基本架构设计。

2.7.1 实现【任务 2-1】

本书的实践贯穿案例是"GIFT-EMS 礼记"项目,该项目是对《Java EE 轻量级框架应用与开发——S2SH》一书中贯穿任务的延续。"GIFT-EMS 礼记"系统是一个针对新潮礼物的 B2C 商城系统,以推荐礼物为核心,收集时下潮流的礼物以及送礼物的方法,为用户呈现热门的礼物攻略,通过"送给 TA"等功能,旨在帮助用户给恋人、家人、朋友和同事制造生日、节日、纪念日惊喜。

在《Java EE 轻量级框架应用与开发——S2SH》中已完成了"GIFT-EMS 礼记"项目的 Web 端,本书贯穿任务将完成 Android 移动端的开发。Android 端的用户购物系统主要提供浏览礼品、查看攻略、购买礼品、生成订单和送礼等功能,功能模块如表 2-6 所示。

表 2-6 Android 端功能列表

模 块	功 能 描 述
礼品浏览	用户通过礼品列表进入礼品详情界面,同一个礼品有多个款式,不同的款式价格不同
个人中心	包括修改密码、我的订单、安全退出等功能点

续表

模 块	功 能 描 述
支付	用户选购完礼品后,进行购买,生成订单然后进行支付
我的订单	可以查看全部订单、已付款订单、已完成订单、待付款订单等,并且可以对订单进行取消、确认收货等操作
攻略	用户可以浏览并查看攻略详情

注意

本项目来源于企业中的真实产品,模块功能较多,并存在多处本书中没有涉及的技术,限于篇幅本书只展示核心功能的实现过程。

2.7.2 实现【任务 2-2】

本任务完成"GIFT-EMS 礼记"项目的创建,并编写实体类、Application 类等基础架构。

1. 创建项目

新建 Android 项目 giftems,包名为 com.qst.giftems。

2. 实体类

针对每个业务实体,需要声明一个实体类,Android 端与服务器进行数据交互时,将以实体类形式封装数据,并以 json 格式传输。根据系统的需求,实体类如表 2-7 所示。

表 2-7 POJO 类列表

类 名	功 能 描 述
User	用户,本系统中所有业务类都是围绕着 User 进行设计的
Gift	礼品,用户可以选择不同的礼品,加入购物车进行购买;也可以购买礼品后,送礼给某一个用户
GiftType	礼品类型,礼品类型有多种,例如鲜花、红酒、宠物等
GiftStyle	礼品款式,每一个礼品都有多个款式,不同的款式价格可以不同,可自行设置
Order	订单,订单分为多个状态,例如交易成功、交易失败、待支付、已支付、待发货、已发货、已收货等状态
OrderItem	订单明细,一个订单对象对应多个订单明细对象,每个订单明细对象又关联着一个礼品对象
ShoppingBagItem	购物袋条目
Strategy	送礼攻略,用于向用户展示的有关各种送礼方式的类
UserAddress	用户的收件人,用于用户购买礼品时指定快递地址
IndexPage	Android APP 首页配置
IndexPageResource	Android APP 首页的图片等资源
ContactEntity	联系人,用于将通讯录中的联系人作为收礼人

系统中主要实体类之间的关系如图 2-24 所示。

在 com.qst.giftems.model 包下创建各个实体类,代码如下所示。

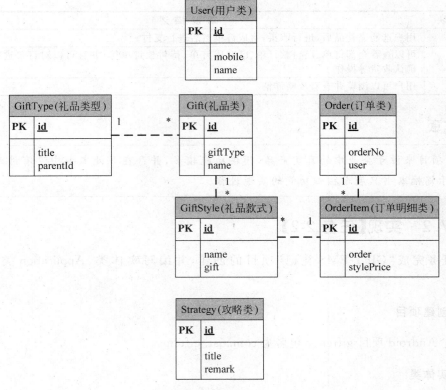

图 2-24 类关系图

【任务 2-2】 User.java

```
/** 用户 */
public class User {
    public int id;
    public String name;            /** 用户名 */
    public String mobile;          /** 手机 */
    public String mobileToken;     /** 登录令牌 */
    public String realName;        /** 真实姓名 */
    public String sex;             /** 性别 */
    public String birthday;        /** 生日 */
    public String work;            /** 工作 */
    public Integer bindStatus;     /** 手机绑定状态 */
}
```

【任务 2-2】 Gift.java

```
/** 礼品 */
public class Gift {
    public int id;
    public String name;            /** 礼品名称 */
    public String remark;          /** 礼品简介 */
    public int likes;              /** 礼品被喜欢的次数 */
    public int sales;              /** 累计销量 */
    public boolean collected;      /** 是否被当前用户收藏 */
```

```
    public ArrayList<GiftStyle> styles = new ArrayList<GiftStyle>();/** 款式列表 */
}
```

【任务 2-2】 GiftStyle.java

```
/** 礼品款式 */
public class GiftStyle {
    public int id;
    public String name;              /** 款式名称 */
    public double price;             /** 原价 */
    public double discount;          /** 折后价 */
    public String pic1;              /** 图片1 */
    public String pic2;              /** 图片2 */
    public String pic3;              /** 图片3 */
    public String pic4;              /** 图片4 */
    public String pic5;              /** 图片5 */
    public String remark;            /** 介绍 */
    public int orderNumber;          /** 顺序号,显示同一礼品的多个款式时排序用 */
    public Gift gift;                /** 对应礼品 */
}
```

【任务 2-2】 GiftType.java

```
/** 礼品类型 */
public class GiftType implements Comparable<GiftType> {
    public int id;
    public String name;              /** 名称 */
    public String pic;               /** 图片 */
    public int orderNum;             /** 顺序号 */
    @Override
    public int compareTo(GiftType another) {
        return orderNum < another.orderNum ? -1
                : (orderNum == another.orderNum ? 0 : 1);}
}
```

【任务 2-2】 Order.java

```
/** 用户订单 */
public class Order {
    /** 状态:"交易成功",前台:确认收货后,状态改变 */
    public static final int PAY_SUCCESS = 1;
    /**
     * 状态:"交易关闭",后台:退货后,交易关闭或前台订单超时后,交易关闭或前台申请退货,后
台操作"退货"按钮
     * 订单待支付时,用户也可以选择管理该订单进行关闭
     */
    public static final int PAY_CLOSED = 2;
    /** 状态:"买家待支付",前台:用户提交订单后,修改至"买家待支付"状态 */
    public static final int PAY_DAIZHIFU = 3;
    /** 状态:"买家已支付",前台:支付成功后,显示"买家已支付",后台:根据"买家已支付"的状
态,修改"卖家代发货"状态 */
    public static final int PAY_YIZHIFU = 4;
    /** 状态:"等待代付",前台:用户操作"代付" */
```

```java
    public static final int PAY_DENGDAIDAIFU = 5;
    /** 状态:"代付完成",前台:代付成功后,显示"代付完成",后台：根据买家支付情况,修改"卖家代发货"状态 */
    public static final int PAY_DAIFUWANCHENG = 6;
    /** 状态:"卖家待发货",后台:卖家发货后,单击"发货"按钮后状态变为"已发货" */
    public static final int PAY_DAIFAHUO = 7;
    /** 状态:"卖家已发货",前台:用户单击"确认收货"按钮后,修改 PAY_SUCCESS 状态 */
    public static final int PAY_YIFAHUO = 8;
    /** 状态:"退货中",前台:单击"退款"按钮,并修改订单状态为 PAY_TUIHUO 状态,后台:受理退货,查询待退款状态的订单,修改状态为"交易关闭" */
    public static final int PAY_TUIHUO = 9;
    public Integer id;                      /** 订单 ID */
    /** 订单编号(yyyyMMddHHmmssSSS + 999 内随机数)随机数默认为 3 位,不够的话前面补 0,这样确定为 20 位 */
    public String orderNo;
    public String deliveryNo;               /** 快递单号:后台操作发货时填写的快递单号 */
    public String deliveryName;             /** 快递公司名称 */
    public String transNo;                  /** 支付交易号 */
    public Integer userId;                  /** 付款人 ID(user) */
    public String userName;                 /** 付款人用户名(user) */
    public String receiverName;             /** 收礼人姓名 */
    public String receiverPhone;            /** 收礼人电话 */
    public String receiverAddress;          /** 收礼人地址 */
    public String arriveDate;               /** 指定送达日期 */
    public Float price;                     /** 订单金额(人民币) */
    public String payTime;                  /** 付款时间 */
    public String createTime;               /** 订单生成时间 */
    public String expiredTime;  /** 订单过期时间(订单生成之后 7 天为缓冲付款时间,过期不
                                    可再付款) */
    public Integer status;                  /** 订单状态 */
    public String orderInfo;                /** 订单状态信息(对应订单状态,为何未付款、付款失败
                                                的原因等) */
    public String attachInfo;               /** 订单附加信息 */
    /**
     * 订单删除 0:正常,1:已删除(用户回收站),2:彻底删除(用户看不到),3:永久隐藏(除非数据库恢复)
     * 对于订单的删除是用户行为,后台对用户彻底删除的订单进行清理(建议不删除)
     */
    public Integer deleted;
    public Integer couponItemId;            /** 优惠券 ID */
    public List<OrderItem> items = new ArrayList<OrderItem>();      /** 订单 Item 集合 */
    public String specifyTime;              /** 指定日期送达时间 */
    public String specifyFlag;              /** 指定日期送达标志位 */
    public String leaveWords;               /** 文字留言 */
    public String leaveSound;               /** 留言录音 */
    private String statusName;              /** 订单状态名称 */
    public String getStatusName() {
        if (statusName == null || statusName.length() == 0) {
            switch (status) {
                case PAY_SUCCESS:
                    statusName = "交易成功";
                    break;
                case PAY_CLOSED:
```

```
                statusName = "交易关闭";
                break;
            case PAY_DAIZHIFU:
                statusName = "等待支付";
                break;
            case PAY_YIZHIFU:
                statusName = "已支付";
                break;
            case PAY_DENGDAIDAIFU:
                statusName = "等待代付";
                break;
            case PAY_DAIFUWANCHENG:
                statusName = "代付完成";
                break;
            case PAY_DAIFAHUO:
                statusName = "等待发货";
                break;
            case PAY_YIFAHUO:
                statusName = "已发货";
                break;
            case PAY_TUIHUO:
                statusName = "退货中";
                break;
        } }
        return statusName; }
}
```

Order 类的 status 属性表示订单的状态, 即订单在从生成到交易完毕的过程中所处的步骤, Order 类中定义了 9 个常量来表示相应的状态。

【任务 2-2】 ContactEntity.java

```
/**通讯录*/
public class ContactEntity {
    private String name;              /**姓名*/
    private String phone;             /**电话*/
    …//省略 getter/setter 方法
}
```

【任务 2-2】 OrderItem.java

```
/**订单项*/
public class OrderItem {
    public Integer id;
    public Integer orderId;           /**所属订单 ID*/
    public Integer giftId;            /**礼品编号*/
    public String giftName;           /**礼品名称*/
    public Integer giftCount;         /**购买礼品数量*/
    public Integer styleId;           /**款式 ID*/
```

```java
    public String styleName;          /** 款式名称 */
    public float discount;            /** 款式折扣价格 */
    public float stylePrice;          /** 款式价格 */
    public String remark;             /** 备注 */
    public String stylePic1;          /** 对应礼品款式的图片 1 */
    public Order order;               /** 对应订单 */
}
```

【任务 2-2】 ShoppingBagItem.java

```java
/** 购物车条目 */
public class ShoppingBagItem {
    public int giftStyleId;           /** 礼品款式 ID */
    public int quantity;              /** 数量 */
    public GiftStyle giftStyle;       /** 礼品款式 */
}
```

【任务 2-2】 Strategy.java

```java
/** 礼品攻略 */
public class Strategy implements Comparable<Strategy> {
    public int id;
    public String title;              /** 攻略名称 */
    public String remark;             /** 说明 */
    public String pic1;               /** 图片 1 */
    public String pic2;               /** 图片 2 */
    public String pic3;               /** 图片 3 */
    public String pic4;               /** 图片 4 */
    public String pic5;               /** 图片 5 */
    public String content;            /** 内容 */
    public String createdTime;        /** 创建时间 */
    @Override
    public int compareTo(Strategy another) {
        return createdTime.compareTo(another.createdTime);}
}
```

【任务 2-2】 UserAddress.java

```java
/** 收件人 */
public class UserAddress {
    public int id;
    public int userId;                /** 对应用户 ID */
    public String name;               /** 姓名 */
    public String mobile;             /** 手机 */
    public int provinceId;            /** 省 ID */
    public int cityId;                /** 市 ID */
    public int areaId;                /** 区 ID */
    public String address;            /** 地址 */
    public String postcode;           /** 邮编 */
```

第2章 Activity和Application

```java
    public String info;              /** 信息 */
    public String provinceName;      /** 省名 */
    public String cityName;          /** 市名 */
    public String areaName;          /** 区名 */
}
```

【任务 2-2】 **UserCalendar.java**

```java
/** 用户日历 */
public class UserCalendar {
    public int id;
    public int userId;               /** 用户 ID */
    public String title;             /** 备注 */
    public String actionTime;        /** 日程日期 yyyy-MM-dd HH:mm:ss */
    public int remindTime1;          /** 提醒时间 1 提前的分钟数 */
    public int remindTime2;          /** 提醒时间 2 提前的分钟数 */
    public int remindTime3;          /** 提醒时间 3 提前的分钟数 */
    public int remindTime4;          /** 提醒时间 4 提前的分钟数 */
    public int remindTime5;          /** 提醒时间 5 提前的分钟数 */
}
```

【任务 2-2】 **IndexPageResource.java**

```java
/** 首页资源 */
public class IndexPageResource {
    public int id;
    public String picUrl;            /** 图片地址 */
    public int type;                 /** 配置类型 1:pc 2:android 3:ios */
    public int orderNum;             /** 每种类型图片的顺序 */
    public int status;               /** 0:未发布,1:已发布 */
    public int place;                /** 位置 1:礼物推荐上部,2.礼物推荐下部左侧,3.礼物
                                         推荐下部右侧,4:礼品中心,5:礼品攻略 */
}
```

【任务 2-2】 **IndexPage.java**

```java
/** 首页资源 */
public class IndexPage {
    static final String SERVER = "http://localhost:8080/giftems";   /** 服务器地址 */
    public String[] recommendTop;              /** 礼物推荐上部轮换内容 */
    public String recommendBottomLeft;         /** 礼物推荐下部左侧图片 */
    public String recommendBottomRight;        /** 礼物推荐下部右侧图片 */
    public String giftCommon;                  /** 普通礼品图片 */
    public String strategy;                    /** 送礼攻略图片 */
    public IndexPage(ArrayList<IndexPageResource> settings) {
        ArrayList<String> recommendTopList = new ArrayList<String>();
        for (IndexPageResource s : settings) {
            switch (s.place) {
                case 1:
                    recommendTopList.add(SERVER + s.picUrl);
                    break;
                case 2:
                    recommendBottomLeft = SERVER + s.picUrl;
```

```
                    break;
                case 3:
                    recommendBottomRight = SERVER + s.picUrl;
                    break;
                case 4:
                    giftCommon = SERVER + s.picUrl;
                    break;
                case 5:
                    strategy = SERVER + s.picUrl;
                    break;
            }
        }
        recommendTop = recommendTopList.toArray(new String[recommendTopList.size()]);}
}
```

3. Application

创建完成实体类后,需要在 com.qst.giftems 包下编写应用程序的 Application 类。在 Android 应用程序中,如果需要存储一些能够在整个应用程序中访问的数据,通常不会采用静态常量的方式,而应该保存在 Application 对象中。在本应用程序中,用户登录后需要存储登录成功的 User 对象,以便后续的多个业务使用,因此编写 App 类继承 Application,并添加 User 属性,代码如下所示。

【任务 2-2】 App.java

```
public class App extends Application {
    public User user;      /** 当前登录用户 */
}
```

App 类编写完成后,还需要在 AndroidManifest.xml 中进行配置,使用 App 类作为整个应用的 Application,代码如下。

【任务 2-2】 AndroidManifest.xml

```xml
...
<application
        android:name = "com.qst.giftems.App"
        android:allowBackup = "true"
        android:icon = "@drawable/logo1"
        android:label = "@string/app_name"
        android:theme = "@style/AppTheme" >
    <activity
            android:name = "com.qst.giftems.MainActivity"
            android:label = "@string/app_name"
            android:screenOrientation = "portrait" >
        <intent-filter>
            <action android:name = "android.intent.action.MAIN" />
            <category android:name = "android.intent.category.LAUNCHER" />
        </intent-filter>
    </activity>
    ...
</application>
...
```

第2章 Activity和Application

2.7.3 实现【任务2-3】

本任务编写项目中 Activity、按钮、文本输入框等控件所使用的 XML 布局文件,这些文件都位于项目 res/drawable 目录下。

1. Activity 背景

首先,项目中各 Activity 需要一个共用的背景,以便统一修改,代码如下所示。

【任务2-3】 background_body.xml

```xml
<?xml version = "1.0" encoding = "utf-8"?>
<layer-list xmlns:android = "http://schemas.android.com/apk/res/android">
    <item>
        <shape android:shape = "rectangle">
            <solid android:color = "@color/background" />
        </shape>
    </item>
</layer-list>
```

上述代码中,首先声明<layer-list>元素,其中包含一个<item>元素,在<item>使用<shape>元素定义了一个图形,其子元素<solid>使用颜色@color/background 进行填充。

2. 按钮背景色和文字颜色

项目中各按钮需要统一的样式,在单击按钮时需要切换背景色和文字颜色,使用户能够更加直观地看到效果,按钮的背景文件代码如下所示。

【任务2-3】 background_button.xml

```xml
<?xml version = "1.0" encoding = "utf-8"?>
<selector xmlns:android = "http://schemas.android.com/apk/res/android">
    <item android:state_pressed = "true">
        <shape>
            <solid android:color = "@color/selected"/>
            <corners
                android:bottomLeftRadius = "8dp"
                android:bottomRightRadius = "8dp"
                android:topLeftRadius = "8dp"
                android:topRightRadius = "8dp" />
            <stroke
                android:width = "1dp"
                android:color = "@color/title_bottom" />
        </shape>
    </item>
    <item>
        <shape>
            <solid android:color = "#FFFFFF"/>
            <corners
                android:bottomLeftRadius = "8dp"
                android:bottomRightRadius = "8dp"
                android:topLeftRadius = "8dp"
```

```
                android:topRightRadius = "8dp" />
            <stroke
                android:width = "1dp"
                android:color = "@color/title_bottom" />
        </shape>
    </item>
</selector>
```

在按钮背景文件中,首先定义了<selector>根元素;其中包含两个<item>元素:第一个<item>元素中的属性"android:state_pressed="true""代表按钮按下时的背景;第二个<item>元素未声明任何属性代表其他状态下的背景。在两个<item>元素中,都通过<solid>元素来定义背景色,<corners>元素来定义 4 个角的弧度,<stroke>元素来定义边框的线宽和颜色。

除此之外,还需要在 XML 文件中声明按钮上的文字在不同状态下的颜色,代码如下所示。

【任务 2-3】 textcolor_button.xml

```
<?xml version = "1.0" encoding = "utf-8"?>
<selector xmlns:android = "http://schemas.android.com/apk/res/android">
    <item android:state_pressed = "false" android:color = "#555555"></item>
    <item android:state_pressed = "true" android:color = "#FFFFFF"></item>
</selector>
```

上述代码中,<item>元素的 android:state_pressed 属性值为 true 或 false 代表两个不同的状态,通过 android:color 属性来设定不同状态下的文字颜色。

在布局文件或样式中,分别指定 Button 控件的 textColor 和 background 属性值为 @drawable/textcolor_button 和@drawable/background_button,从而使用前面定义的选择器,代码如下所示。

```
<Button
    android:textColor = "@drawable/textcolor_button"
    android:background = "@drawable/background_button"
    ... />
```

运行结果如图 2-25 所示,其中"确定"按钮是单击后的效果,"取消"按钮是未单击效果。

图 2-25 自定义背景色和文字颜色的按钮

3. 文本输入框背景

与按钮类似,项目中输入框也需要定义统一的样式,当在输入框获取焦点时需要切换背景。输入框的背景文件代码如下所示。

第2章 Activity和Application

【任务2-3】 background_edit.xml

```xml
<?xml version = "1.0" encoding = "utf-8"?>
<selector xmlns:android = "http://schemas.android.com/apk/res/android">
    <item android:state_focused = "true">
        <shape android:shape = "rectangle">
            <solid android:color = "#FFFFFF" />
            <corners
                android:bottomLeftRadius = "3dip"
                android:bottomRightRadius = "3dip"
                android:topLeftRadius = "3dip"
                android:topRightRadius = "3dip" />
            <stroke
                android:width = "1.5dp"
                android:color = "#0080FF" />
        </shape>
    </item>
    <item>
        <shape android:shape = "rectangle">
            <solid android:color = "#FFFFFF" />
            <corners
                android:bottomLeftRadius = "3dip"
                android:bottomRightRadius = "3dip"
                android:topLeftRadius = "3dip"
                android:topRightRadius = "3dip" />
            <stroke
                android:width = "1px"
                android:color = "@color/title_bottom" />
        </shape>
    </item>
</selector>
```

与按钮的背景文件相似,上述文件中分别定义了在获取焦点后和其他状态下的填充颜色、4个角的弧度以及边框的线宽和颜色。

在布局文件或样式中,指定EditText控件的background属性值为@drawable/background_edit,从而使用前面定义的编辑框背景,代码如下所示。

```xml
<EditText
    android:layout_width = "wrap_content"
    android:layout_height = "wrap_content"
    android:background = "@drawable/background_edit" />
```

运行结果如图2-26所示,其中第一个EditText是获取焦点后的效果,第二个EditText为失去焦点的效果。

图2-26 自定义背景和边框的输入框

2.7.4 实现【任务 2-4】

本任务完成项目中使用的统一样式文件，在 res/values 目录下添加 styles.xml 文件，代码如下所示。

【任务 2-4】 styles.xml

```xml
<?xml version = "1.0" encoding = "utf-8"?>
<resources xmlns:android = "http://schemas.android.com/apk/res/android">
    <style name = "AppBaseTheme"
           parent = "android:Theme.Holo.NoActionBar"></style>
    <style name = "AppTheme" parent = "AppBaseTheme"></style>
    <style name = "text1">
        <item name = "android:textColor">@color/gray</item>
        <item name = "android:gravity">center</item>
        <item name = "android:textStyle">bold</item>
        <item name = "android:textSize">14sp</item>
    </style>
    <style name = "text3">
        <item name = "android:textColor">@color/selected</item>
        <item name = "android:textStyle">bold</item>
        <item name = "android:gravity">center</item>
        <item name = "android:textSize">14sp</item>
    </style>
    <style name = "button">
        <item name = "android:gravity">center</item>
        <item name = "android:textStyle">bold</item>
        <item name = "android:textSize">14sp</item>
        <item name = "android:textColor">@drawable/textcolor_button</item>
        <item name = "android:background">@drawable/background_button</item>
    </style>
    <style name = "item">
        <item name = "android:textSize">14sp</item>
        <item name = "android:textColor">@drawable/textcolor_button</item>
        <item name = "android:background">@drawable/background_item</item>
    </style>
</resources>
```

上述样式文件中，AppBaseTheme 继承自 android:Theme.Holo.NoActionBar，然后 AppTheme 又继承 AppBaseTheme。另外，还自定义了 text1 和 text3 两个字体样式、按钮 button 样式和下拉菜单项 item 样式。

本章总结

小结

- Activity 是 Android 系统最重要的组件，是 Android 程序开发的入口点，深刻领会 Activity 编程的步骤对于 Android 开发非常重要。
- Activity 有运行、暂停、停止和销毁 4 种状态。

- 资源管理是 Android 编程的一大亮点,体现了 MVC 编程的优势,对于提高程序的可读性以及可靠性提供了有效的手段。
- Android 开发中常用的资源主要包括文本字符串(strings)、颜色(colors)、数组(arrays)、动画(anim)、布局(layout)、图像和图标(drawable)、音频视频(media)和其他应用程序使用的组件。
- AndroidManifest.xml 清单文件是整个 Android 应用程序的全局描述配置文件,也是每一个 Android 应用程序必须有的且放在根目录下的文件。
- Android 应用程序从高到低划分了 5 个优先级:前台进程、可见进程、服务进程、后台进程和空进程。
- 程序生存周期在认识 Android 应用的运行机理、从何处入手写编写代码等方面提供了一个路径。
- Application 类代表当前运行的应用程序,应用程序启动时,系统会自动创建对应 Application 类的实例,并一直伴随应用程序的生命周期,而且始终维持一个实例。

Q&A

1. 问题:简述资源的种类、访问方法以及不同的资源在程序编译后的呈现形式。

回答:资源可以分成两大类:一类是能够写入 R.java 类的资源,这类资源在程序编译后被一起编译到了要发行的应用程序中;另一类是原生资源,这类资源一般体积较大,不宜编入程序中,发行时需要与应用程序同时提供给用户。不同类别的资源访问方式也不同。

2. 简述 Application 的应用场景。

回答:Application 类是 Android 框架的一个系统组件,当 Android 应用程序启动时,系统会自动创建一个 Application 对象来存储系统的一些全局信息,如果需要也可以创建自己的 Application 对象。Application 类采用单例(Singleton)模式设计,其生命周期就等于整个应用程序的生命周期,因为 Application 对象是全局和单例的,所以在不同的 Activity、Service 等组件的对象获得的 Application 对象都是同一个对象。

章节练习

习题

1. 以下有关 Android 系统进程优先级的说法,不正确的是_____。
    A. 前台进程是 Android 系统中最重要的进程
    B. 空进程在系统资源紧张时会被首先清除
    C. 服务进程没有用户界面并且在后台长期运行
    D. Android 系统中一般存在数量较多的可见进程
2. 以下有关 Activity 生命周期的描述,不正确的是_____。
    A. Activity 的状态之间是可以相互转换的
    B. Activity 的生命周期是从 Activity 建立到销毁的全部过程,开始于 onCreate(),

结束于 onDestroy()
   C. 活动生命周期是 Activity 在屏幕的最上层,并能够与用户交互的阶段
   D. onPause()函数在 Android 系统因资源不足终止 Activity 前调用
3. 对 Android 项目工程里的文件,下面描述错误的是_____。
   A. res 目录存放程序中需要使用的资源文件,在打包过程中 android 的工具会对这些文件做对应的处理
   B. R.java 文件是自动生成而不需要开发者维护的,在 res 文件夹中内容发生任何变化,R.java 文件都会同步更新
   C. 在 Assets 目录下存放的文件,在打包过程中将会经过编译后打包在 APK 中
   D. AndroidManifest.xml 是程序的配置文件,程序中用到的所有 Activity、Service、BroadcastIntentReceiver 和 ContentProvider 都必须在这里进行声明
4. 简述 Activity 生命周期中的各个方法。
5. 简述 Application 的作用。
6. 简述 AndroidManifest.xml 文件中主要包括哪些信息。

上机练习

训练目标:创建 Activity。

培养能力	掌握 Activity 生命周期中的各个方法,并熟练创建 Activity		
掌握程度	★★★★★	难度	中
代码行数	100	实施方式	编码强化
结束条件	搭建环境并测试		
参考训练内容			
(1) 创建一个 Activity,并重写其生命周期的 7 个方法; (2) 运行 Activity,测试 Activity 不同状态下的输出结果			

第 3 章

UI编程基础

本章任务是完成"GIFT-EMS 礼记"的主界面及功能 Activity：
- 【任务 3-1】 编写主界面 Activity。
- 【任务 3-2】 编写各个业务 Activity 的父类 BaseActivity。
- 【任务 3-3】 编写"GIFT-EMS 礼记"的辅助功能对应的 Activity。

知 识 点	Listen（听）	Know（懂）	Do（做）	Revise（复习）	Master（精通）
UI 元素	★	★			
页面布局	★	★	★	★	★

知 识 点	Listen(听)	Know(懂)	Do(做)	Revise(复习)	Master(精通)
事件处理	★	★	★	★	
Widget 简单组件	★	★	★	★	★
Dialog 使用	★	★	★	★	

3.1 Android UI 元素

用户界面(User Interface,UI)设计是指对软件人机交互、操作逻辑、界面美观的整体设计。良好的 UI 设计不仅让软件变得更加人性化,还让软件的操作变得舒适、简单、自由,充分体现了软件的定位和特点。Android 借鉴了 Java 中的 UI 设计思想,包括事件响应机制和布局管理,提供了丰富的可视化用户界面组件,例如菜单、对话框、按钮和文本框等。

Android 中界面元素主要由以下几个部分构成。

- 视图(View):视图是所有可视界面元素(通常称为控件或小组件)的基类,所有 UI 控件都是由 View 类派生而来的。
- 视图容器(ViewGroup):视图容器是视图类的扩展,其中包含多个子视图。通过扩展 ViewGroup 类,可以创建由多个相互连接的子视图所组成的复合控件,还可以创建布局管理器,从而实现 Activity 中的布局。
- 布局管理(Layout):布局管理器是由 ViewGroup 派生而来,用于管理组件的布局格式,组织界面中组件的呈现方式。
- Activity:用于为用户呈现窗口或屏幕,当程序需要显示一个 UI 界面时,需要为 Activity 分配一个视图(通常是一个布局或 Fragment)。
- Fragment:Fragment 是 Android 3.0 引入的新 API,代表了 Activity 的子模块,即 Activity 片段(Fragment 本身就是片段的意思)。Fragment 可用于 UI 的各个部分,特别适合针对不同屏幕尺寸,优化 UI 布局以及创建可重用的 UI 元素。每个 Fragment 都包含自己的 UI 布局,并接收相应的输入事件,但使用时必须与 Activity 紧密绑定在一起(Fragment 必须嵌入到 Activity 中)。

因此,一个复杂的 Android 界面设计往往需要不同的组件组合才能实现,有时需要对这些标准视图进行扩展或者修改,从而提供更好的用户体验。

3.1.1 视图

View 视图组件是用户界面的基础元素,View 对象是 Android 屏幕上一个特定的矩形区域的布局和内容属性的数据载体,通过 View 对象可实现布局、绘图、焦点变换、滚动条、屏幕区域的按键、用户交互等功能。Android 应用的绝大部分 UI 组件都放在 android.widget 包及其子包中,所有这些 UI 组件都继承了 View 类。View 的常见子类及功能如表 3-1 所示,本章将对这些 View 组件进行重点讲解。

表 3-1 View 类的主要子类

类　　名	功 能 描 述	类　　名	功 能 描 述
TextView	文本视图	DigitalClock	数字时钟
EditText	编辑文本框	AnalogClock	模拟时钟
Button	按钮	ProgessBar	进度条
Checkbox	复选框	RatingBar	评分条
RadioGroup	单选按钮组	SeekBar	搜索条
Spinner	下拉列表	GridView	网格视图
AutoCompleteTextView	自动完成文本框	ListView	列表视图
DataPicker	日期选择器	ScrollView	滚动视图
TimePicker	时间选择器		

3.1.2 视图容器

View 类还有一个非常重要的 ViewGroup 子类,该类通常作为其他组件的容器使用。View 组件可以添加到 ViewGroup 中,也可以将一个 ViewGroup 添加到另一个 ViewGroup 中。Android 中的所有 UI 组件都是建立在 View、ViewGroup 基础之上,Android 采用了"组合器"模式来设计 View 和 ViewGroup;其中 ViewGroup 是 View 的子类,因此 ViewGroup 可以当成 View 来使用。对于一个 Android 应用的图形 UI 而言,ViewGroup 又可以作为容器来盛装其他组件;ViewGroup 不仅可以包含普通的 View 组件,还可以包含其他 ViewGroup 组件。Android 图形 UI 的组件层次如图 3-1 所示。

图 3-1 UI 组件层次图

图 3-1 来自 Android 开发文档,对于每个 Android 程序员而言,Android 提供的官方文档需要仔细阅读。

ViewGroup 类提供的主要方法如表 3-2 所示。

表 3-2 ViewGroup 类的方法功能

方　　法	功 能 描 述
ViewGroup()	构造方法
void addView(View child)	用于添加子视图,以 View 作为参数,将该 View 增加到当前视图组中

续表

方　法	功　能　描　述
removeView(View view)	将指定的 View 从视图组中移除
updateViewLayout(View view, ViewGroup.LayoutParams params)	用于更新某个 View 的布局
void bringChildToFront(View child)	将参数所指定的视图移动到所有视图之前显示
boolean clearChildFocus(View child)	清除参数所指定的视图的焦点
boolean dispatchKeyEvent(KeyEvent event)	将参数所指定的键盘事件分发给当前焦点路径的视图。当分发事件时，按照焦点路径来查找合适的视图。若本视图为焦点，则将键盘事件发送给自己；否则发送给焦点视图
boolean dispatchPopulateAccessibilityEvent(AccessibilityEvent event)	将参数所指定的事件分发给当前焦点路径的视图
boolean dispatchSetSelected(boolean selected)	为所有的子视图调用 setSelected()方法

注意

ViewGroup 继承了 View 类，虽然可以当成普通的 View 来使用，但习惯上将 ViewGroup 当容器来使用。由于 ViewGroup 是一个抽象类，在实际应用中通常使用 ViewGroup 的子类作为容器，例如各种布局管理器。

1. ViewGroup 继承结构

ViewGroup 的继承者大部分位于 android.widget 包中，其直接子类包括 AdapterView、AbsoluteLayout、FrameLayout、LinearLayout 和 RelativeLayout 等类。以上直接子类又分别具有子类，ViewGroup 继承者的体系结构如图 3-2 所示。

图 3-2　ViewGroup 继承者的体系结构

如图 3-2 所示，ViewGroup 直接子类均可作为容器来使用，这些类为子类提供不同的布局方法，用于设置子类之间的位置和尺寸关系。ViewGroup 类的间接子类中，有些不能作为容器来使用，仅能当作普通的组件来使用。

2. 布局参数类

在 Android 布局文件中，每个组件所能使用的 XML 属性有以下三类：组件本身的

XML 属性、组件祖先类的 XML 属性、组件所属容器的布局参数。

其中,布局参数是包含该组件的容器(例如 ViewGroup 子类)所提供的参数。在 Android 中,ViewGroup 子类都有一个相应的{XXX}.LayoutParams 静态子类,用于设置子类所使用的布局方式。这些子类继承关系和 ViewGroup 子类的继承关系具有相似性。

ViewGroup 容器使用 ViewGroup.LayoutParams 和 ViewGroup.MarginLayoutParams 两个内部类来控制子组件在其中的分布位置,这两个内部类中都提供了一些 XML 属性,ViewGroup 容器中的子组件通过指定 XML 属性来控制组件的位置,如表 3-3 所示。

表 3-3　ViewGroup 子元素支持的属性

XML 属性	功 能 描 述
android:layout_width	设定该组件的子组件布局的宽度
android:layout_height	设定该组件的子组件布局的高度

android:layout_height 和 android:layout_width 属性都支持以下三个属性值:

(1) fill_parent 属性用于指定子组件的高度、宽度与父容器的高度、宽度相同;

(2) match_parent 与 fill_parent 的功能完全相同,从 Android 2.2 开始推荐使用该属性值来代替 fill_parent;

(3) wrap_content 属性用于指定子组件的大小恰好能包裹其内容即可。

在实际应用中,除了为组件指定高度、宽度,还需要设置布局的高度、宽度,这是由 Android 的布局机制决定的。Android 组件的大小不仅由实际的宽度、高度控制,还由布局的高度、宽度控制。例如一个组件的宽度为 30px,如果将其布局宽度设置为 match_parent,那么该组件的宽度将会被"拉宽"并占满其所在的父容器;如果将其布局宽度设为 wrap_content,那么该组件的宽度才会是 30px。

ViewGroup.MarginLayoutParams 用于控制子组件周围的页边距(即组件四周的留白),所支持的 XML 属性如表 3-4 所示。

表 3-4　MarginLayoutParams 支持的属性

XML 属性	功 能 描 述
android:layout_marginTop	指定该子组件上面的页边距
android:layout_marginRight	指定该子组件右面的页边距
android:layout_marginBottom	指定该子组件下面的页边距
android:layout_marginLeft	指定该子组件左面的页边距

由于 LayoutParams 也具有继承关系,因此 LinearLayout 的子类除了可以使用 LinearLayout.LayoutParams 所提供的 XML 属性外,还可以使用其祖先类 ViewGroup.LayoutParams 的 XML 属性。

3.1.3 布局管理

针对不同的手机屏幕(如手机屏幕的分辨率不同或屏幕尺寸不同等情况),当在程序中手动控制每个组件的大小和位置时,将会给编程带来巨大的困难。为了解决这个问题,Android 提供了布局管理器,使得 Android 各类组件(如按钮、文本等组件)能够适应屏幕的变化。布局管理器可以根据运行平台来调整组件的大小,开发者只须为容器选择合适的布局管理器即可。

Android 的布局管理器本身是一种 UI 组件,所有的布局管理器都是 ViewGroup 的子类,Android 布局管理器类之间的关系如图 3-3 所示。

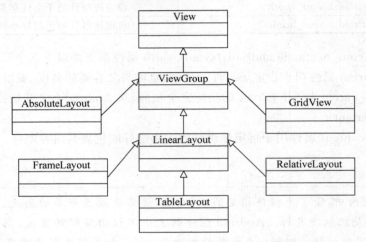

图 3-3 Android 布局管理器类之间的关系

所有布局都可以作为容器来使用,通过调用 addView()方法向布局管理器中添加组件。此外,布局管理器还继承了 View 类,在实际编程过程中可以把布局管理器作为普通的 UI 组件嵌套到其他布局管理器中。

Android 提供了多种布局,常用的布局有以下几种。

- **LinearLayout 线性布局**:该布局中子元素之间呈线性排列,即在水平或垂直方向上的顺序排列。
- **RelativeLayout 相对布局**:该布局是一种根据相对位置排列元素的布局方式,允许子元素指定相对于其他元素或父元素的位置(一般通过 ID 指定)。在线性布局中排列子元素时,不需要特殊指定参照物,而相对布局中的子元素必须指定其参照物,只有指定参照物之后,才能定义该元素的相对位置。
- **TableLayout 表格布局**:该布局将子元素的位置分配到表格的行或列中,即按照表格形式的顺序排列。一个表格布局中有多个"表格行",而表格行中又包含多个"表格单元"。表格布局并不是真正意义上的表格,只是按照表格的方式组织元素的布局,元素之间并没有实际表格中的分界线。
- **AbsoluteLayout 绝对布局**:按照绝对坐标对元素进行布局。与相对布局不同,绝对布局不需要指定参照物,而是使用整个手机界面作为坐标系,通过坐标系的水平偏移量和垂直偏移量来确定其唯一位置。

3.1.4 Fragment

Fragment 允许将 Activity 拆分成多个完全独立的可重用的组件,每个组件具有自己的生命周期和 UI 布局。Fragment 最大的优点就是灵活地为不同大小屏幕的设备创建 UI 界面,例如小屏幕的智能手机和大屏幕的平板电脑。

每个 Fragment 都是一个独立的模块,并与所绑定的 Activity 紧密地联系在一起。一个 Fragment 可以被多个 Activity 所共用,一个界面可以有多个 UI 模块,对于像平板电脑这样的设备,Fragment 展现了很好的适应性和动态创建 UI 的能力,在一个 Activity 中可以添加、删除、更换 Fragment。Fragment 为不同型号、尺寸、分辨率的设备提供了统一的 UI 优化方案。

有关 Fragment 的生命周期及详细使用方法参见第 4 章。

3.2 界面布局

Android 中提供了以下两种创建布局的方式。

(1) **在 XML 布局文件中声明**:首先将需要显示的组件在布局文件中进行声明,然后在程序中通过 setContentView(R.layout.XXX)方法将布局呈现在 Activity 中。推荐使用此种方式,前面的程序也一直使用此种方式。

(2) **在程序中直接实例化布局及其组件**:此种方式并不提倡使用,除非界面中的组件及布局需要动态改变。

常见的 Android 布局包括 LinearLayout、RelativeLayout、TableLayout 和 AbsoluteLayout 等多种布局。

3.2.1 线性布局

LinearLayout 是一种线性排列的布局,布局中的组件按照垂直或者水平方向进行排列,排列方向由 android:orientation 属性进行控制,其属性值包括垂直(vertical)和水平(horizontal)两种。LinearLayout 对应的类为 android.widget.LinearLayout。

LinearLayout 常用的 XML 属性及相关方法的说明如表 3-5 所示。

表 3-5 LinearLayout 常用的 XML 属性及对应方法

XML 属性	对应方法	功 能 描 述
android:divider	setDividerDrawable()	设置垂直布局时两个按钮之间的分隔条
android:gravity	setGravity()	设置布局管理器内组件的对齐方式。该属性支持 top、bottom、left、right、center_vertical、fill_vertical、center_horizontal、fill_horizontal、center、fill、clip_vertical、clip_horizontal、start、end 等属性值,也可以指定多种对齐方式的组合,例如,left|center_vertical 代表出现在屏幕左边,且垂直居中

续表

XML 属性	对应方法	功能描述
android:orientation	setOrientation()	设置布局管理器内组件的排列方式,参数可以为 horizontal(水平排列)或 vertical(垂直排列、默认值)

此外,LinearLayout 中包含的所有子元素的位置都受 LinearLayout.LayoutParams 控制,LinearLayout 包含的子元素可以额外指定属性,如表 3-6 所示。

表 3-6 LinearLayout 子元素常用 XML 属性及说明

XML 属性	功能描述
android:layout_gravity	指定子元素在 LinearLayout 中的对齐方式
android:layout_weight	指定子元素在 LinearLayout 中所占的比重

注意

线性布局不会换行,当组件顺序排列到屏幕边缘时,剩余的组件不会被显示出来。

在项目的 res/layout 目录下创建一个线性布局文件 linearlayout.xml,用于演示 LinearLayout 的用法,其布局代码如下所示。

【代码 3-1】 linearlayout.xml

```xml
<?xml version = "1.0" encoding = "utf-8"?>
<LinearLayout xmlns:android = "http://schemas.android.com/apk/res/android"
    xmlns:tools = "http://schemas.android.com/tools"
    android:layout_width = "match_parent"
    android:layout_height = "match_parent"
    android:orientation = "vertical"
    android:gravity = "center_horizontal">
    <TextView
        android:layout_width = "wrap_content"
        android:layout_height = "wrap_content"
        android:text = "txtView1"
        android:textColor = "#000000"
        android:textSize = "20sp"/>
    <TextView
        android:layout_width = "wrap_content"
        android:layout_height = "wrap_content"
        android:text = "txtView2"
        android:textColor = "#000000"
        android:textSize = "20sp"/>
    …//剩余 3 个 TextView 与前两个类似,此处省略
</LinearLayout>
```

上述代码中,页面布局相对比较简单,仅定义了一个线性布局,并在布局中定义了 5 个 TextView;在定义线性布局时默认采用垂直排列方式,且所有组件在容器的顶部居中对齐。

在 LayoutActivity 中使用 linearlayout.xml 布局,代码如下所示。

【代码 3-2】 LayoutActivity.java

```
public class LayoutActivity extends AppCompatActivity {
    @Override
    public void onCreate(Bundle savedInstanceState) {
        super.onCreate(savedInstanceState);
        setContentView(R.layout.linearlayout);}
}
```

上述代码中，调用 setContentView( )方法将布局设置到屏幕中，运行结果如图 3-4 所示。

在上述布局文件中，将 LinearLayout 属性修改为 android:gravity="center"，即垂直方向居中，再次运行 LayoutActivity，结果如图 3-5 所示。

图 3-4 线性布局

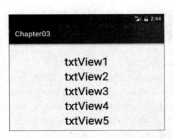
图 3-5 属性修改后线性布局

3.2.2 表格布局

TableLayout 类似表格形式，以行和列的方式来布局子组件。TableLayout 继承了 LinearLayout，因此其本质上依然是线性布局。TableLayout 并不需要明确地声明所包含的行数和列数，而是通过 TableRow 及其子元素来控制表格的行数和列数。

通常情况下，TableLayout 的行数由开发人员直接指定，即 TableRow 对象（或 View 控件）的个数；TableLayout 的列数等于含有最多子元素的 TableRow 所包含的元素个数，例如第一个 TableRow 中含两个子元素，第二个 TableRow 含 3 个，第三个 TableRow 含 4 个，则该 TableLayout 的列数为 4。

在 TableLayout 布局中，某列的宽度是由该列中最宽的那个单元格决定的，整个表格布局的宽度则取决于父容器的宽度（默认总是占满父容器本身）。

在表格布局器中，可以通过以下 3 种方式对单元格进行设置。

（1）**Shrinkable**：如果某个列被设置为 Shrinkable，那么该列中所有单元格的宽度都可以被收缩，以保证表格能适应父容器的宽度。

（2）**Stretchable**：如果某个列被设置为 Stretchable，那么该列中所有单元格的宽度都可以被拉伸，以保证组件能够完全填满表格的空余空间。

（3）**Collapsed**：如果某个列被设置为 Collapsed，那么该列中的所有单元格都会被隐藏。

TableLayout 可设置的属性包括全局属性和单元格属性，全局属性也被称为列属性。TableLayout 常用的全局 XML 属性及相关方法如表 3-7 所示。

表 3-7 TableLayout 常用 XML 属性及对应方法

XML 属性	对应方法	功能描述
android:shrinkColumns	setShrinkAllColumns(boolean)	设置可收缩的列。当该列子控件的内容太多，已经挤满所在行时，子控件的内容将往列方向显示，多个列之间用逗号隔开
android:stretchColumns	setStretchAllColumns(boolean)	设置可伸展的列。该列可以横向伸展，最多可占据一整行，多个列之间用逗号隔开
android:collapseColumns	setColumnCollapsed(int,boolean)	设置要隐藏的列，多个列之间用逗号隔开

下述代码演示了表格的全局属性的使用，代码如下所示。

【示例】 全局属性的设置

```
<?xml version = "1.0" encoding = "utf - 8"?>
<TableLayout xmlns:android = "http://schemas.android.com/apk/res/android"
    android:layout_width = "match_parent"
    android:layout_height = "match_parent"
    android:stretchColumns = "0"
    android:collapseColumns = " * "
    android:shrinkColumns = "1,2" >
</TableLayout>
```

其中：android:stretchColumns = "0" 表示第 0 列可伸展；android:shrinkColumns = "1,2" 表示第 1、2 列皆可收缩；android:collapseColumns = " * " 表示隐藏所有行。

注意

> 列可以同时具备 stretchColumns 和 shrinkColumns 属性；当该列的内容较多时，将以"多行"方式显示其内容(此处所指的"多行"不是真正的多行，而是系统根据需要自动调节该行的 layout_height)。

TableRow.LayoutParams 常用的单元格 XML 属性及方法如表 3-8 所示，通常对 TableRow 的子元素进行修饰。

表 3-8 TableRow.LayoutParams 的单元格 XML 属性及对应方法

XML 属性	功能描述
android:layout_column	指定该单元格在第几列显示
android:layout_span	指定该单元格占据的列数(未指定时，默认为 1)

下述代码演示了表格属性的设置，代码如下所示。

【示例】 对表格属性进行设置

```
<TableLayout xmlns:android = "http://schemas.android.com/apk/res/android" …>
  <TableRow>
        <Button android:layout_span = "2" />
```

```
            <Button android:layout_column = "1"/>
        </TableRow>
</TableLayout>
```

其中：android:layout_span="2"表示该控件占据 2 列；android:layout_column="1"表示该控件显示在第 1 列。

注意

> 由于 TableLayout 继承了 LinearLayout，因此完全支持 LinearLayout 所支持的全部 XML 属性。

下述代码用于演示 TableLayout 的基本使用。在 res/layout 目录下创建一个表格布局文件 tablelayout.xml，代码如下所示。

【代码 3-3】 tablelayout.xml

```
<?xml version = "1.0" encoding = "utf-8"?>
<TableLayout xmlns:android = "http://schemas.android.com/apk/res/android"
    android:id = "@ + id/MorePageTableLayout_Favorite"
    android:layout_width = "fill_parent"
    android:layout_height = "wrap_content"
    android:collapseColumns = "2"
    android:shrinkColumns = "0"
    android:stretchColumns = "0" >
    <TableRow
        android:id = "@ + id/more_page_row1"
        android:layout_width = "fill_parent"
        android:layout_marginLeft = "2.0dip"
        android:layout_marginRight = "2.0dip"
        android:paddingBottom = "16.0dip"
        android:paddingTop = "8.0dip" >
        <TextView
            android:layout_width = "wrap_content"
            android:layout_height = "fill_parent"
            android:drawablePadding = "10.0dip"
            android:gravity = "center_vertical"
            android:includeFontPadding = "false"
            android:paddingLeft = "17.0dip"
            android:text = "账号管理"
            android:textColor = "#ff333333"
            android:textSize = "16.0sp" />
        <ImageView
            android:layout_width = "wrap_content"
            android:layout_height = "fill_parent"
            android:layout_gravity = "right"
            android:gravity = "center_vertical"
            android:paddingRight = "20.0dip"
            android:src = "@drawable/item_arrow" />
    </TableRow>
    …//省略"搜索商品"和"浏览记录"的<TableRow>
</TableLayout>
```

上述代码中,使用< TableLayout >元素定义了表格布局,该元素的 android:collapseColumns 属性用于指明表格的列数,此处设置表格的列数为 2。android:stretchColumns 属性用于指明表格的伸展列,将指定列进行拉伸以填满剩余的空间;注意列号从 0 开始,此处 0 表示第 1 列为伸展列。使用< TableRow >元素定义了表格中的行,其他组件都放在该元素内。

在 LayoutActivity 中,使用 tablelayout.xml 布局,相关代码如下所示。

```
setContentView(R.layout.tablelayout);
```

运行结果如图 3-6 所示。

将< TableLayout >元素中的 android:stretchColumns = "0" 删除,即不指定伸展列时,运行结果如图 3-7 所示。

图 3-6　第一列为延伸列　　　　　　图 3-7　普通的表格布局

> Android 的表格布局跟 HTML 中的表格布局非常类似,TableRow 相当于 HTML 表格的< tr >标记。

3.2.3　相对布局

RelativeLayout 是一组相对排列的布局方式,在相对布局容器中,子组件的位置总是相对于兄弟组件或父容器,例如一个组件在另一个组件的左边、右边、上面或下面等位置。在相对布局容器中,当 A 组件的位置由 B 组件来决定时,Android 要求先定义 B 组件,再定义 A 组件。

RelataiveLayout 位于 android.widget 包中,其常用 XML 属性及方法如表 3-9 所示。

表 3-9　RelativeLayout 常用 XML 属性及方法

XML 属性	对应方法	功能描述
android:gravity	setGravity()	设置布局管理器内组件的对齐方式。该属性支持包括 top、bottom、left、right、center_vertical、fill_vertical、center_horizontal、fill_horizontal、center、fill、clip_vertical、clip_horizontal、start 和 end。也可以同时指定多种对齐方式的组合,例如 left|center_vertical 代表出现在屏幕左边且垂直居中
android:ignoreGravity	setIgnoreGravity()	设置特定的组件不受 gravity 属性的影响

为了控制该布局容器中各个子组件的布局分布，RelativeLayout 提供了一个内部类——RelativeLayout.LayoutParams，该类提供了大量的 XML 属性来控制 RelativeLayout 布局中子组件的位置分布，如表 3-10 所示，表中所列属性取值只能为 true 或 false。

表 3-10　RelativeLayout.LayoutParams 的 XML 属性及说明（一）

XML 属性	功 能 描 述
android:layout_alignParentLeft	指定该组件是否与布局容器左对齐
android:layout_alignParentTop	指定该组件是否与布局容器顶端对齐
android:layout_alignParentRight	指定该组件是否与布局容器右对齐
android:layout_alignParentBottom	指定该组件是否与布局容器底端对齐
android:layout_centerInParent	指定该组件是否位于布局容器的中央位置
android:layout_centerHorizontal	指定该组件是否位于布局容器的水平居中
android:layout_centerVertical	指定该组件是否位于布局容器的垂直居中

RelativeLayout.LayoutParams 中另外一部分的属性值可以是其他 UI 组件的 ID 值，表示当前组件与指定 ID 组件的相对位置，如表 3-11 所示。

表 3-11　RelativeLayout.LayoutParams 的 XML 属性及说明（二）

XML 属性	功 能 描 述
android:layout_toLeftOf	控制该组件位于指定 ID 组件的左侧
android:layout_toRightOf	控制该组件位于指定 ID 组件的右侧
android:layout_above	控制该组件位于指定 ID 组件的上方
android:layout_below	控制该组件位于指定 ID 组件的下方
android:layout_alignLeft	控制该组件与指定 ID 组件的左边界进行对齐
android:layout_alignTop	控制该组件与指定 ID 组件的上边界进行对齐
android:layout_alignRight	控制该组件与指定 ID 组件的右边界进行对齐
android:layout_alignBottom	控制该组件与指定 ID 组件的下边界进行对齐

此外，RelativeLayout.LayoutParams 还继承了 android.view.ViewGroup.MarginLayoutParams 类，该类用于定义组件边缘的空白，具有 android:layout_marginTop、android:layout_marginLeft、android:layout_marginBottom、android:layout_marginRight 四个 XML 属性，分别表示上、左、下、右 4 个方向的边缘空白。

下面通过对"东西南北中"的布局来演示 RelativeLayout 的使用。

【代码 3-4】　relativelayout.xml

```
<?xml version = "1.0" encoding = "utf-8"?>
<RelativeLayout xmlns:android = "http://schemas.android.com/apk/res/android"
    android:layout_width = "match_parent"
    android:layout_height = "match_parent" >
    <TextView
        android:id = "@+id/middle"
        android:layout_width = "wrap_content"
        android:layout_height = "wrap_content"
        android:layout_centerInParent = "true"
        android:text = "中" />
```

```xml
<TextView
    android:id = "@ + id/west"
    android:layout_width = "wrap_content"
    android:layout_height = "wrap_content"
    android:layout_toLeftOf = "@id/middle"
    android:text = "西" />
<TextView
    android:id = "@ + id/east"
    android:layout_width = "wrap_content"
    android:layout_height = "wrap_content"
    android:layout_toRightOf = "@id/middle"
    android:text = "东" />
<TextView
    android:id = "@ + id/north"
    android:layout_width = "wrap_content"
    android:layout_height = "wrap_content"
    android:layout_above = "@id/middle"
    android:text = "北" />
<TextView
    android:id = "@ + id/south"
    android:layout_width = "wrap_content"
    android:layout_height = "wrap_content"
    android:layout_below = "@id/middle"
    android:text = "南" />
</RelativeLayout>
```

上述代码使用< RelativeLayout >元素定义了一个相对布局,该布局中含有 5 个文本,分别位于"东、西、南、北、中"。由于相对布局中的组件总是由其他组件来决定分布的位置,因此在设计过程中首先把"中"元素放到布局容器的中间,然后以该组件为中心,依次将"东、西、南、北"4 个组件分布到四周,这样就形成了"上北、下南、左西、右东"的布局效果。文本的摆放位置具体如下:

- "西"位于文本"中"的左边,即通过 layout_toLeftOf 属性进行设置;
- "东"位于文本"中"的右边,即通过 layout_toRightOf 属性进行设置;
- "北"位于文本"中"的上边,即通过 layout_above 属性进行设置;
- "南"位于文本"中"的下边,即通过 layout_below 属性进行设置。

在 LayoutActivity 中,使用 relativelayout.xml 布局,相关代码如下所示。

```
setContentView(R.layout.relativelayout);
```

运行结果如图 3-8 所示。在图 3-8 中,以"中"为中心,所显示的"上北、下南、左西、右东"偏离了预想的位置;如果希望"东、西、南、北"4 个元素以"中"为中心,分别位于正东、正西、正南、正北"4 个方位,需要在布局文件中通过 android:layout_alignXXX 属性来指定具体的对齐方式:

- "中"的"正西",由于两个文字的高度相同,可以通过 android:layout_alignTop 属性进行设置;
- "中"的"正东",通过 android:layout_alignTop 属性进行设置;
- "中"的"正南",由于两个文字的长度相同,可以通过 android:layout_alignLeft 属性进行设置；
- "中"的"正北"，通过 android:layout_alignLeft 属性进行设置。

对 relativelayout.xml 布局文件进行改进，改进后的代码如下所示。

【代码 3-5】 relativelayout.xml

```
<RelativeLayout xmlns:android = "http://schemas.android.com/apk/res/android" …>
    <TextView
        android:id = "@ + id/middle"
        android:layout_width = "wrap_content"
        android:layout_height = "wrap_content"
        android:layout_centerInParent = "true"
        android:text = "中" />
    <TextView
        android:id = "@ + id/west"
        android:layout_width = "wrap_content"
        android:layout_height = "wrap_content"
        android:layout_alignTop = "@id/middle"
        android:layout_toLeftOf = "@id/middle"
        android:text = "西" />
    …省略部分代码
</RelativeLayout>
```

> **注意**
>
> 如果五个方位的字符串长度不同，则需要选择居中对齐的方式，例如使用 android:layout_centerHorizontal＝"true"来实现。

运行上述代码结果如图 3-9 所示。图 3-9 显示了传统意义上的"上北、下南、左西、右东"各个方位，但 5 个方位之间过于紧凑，需要调整一下元素之间的间距，因此可以使用 android:layout_margin 等属性来设置元素的边缘空白，例如将边缘空白设置为 10dp。

对 relativelayout.xml 布局文件进一步改进，改进后的代码如下所示。

【代码 3-6】 relativelayout.xml

```
<RelativeLayout xmlns:android = "http://schemas.android.com/apk/res/android" …
    android:layout_height = "match_parent" >
    <TextView
        android:id = "@ + id/middle"
        android:layout_width = "wrap_content"
        android:layout_height = "wrap_content"
        android:layout_centerInParent = "true"
        android:layout_margin = "10dp"
        android:text = "中" />
    …省略部分代码
</RelativeLayout>
```

运行上述代码，结果如图 3-8～图 3-10 所示。

图 3-8　相对布局　　　　　图 3-9　相对布局（对齐）　　　图 3-10　相对布局（页边距）

> 上述效果除了可以通过设置"中"之外，还可以通过设置其他 4 个方位对象的 android:layout_marginXX 来实现，此处不再赘述，请读者验证之。

3.2.4　绝对布局

AbsoluteLayout 通过指定组件的确切 X、Y 坐标来确定组件的位置。下述代码用于演示 AbsoluteLayout 的使用。

【代码 3-7】　absolutelayout.xml

```xml
<?xml version = "1.0" encoding = "utf-8"?>
<LinearLayout xmlns:android = "http://schemas.android.com/apk/res/android"
    android:orientation = "vertical" android:layout_width = "match_parent"
    android:layout_height = "match_parent">
    <AbsoluteLayout android:id = "@+id/AbsoluteLayout01"
        android:layout_width = "wrap_content"
        android:layout_height = "wrap_content">
        <Button android:text = "A" android:id = "@+id/Button01"
            android:layout_width = "wrap_content"
            android:layout_height = "wrap_content"
            android:layout_x = "10dp" android:layout_y = "20dp"></Button>
        <Button android:text = "B" android:id = "@+id/Button02"
            android:layout_width = "wrap_content"
            android:layout_height = "wrap_content"
            android:layout_x = "100dp" android:layout_y = "20dp"></Button>
        <Button android:text = "C" android:id = "@+id/Button03"
            android:layout_width = "wrap_content"
            android:layout_height = "wrap_content"
            android:layout_x = "10dp" android:layout_y = "80dp"></Button>
        <Button android:text = "D" android:id = "@+id/Button04"
            android:layout_width = "wrap_content"
            android:layout_height = "wrap_content"
            android:layout_x = "100dp" android:layout_y = "80dp"></Button>
    </AbsoluteLayout>
</LinearLayout>
```

上述代码使用<AbsoluteLayout>元素来定义绝对布局，该布局中有 4 个按钮，每个按钮的位置都是通过 X、Y 轴坐标进行指定的，其中 layout_x 属性用于指定元素的 X 轴坐标，

layout_y 属性用于指定元素的 Y 轴的坐标。

在 LayoutActivity 中使用 absolutelayout.xml 进行布局，相关代码如下所示。

```
setContentView(R.layout.absolutelayout);
```

运行结果如图 3-11 所示。

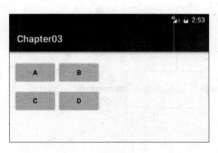

图 3-11　绝对布局

3.3　事件处理

当用户在程序界面上执行各种操作时，应用程序必须为用户提供响应动作，通过响应动作来完成事件处理。在图形界面（UI）的开发中，有两个非常重要的内容：一个是控件的布局；另一个就是控件的事件处理。其中，控件的布局已经在 3.2 节进行了简要介绍，本节主要对事件处理进行介绍。

Android 提供了两种方式的事件处理：基于回调的事件处理和基于监听的事件处理。Android 系统充分利用这两种事件处理方式的优点，允许开发人员采用自己熟悉的事件处理方式为用户的操作提供响应动作，从而可以开发出界面友好、人机交互效果好的 Android 应用程序。

3.3.1　基于监听的事件处理

基于监听的事件处理方式和 Java Swing/AWT 的事件处理方式几乎完全相同，如果开发者具有 Java Swing 方面的编程经验，则更容易上手。

Android 系统中引用了 Java 事件处理机制，包括事件、事件源和事件监听器三个事件模型。

- 事件（Event）：是一个描述事件源状态改变的对象，事件对象不是通过 new 运算符创建的，而是在用户触发事件时由系统生成的对象。事件包括键盘事件、触摸事件等，一般作为事件处理方法的参数，以便从中获取事件的相关信息。
- 事件源（Event Source）：触发事件的对象，事件源通常是 UI 组件，例如单击按钮时按钮就是事件源。
- 事件监听器（Event Listener）：当触发事件时，事件监听器用于对该事件进行响应和处理。监听器需要实现监听接口中所定义的事件处理方法。

当用户按下一个按钮或单击某个菜单选项时，这些操作就会触发一个响应事件，该事件就会调用在事件源上注册的事件监听器,事件监听器调用相应的事件处理程序并完成相应的事件处理,基于监听的事件处理流程如图 3-12 所示。

图 3-12　基于监听的事件处理流程

Android 的事件处理机制是一种委派式事件处理机制,该处理方式类似于人类社会的分工协作,例如某个企业(事件源)进行货物采购(事件)时,企业通常不会自己运输物品,而是找特定的物流公司来运输;如果发生了火灾(事件),则会委派给消防局(事件监听器)来处理;而消防局或物流公司也会同时监听多个企业的火灾事件或货物运输事件。委派式的处理方式将事件源和事件监听器分离,从而提供更好的程序模型,有利于提高程序的可维护性和代码的健壮性。

在 Android 应用程序中,所有的组件都可以针对特定的事件指定一个事件监听器,每个事件监听器可以监听一个或多个事件源。同一个事件源上也可能发生多个事件,例如,在按钮上可能发生单击、获取焦点等事件,委派式事件处理将事件源上所有可能发生的事件分别委派给不同的事件监听器来处理,同时也可以让同一类事件都使用同一个事件监听器来处理。

Android 中常用的事件监听器如表 3-12 所示,这些事件监听器以内部接口的形式定义在 android.view.View 中。

表 3-12　Android 中的事件监听器

事件监听器接口	事　　件	功　能　描　述
OnClickListener	单击事件	当用户单击某个组件或者方向键时触发该事件
OnFocusChangeListener	焦点事件	当组件获得或者失去焦点时触发该事件
OnKeyListener	按键事件	当用户按下或者释放设备上的某个按键时触发该事件
OnTouchListener	触摸事件	当设备具有触摸屏功能,在触碰屏幕时触发该事件
OnCreateContextMenuListener	创建上下文菜单事件	当创建上下文菜单时触发该事件
OnCheckedChangeListener	选项改变事件	当选择改变时触发该事件

由此可知,事件监听器本质上是一个实现了特定接口的 Java 对象。在程序中实现事件监听器,通常有以下几种形式:

● **Activity 本身作为事件监听器**:通过 Activity 实现监听器接口,并实现事件处理方法;

- **匿名内部类形式**：使用匿名内部类创建事件监听器对象；
- **内部类或外部类形式**：将事件监听类定义为当前类的内部类或普通的外部类；
- **绑定标签**：在布局文件中为指定标签绑定事件处理方法。

通常实现基于监听的事件处理步骤如下：①创建事件监听器；②在事件处理方法中编写事件处理代码；③在相应的组件上注册监听器。

1. Activity 本身作为事件监听器

通过 Activity 实现监听器接口，并实现该接口中对应的事件处理方法。下述代码演示了在 Button 按钮上绑定单击事件，当单击按钮时改变文字的内容，布局代码如下所示。

【代码 3-8】 event_btn.xml

```xml
<?xml version = "1.0" encoding = "utf - 8"?>
<LinearLayout xmlns:android = "http://schemas.android.com/apk/res/android"
    android:layout_width = "match_parent"
    android:layout_height = "match_parent"
    android:gravity = "center_horizontal"
    android:orientation = "vertical" >
    <EditText
        android:id = "@ + id/showTxt"
        android:layout_width = "match_parent"
        android:layout_height = "wrap_content"
        android:editable = "false" />
    <Button
        android:id = "@ + id/clickBtn"
        android:layout_width = "wrap_content"
        android:layout_height = "wrap_content"
        android:text = "单击我" />
</LinearLayout>
```

上述代码定义了 Button 和 EditText 两个组件，主要用于实现 Button 的单击事件。实现监听和事件处理的 Activity 代码如下所示。

【代码 3-9】 EventBtnActivity.java

```java
//1.实现事件监听器接口
public class EventBtnActivity extends AppCompatActivity
        implements OnClickListener{
    private Button clickBtn;
    //单击 Button
    private TextView showTxt;
    //文字显示
    @Override
    protected void onCreate(Bundle savedInstanceState) {
        super.onCreate(savedInstanceState);
        setContentView(R.layout.event_btn);
        //初始化组件
        showTxt = (TextView) findViewById(R.id.showTxt);
        clickBtn = (Button)findViewById(R.id.clickBtn);
        //2.直接使用 Activity 作为事件监听器
        clickBtn.setOnClickListener(this);}
```

```
//3. 在事件处理方法中编写事件处理代码
@Override
public void onClick(View v) {
    //实现事件处理方法
    showTxt.setText("btn 按钮被单击了!");}
}
```

运行上述代码，当单击"单击我"按钮时，TextView 文本内容将发生改变，效果如图 3-13 所示。

上述代码中，定义的 EventBtnActivity 继承了 Activity，并实现了 OnClickListener 接口，此时 Activity 对象允许作为事件监听器进行使用。代码"clickBtn.setOnClickListener(this);"用于为 clickBtn

图 3-13　按钮单击效果

按钮注册事件监听器。当单击 Button 按钮时，触发鼠标单击事件并调用 onClick()事件处理方法，TextView 文本内容变成"btn 按钮被单击了!"。从上面程序中可以看出，基于监听的事件处理模型的编程步骤如下。

（1）获取所要触发事件的事件源控件，例如本例中的 clickBtn 对象。

（2）实现事件监听器类，本例中的监听器类是 Activity 对象本身（实现了 OnClickListener 接口）。

（3）调用事件源的 setXxxListener()方法，将事件监听器注册给事件源对象；当事件源上发生指定事件时，Android 会触发事件监听器，由事件监听器调用相应的方法来处理事件。

2. 匿名内部类形式

Activity 的主要任务是完成界面的初始化工作，而代码 3-9 中使用 Activity 本身作为监听器类，并在 Activity 类中定义事件处理方法，易造成程序结构混乱。大部分情况下，事件监听器只是临时使用一次，所以匿名内部类形式的事件监听器更合适。将代码 3-9 改为匿名内部类形式，代码如下所示。

【代码 3-10】 AnonymousBtnActivity.java

```
//实现事件监听器接口
public class AnonymousBtnActivity extends AppCompatActivity{
    private Button clickBtn;          //单击 Button
    private TextView showTxt;         //文字显示
    @Override
    protected void onCreate(Bundle savedInstanceState) {
        super.onCreate(savedInstanceState);
        setContentView(R.layout.event_btn);
        //初始化组件
        showTxt = (TextView) findViewById(R.id.showTxt);
        clickBtn = (Button)findViewById(R.id.clickBtn);
        //使用匿名内部类创建一个监听器
        clickBtn.setOnClickListener(new OnClickListener() {
            @Override
```

```
            public void onClick(View v) {
                //实现事件处理方法
                showTxt.setText("btn 按钮被单击了!");}});}
}
```

上述代码中粗体部分使用匿名内部类创建了一个事件监听器对象,界面效果与图 3-13 一致,此处不再演示。

3. 内部类、外部类形式

所谓的"内部类"形式,是指将事件监听器定义成当前类的内部类。下述代码演示使用内部类的方式实现事件监听。

【代码 3-11】 InnerClassBtnActivity.java

```
//实现事件监听器接口
public class InnerClassBtnActivity extends AppCompatActivity{
    private Button clickBtn;
    //单击 Button
    private TextView showTxt;
    //文字显示
    @Override
    protected void onCreate(Bundle savedInstanceState) {
        super.onCreate(savedInstanceState);
        setContentView(R.layout.event_btn);
        //初始化组件
        showTxt = (TextView) findViewById(R.id.showTxt);
        clickBtn = (Button)findViewById(R.id.clickBtn);
        //直接使用 Activity 作为事件监听器
        clickBtn.setOnClickListener(new ClickListener());}
    //内部类方式定义一个事件监听器
    class ClickListener implements OnClickListener{
        @Override
        public void onClick(View v) {
            //实现事件处理方法
            showTxt.setText("btn 按钮被单击了!");}}
}
```

用内部类有以下优点:

(1) 可以在当前类中复用内部监听器类;

(2) 由于监听器是当前类的内部类,所以可以访问当前类的所有界面组件。

> **注意**
>
> 外部监听器的定义方式和内部类的定义方式相似,由于使用外部类事件监听器的形式比较少见,故此处不再赘述。

4. 绑定标签

Android 还有一种更简单的绑定事件的方式,在界面布局文件中直接为指定标签绑定事件处理方法。对于大多数 Android 界面的组件标签而言,基本都支持 onClick 事件属性,相应的属性值就是一个类似 xxxMethod 形式的方法名称。

【代码 3-12】 event_tag.xml

```xml
<LinearLayout xmlns:android="http://schemas.android.com/apk/res/android" …>
    <EditText
        android:id = "@+id/showTxt"
        android:layout_width = "match_parent"
        android:layout_height = "wrap_content"
        android:editable = "false" />
    <Button
        android:id = "@+id/clickBtn"
        android:layout_width = "wrap_content"
        android:layout_height = "wrap_content"
        android:onClick = "clickMe"
        android:text = "单击我" />
</LinearLayout>
```

上述代码中,粗体部分用于为 clickBtn 按钮绑定一个事件处理方法:clickMe。此时需要开发者在相应的 Activity 中定义一个名为 clickMe 的方法,该方法用于负责处理按钮上的单击事件,代码如下所示。

【代码 3-13】 BindTagActivity.java

```java
//实现事件监听器接口
public class BindTagActivity extends AppCompatActivity{
    private Button clickBtn;                    //单击 Button
    private TextView showTxt;                   //文字显示
    @Override
    protected void onCreate(Bundle savedInstanceState) {
        super.onCreate(savedInstanceState);
        setContentView(R.layout.event_btn);
        //初始化组件
        showTxt = (TextView) findViewById(R.id.showTxt);
        clickBtn = (Button)findViewById(R.id.clickBtn);}
    public void clickMe(View v){
        //实现事件处理方法
        showTxt.setText("btn 按钮被单击了!");}
}
```

上述代码中,粗体部分定义了 clickMe()方法,其中有一个 View 类型的参数,方法的返回类型为 void。运行上述代码,界面效果与前面的图 3-13 一致。

3.3.2 基于回调机制的事件处理

在 Android 平台中,每个 View 都拥有事件处理的回调方法,开发人员通过重写这些回调方法来实现所需要响应的事件。当某个事件没有被任何一个 View 控件处理时,便会调用 Activity 中相应的回调方法进行处理。从代码实现的角度来看,基于回调的事件处理模型要比基于监听的事件处理模型更为简单。从 3.2 节内容可知,事件监听机制是一种委托式的事件处理,而回调机制则恰好与之相反;对于基于回调的事件处理模型而言,事件源和事件监听器是统一的,当用户在 GUI 组件上触发某个事件时,组件自身的方法将会负责处

第3章 UI编程基础

理该事件。

为了实现回调机制的事件处理，Android 为所有的 GUI 组件都提供了事件处理的回调方法，例如 View 中提供了 onKeyDown()、onKeyUp()、onTouchEvent()、onTrackBallEvent() 和 onFocusChanged() 等事件回调方法。

1. onKeyDown()

onKeyDown() 方法是 KeyEvent.Callback 接口中的抽象方法，所有的 View 都实现了该接口并重写了 onKeyDown() 方法，onKeyDown() 方法用来捕捉手机键盘被按下的事件，方法的签名如下所示。

【语法】

```
public boolean onKeyDown(int keyCode, KeyEvent event)
```

其中：

- 参数 keyCode 表示被按下的键值（即键盘码），手机键盘中每个按钮都有一个单独的键盘码，在应用程序中可通过键盘码的值来判断用户按下的是哪个键。在 KeyEvent 类中定义了许多常量来表示不同的 keyCode，如表 3-13 所示。
- 参数 event 用于封装按键事件的对象，其中包含了触发事件的详细信息，例如事件的状态、事件的类型以及事件触发的时间等。当用户按下某个键时，系统自动将事件封装成 KeyEvent 对象供应用程序使用。

表 3-13　keyCode 的部分值

常　量　名	功能描述	常　量　名	功能描述
KEYCODE_CALL	拨号键	KEYCODE_POWER	电源键
KEYCODE_ENDCALL	挂机键	KEYCODE_NOTIFICATION	通知键
KEYCODE_HOME	按键 Home	KEYCODE_MUTE	话筒静音键
KEYCODE_MENU	菜单键	KEYCODE_VOLUME_MUTE	扬声器静音键
KEYCODE_BACK	返回键	KEYCODE_VOLUME_UP	音量增加键
KEYCODE_SEARCH	搜索键	KEYCODE_VOLUME_DOWN	音量减小键
KEYCODE_CAMERA	拍照键	KEYCODE_CALL	拨号键
KEYCODE_FOCUS	拍照对焦键	KEYCODE_ENDCALL	挂机键

onKeyDown() 方法的返回值为 boolean 类型。当方法返回 true 时，表示已经完整地处理了该事件，并不希望其他的回调方法再次进行处理；当方法返回 false 时，表示没有完全处理完该事件，其他回调方法可以继续对该事件进行处理，例如 Activity 中的回调方法。

注意

> 表 3-13 中是常见的手机键盘中的 keyCode 值，此外，keyCode 还包括控制键、组合键、基本键、符号键、小键盘键和功能键等，读者可以参见 KeyEvent 代码。

下述代码通过一个简单例子来演示 onKeyDown() 方法的使用。通过用户的按键来捕获手机键盘事件，并根据按键情况来显示相关信息，代码如下所示。

【代码 3-14】 keydown_btn.xml

```xml
<LinearLayout xmlns:android="http://schemas.android.com/apk/res/android" …>
    <EditText
        android:id="@+id/showTxt"
        android:layout_width="match_parent"
        android:layout_height="wrap_content"
        android:editable="false" />
</LinearLayout>
```

上述代码定义了一个 EditText 文本框,用于显示用户单击按键时不同的文本。在相应的 Activity 中实现 onKeyDown 事件监听,代码如下所示。

【代码 3-15】 KeyDownActivity.java

```java
public class KeyDownActivity extends AppCompatActivity {
    //自定义的 Button
    EditText showText;
    public void onCreate(Bundle savedInstanceState) { //重写的 onCreate 方法
        super.onCreate(savedInstanceState);
        setContentView(R.layout.keydown_btn);
        showText = (EditText) findViewById(R.id.showTxt);}
    public boolean onKeyDown(int keyCode, KeyEvent event) {
        //重写的键盘按下监听
        switch (keyCode) {
        case KeyEvent.KEYCODE_BACK:
            showText.setText("单击了【回退键】");
            break;
        case KeyEvent.KEYCODE_0:
            showText.setText("0 键");
            break;
        case KeyEvent.KEYCODE_A:
            showText.setText("A 键");
            break;
        case KeyEvent.KEYCODE_VOLUME_DOWN:
            showText.setText("音量-");
            break;
        case KeyEvent.KEYCODE_VOLUME_UP:
            showText.setText("音量+");
            break;
        default:
            break;}
        return super.onKeyDown(keyCode, event);}
}
```

上述代码中,粗体部分为重写的 Activity 中的 onKeyDown()方法,在该方法中实现键盘按下的事件处理,方法返回之前调用父类的同名方法并返回处理结果。当单击手机键盘上的回退键、0 键、A 键、音量-或音量+键时,在 showText 文本框中显示对应的文本信息,例如,当在键盘上单击"音量+"时,显示结果如图 3-14 所示。

如果将 onKeyDown()方法中的返回代码"return super.onKeyDown(keyCode, event);"改为"return true;",然后单击回退键时,则回退键的单击事件会被捕获并进行处理,但界面仍然在当前页面,并没有回退效果,如图 3-15 所示。

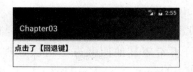

图 3-14　单击音量+效果　　　　图 3-15　单击回退键效果

> 在实际应用中,有时需要对 HOME 等特殊键进行屏蔽处理,可以采用上述方式对这些键进行捕获并处理。

2. onKeyUp()

onKeyUp()方法也是接口 KeyEvent.Callback 中的一个抽象方法,并且所有的 View 都实现了该接口并重写了 onKeyUp()方法,onKeyUp()方法用来捕捉手机键盘按键抬起的事件,方法的签名如下所示。

【语法】

```
public boolean onKeyUp ( int keyCode, KeyEvent event)
```

其中:
- 参数 keyCode 表示触发事件的按键码,需要注意的是,同一个按键在不同型号的手机中的按键码可能不同。
- 参数 event 是一个事件封装类的对象,其含义与 onKeyDown()方法中的完全相同,此处不再赘述。

onKeyUp()方法返回值的含义与 onKeyDown()方法相同,同样通知系统是否希望其他回调方法再次对该事件进行处理。onKeyUp()的使用方式与 onKeyDown()基本相同,只是 onKeyUp()方法在按键抬起时触发调用。如果用户需要对按键被抬起时进行事件处理,可以通过重写该方法来实现。

3. onTouchEvent()

onKeyDown()和 onKeyUp()方法主要针对手机键盘事件的处理,onTouchEvent()方法则主要针对手机屏幕事件的处理。onTouchEvent()方法在 View 类中定义,并且所有的 View 都重写了该方法,应用程序可以通过 onTouchEvent()方法来处理手机屏幕的触摸事件。onTouchEvent()方法的签名如下所示。

【语法】

```
public boolean onTouchEvent (MotionEvent event)
```

其中:
- 参数 event 是一个手机屏幕触摸事件封装类的对象,用于封装件的相关信息,例如触摸的位置、触摸的类型以及触摸的时间等。在用户触摸手机屏幕时由系统创建

event 对象。
- onTouchEvent()方法的返回机制与键盘响应事件的相同,当已经完整地处理了该事件且不希望其他回调方法再次处理时返回 true,否则返回 false。

与 onKeyDown()、onKeyUp()方法不同的是,onTouchEvent()方法可以处理多种事件。一般情况下,屏幕中的按下、抬起和拖动事件均可由 onTouchEvent()方法进行处理,只是每种情况中的动作值有所不同。

- **屏幕被按下**:当屏幕被按下时,会自动调用 onTouchEvent()方法来处理事件,此时 MotionEvent.getAction()的值为 MotionEvent.ACTION_DOWN,如果在应用程序中需要处理屏幕被按下的事件,只需重写该回调方法,并在方法中进行动作的判断即可。
- **触摸动作抬起**:离开屏幕时所触发的事件,该事件需要 onTouchEvent()方法来捕捉,并在该方法中进行动作判断。当 MotionEvent.getAction()的值为 MotionEvent.ACTION_UP 时,表示触发的是屏幕被抬起的事件。
- **在屏幕中拖动**:onTouchEvent()方法还用于处理在屏幕上滑动的事件,根据 MotionEvent.getAction()方法的返回值来判断动作值是否为 MotionEvent.ACTION_MOVE,然后进行相应的处理。

下述代码通过一个简单例子来演示 onTouchEvent()方法的使用。在用户单击的位置绘制一个矩形,然后监测用户触摸的状态,当用户在屏幕上移动手指时,使矩形随之移动,而当用户手指离开手机屏幕时,停止绘制矩形。代码如下所示。

【代码 3-16】 KeyTouchActivity.java

```
public class KeyTouchActivity extends AppCompatActivity {
    TouchView touchView;
    @Override
    public void onCreate(Bundle savedInstanceState) {
        super.onCreate(savedInstanceState);
        touchView = new TouchView(this);        //初始化自定义的View①
        setContentView(touchView);         }    //设置当前显示的用户界面②
    //重写的 onTouchEvent 回调方法
    @Override
    public boolean onTouchEvent(MotionEvent event) {
        switch (event.getAction()) {
            case MotionEvent.ACTION_DOWN:               //手指按下 ③
                touchView.x = (int) event.getX();       //改变 x 坐标
                touchView.y = (int) event.getY() - 52;  //改变 y 坐标
                touchView.postInvalidate();
                //重绘
                break;
            case MotionEvent.ACTION_MOVE:               //手指移动 ④
                touchView.x = (int) event.getX();       //改变 x 坐标
                touchView.y = (int) event.getY() - 52;  //改变 y 坐标
                touchView.postInvalidate();
                //重绘
                break;
            case MotionEvent.ACTION_UP:                 //手指抬起⑤
                touchView.x = -100;                     //改变 x 坐标
```

```
            touchView.y = -100;                       //改变 y 坐标
        //重绘
            touchView.postInvalidate();
            break;}
        return super.onTouchEvent(event);}
//定义 View 的子类⑥
class TouchView extends View {
    Paint paint;                                      //画笔
    int x = 300;                                      //x 坐标
    int y = 300;                                      //y 坐标
    int width = 100;                                  //矩形的宽度
    public TouchView(Context context) {
        super(context);
        paint = new Paint();}                         //初始化画笔⑦
    @Override
    protected void onDraw(Canvas canvas) {            //绘制方法⑧
        canvas.drawColor(Color.WHITE);                //绘制背景色⑨
        canvas.drawRect(x, y, x + width, y + width, paint);   //绘制矩形⑩
        super.onDraw(canvas);}}
}
```

代码解释如下：标号①处创建了一个自定义的 TouchView 对象，标号②处将该 View 的对象设置为当前显示的用户界面。标号③用于判断当前事件是否为屏幕被按下的事件，通过调用 MotionEvent 的 getX() 和 getY() 方法得到事件发生的 x 和 y 坐标，并赋给 TouchView 对象的 x 与 y 成员变量。标号④用于判断是否为屏幕的滑动事件，同样将得到事件发生的位置并赋给 TouchView 对象的 x、y 成员变量。需要注意的是，因为此时手机屏幕并不是全屏模式，所以需要对坐标进行调整。标号⑤用于判断是否为屏幕被抬起的事件，此时将 TouchView 对象的 x、y 成员变量设成 -100，表示并不需要在屏幕中绘制矩形；标号⑥处自定义了 TouchView 类，并重写了 View 类的 onDraw() 绘制方法。在⑦处的构造方法中初始化绘制时需要的画笔，然后在第⑧~⑩行的方法中根据成员变量 x、y 的值来绘制矩形。

图 3-16　矩形绘制

运行上述代码，效果如图 3-16 所示。

当单击屏幕时，会在所单击的位置绘制一个矩形；当手指在屏幕中滑动时，该矩形会随之移动；而当手指离开屏幕时，则取消所绘制的矩形。

注意

自定义的 View 并不会自动刷新，所以每次改变数据模型时都需要手动调用 postInvalidate() 方法进行屏幕的刷新操作。关于自定义 View 的使用方法，将会在后面进行详细介绍，此处只是简单地使用。

由于无法在图书上展示动态效果，需要读者自行验证。

4. onTrackBallEvent()

onTrackBallEvent() 方法是手机中轨迹球的处理方法。所有的 View 同样全部实现了该方法,该方法的签名如下所示。

【语法】

```
public Boolean onTrackballEvent (MotionEvent event)
```

其中:
- 参数 event 为手机轨迹球事件封装类的对象,用于封装触发事件的相关信息,包括事件的类型、触发时间等。一般情况下,该对象会在用户操控轨迹球时由系统创建。
- onTrackBallEvent()方法的返回机制与前面介绍的各个回调方法完全相同,此处不再赘述。

轨迹球与手机键盘有一定的区别,具体如下:

① 某些型号的手机设计出的轨迹球会比只有手机键盘时更美观,可增添用户对手机的整体印象;

② 轨迹球使用更为简单,例如在某些游戏中使用轨迹球控制会更为合理;

③ 使用轨迹球会比键盘更为细化,即滚动轨迹球时,后台表示状态的数值会变得更细微、更精准;

④ onTrackBallEvent()方法的使用与前面介绍过的各个回调方法基本相同,可以在 Activity 中重写该方法,也可以在 View 的子类中重写。

注意

在模拟器运行状态下,可以通过 F6 键打开模拟器的轨迹球,然后通过鼠标的移动来模拟轨迹球事件。

5. onFocusChanged()

前面介绍的各个方法都可以在 View 及 Activity 中重写,接下来介绍的 onFocusChanged() 方法只能在 View 中重写。该方法是焦点改变的回调方法,在某个控件重写了该方法后,当焦点发生变化时,会自动调用该方法来处理焦点改变的事件,该方法的签名如下所示。

【语法】

```
protected void onFocusChanged (
    Boolean gainFocus, int direction, Rect previouslyFocusedRect)
```

其中:
- 参数 gainFocus 表示触发该事件的 View 是否获得了焦点,当该控件获得焦点时 gainFocus 为 true,否则为 false;
- 参数 direction 表示焦点移动的方向,使用数值表示,有兴趣的读者可以重写 View 中的该方法并打印该参数进行观察;
- 参数 previouslyFocusedRect 表示在触发事件的 View 的坐标系中前一个获得焦点的矩形区域,即表示焦点是从哪里来的,如果不可用则为 null。

下述代码通过一个简单例子来演示 onFocusChanged()方法的使用。通过移动上下方向键来观察屏幕中 4 个按钮在获得或失去焦点后按钮上文字的变化情况,代码如下所示。

【代码 3-17】 FocusEventActivity.java

```java
public class FocusEventActivity extends AppCompatActivity {
    //定义 4 个 button ①
    FocusButton focusButton1;
    FocusButton focusButton2;
    FocusButton focusButton3;
    FocusButton focusButton4;
    //声明 myButton04 的引用
    public void onCreate(Bundle savedInstanceState) {
        super.onCreate(savedInstanceState);
        //创建 4 个 FocusButton 对象 ②
        focusButton1 = new FocusButton(this);
        focusButton2 = new FocusButton(this);
        focusButton3 = new FocusButton(this);
        focusButton4 = new FocusButton(this);
        focusButton1.setText("focusButton1");     //设置 focusButton1 上的文字 ③
        focusButton2.setText("focusButton2");     //设置 focusButton2 上的文字
        focusButton3.setText("focusButton3");     //设置 focusButton3 上的文字
        focusButton4.setText("focusButton4");     //设置 focusButton4 上的文字
        LinearLayout linearLayout = new LinearLayout(this); //创建一个线性布局 ④
        linearLayout.setOrientation(LinearLayout.VERTICAL); //设置其布局方式为垂直
        linearLayout.addView(focusButton1);        //将 focusButton1 添加到布局中 ⑤
        linearLayout.addView(focusButton2);        //将 focusButton2 添加到布局中
        linearLayout.addView(focusButton3);        //将 focusButton3 添加到布局中
        linearLayout.addView(focusButton4);        //将 focusButton4 添加到布局中
        setContentView(linearLayout);}             //设置当前的用户界面 ⑥
    //自定义 Button ⑦
    class FocusButton extends Button {
        //自定义 Button
        public FocusButton(Context context) {
            //构造器
            super(context);     }
        @Override
        protected void onFocusChanged(boolean focused, int direction,
                Rect previouslyFocusedRect) {
            String suffix = "(选中)";
            String text = getText().toString();
            //重写的焦点变化方法
            if(focused){
                //获取焦点时,添加(选中)文字
                if(!text.contains(suffix)){
                    this.setText(text + suffix);}
            }else{
                //去掉(选中)文字
                if(text.contains(suffix)){
                    text = text.substring(0,text.length() - suffix.length());
                    this.setText(text);    }}
            super.onFocusChanged(focused, direction, previouslyFocusedRect);}}
}
```

上述代码解释如下:标号①处声明了四个自定义的按钮变量;标号②处初始化标号①所声明的 4 个自定义的按钮控件,然后在标号③处分别设置了各个按钮上的文字;标号④处创建一个线性布局,并设置其布局方式为垂直;标号⑤处用于将 4 个按钮控件依次添加到线性布局中,然后在⑥处将该线性布局设置成当前显示的用户界面;标号⑦处为自定义的 FocusButton 类,在该类中重写了 onFocusChanged() 方法,并在该方法内判断是否获取焦点,如果按钮获取焦点,则为该按钮添加"(选中)"文字,否则取消"(选中)"文字。

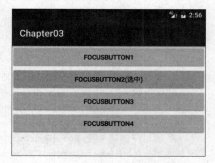

图 3-17 焦点获取

运行上述代码,效果如图 3-17 所示。

读者可以通过上下方向键来控制各个按钮的焦点切换,并观察界面的变化情况。

注意

> 每按下一次按键,会调用两次 onFocusChanged() 方法,一次是某个按钮失去焦点时调用,第二次是另一个按钮获得焦点时调用。同时方向 Direction 的值会根据情况的不同而有所不同。此外,默认情况下 Android 5.0.1 模拟器的方向键可能不起作用,需要读者自己重新设置一下,将 C:\Users\xxuser\.android\avd\android_720p.avd\文件夹下 config.ini 中的 hw.dPad 属性值改为 yes。其中,xxuser 表示当前系统用户,android_720p 表示自定义的模拟器名称。

在介绍 onFocusChanged() 方法时,提到了焦点的概念。焦点描述了按键事件(或者是屏幕事件等)的接收者,每次按键事件都发生在拥有焦点的 View 上。在应用程序中,开发人员可以对焦点进行控制,例如将焦点从一个 View 移动另一个 View 上。下面列出一些与焦点有关的方法,如表 3-14 所示,读者可以进一步进行学习。

表 3-14 常见的焦点相关方法

方　　法	功　能　描　述
setFocusable()	用于设置 View 是否可以拥有焦点
isFocusable()	用于判断 View 是否可以拥有焦点
setNextFocusDownId()	用于设置 View 的焦点向下移动后获得焦点 View 的 ID
hasFocus()	用于判断 View 的父控件是否获得了焦点
requestFocus()	用于尝试让此 View 获得焦点
isFocusableTouchMode()	用于设置 View 是否可以在触摸模式下获得焦点,默认情况下不可用

3.4 Widget 简单组件

本节将介绍 Android 的基本组件。一个易操作、美观的 UI 界面,都是从界面布局开始,然后不断地向布局容器中添加界面组件。掌握这些最基本的用户界面组件,是学好

Android 编程的基础。几乎所有的用户界面组件都定义在 android.widget 包中，如 Button、TextView、EditText、CheckBox、RadioGroup 和 Spinner 等。

3.4.1 Widget 组件通用属性

对 Widget 组件进行 UI 设计时，既可以采用 xml 布局方式也可以采用编写代码的方式。其中 xml 布局文件方式由于简单易用，被广泛使用。Widget 组件几乎都属于 View 类，因此大部分属性在这些组件之间是通用的，如表 3-15 所示。

表 3-15 Widget 组件通用属性

属 性 名 称	功 能 描 述
android:id	设置控件的索引，Java 程序可通过 R.id.<索引>形式来引用该控件
android:layout_height	设置布局高度，使用以下三种方式来指定高度：fill_parent（和父元素相同）、wrap_content（随组件本身的内容调整）、通过指定 px 值来设置高度
android:layout_width	设置布局宽度，也可以采用三种方式：fill_parent、wrap_content、指定 px 值
android:autoLink	设置是否当文本为 URL 链接时，文本显示为可单击的链接。可选值为 none、web、email、phone、map 和 all
android:autoText	如果设置，将自动执行输入值的拼写纠正
android:bufferType	指定 getText() 方式取得的文本类别
android:capitalize	设置英文字母大写类型。需要弹出输入法才能看得到
android:cursorVisible	设定光标为显示/隐藏，默认显示
android:digits	设置允许输入哪些字符，如 1234567890.+-*/%\n()
android:drawableBottom	在 text 的下方输出一个 drawable
android:drawableLeft	在 text 的左边输出一个 drawable
android:drawablePadding	设置 text 与 drawable（图片）的间隔，与 drawableLeft、drawableRight、drawableTop、drawableBottom 一起使用，可设置为负数，单独使用没有效果
android:drawableRight	在 text 的右边输出一个 drawable 对象
android:inputType	设置文本的类型，用于帮助输入法显示合适的键盘类型
android:cropToPadding	是否截取指定区域用空白代替；单独设置无效果，需要与 scrollY 一起使用
android:maxHeight	设置 View 的最大高度

3.4.2 TextView 文本框

TextView 文本框直接继承了 View 类，位于 android.widget 包中。TextView 定义了操作文本框的基本方法，是一个不可编辑的文本框，多用于在屏幕中显示静态字符串。从功能上来看，TextView 实际上是一个文本编辑器，只是 Android 关闭了其文字编辑功能，如果开发者想要定义一个可以编辑内容的文本框，可以使用其子类 EditText 来实现，EditText 允许用户编辑文本框的内容。此外 TextView 还是 Button 的父类，TextView 类及其子类的继承关系如图 3-18 所示。

TextView 提供了大量的 XML 属性，这些属性不仅适用于 TextView，适用于其子类。TextView 所支持的 XML 属性及说明如表 3-16 所示。

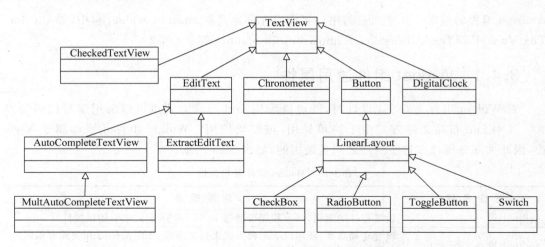

图 3-18 TextView 及其子类的继承关系

表 3-16 TextView 类的 XML 属性及描述

XML 属性	功 能 描 述
android:autoLink	设置是否当文本为 URL 链接(例如：email、电话号码、map)时，文本显示为可单击的链接。可选值有 none、web、email、phone、map 和 all
android:autoText	如果设置,将自动执行输入值的拼写纠正。此处无效,在显示输入法并输入的时候起作用
android:digits	设置允许输入哪些字符。如 1234567890.+-*/%\n()
android:drawableLeft	在 text 的左边输出一个 drawable,如图片
android:drawablePadding	设置 text 与 drawable(图片)的间隔,与 drawableLeft、drawableRight、drawableTop、drawableBottom 一起使用,可设置为负数,单独使用没有效果
android:drawableRight	在 text 的右边输出一个 drawable,如图片
android:drawableTop	在 text 的正上方输出一个 drawable,如图片
android:ellipsize	设置当文字过长时如何显示该控件,取值情况如下：start,省略号显示在开头；End,省略号显示在结尾；middle,省略号显示在中间；marquee,以跑马灯的方式显示(动画横向移动)
android:gravity	设置文本位置,例如 center 表示文本将居中显示
android:hint	文本为空时显示的提示信息,可通过 textColorHint 设置提示信息的颜色。此属性在 TextView 和 EditView 中使用
android:ems	设置 TextView 的宽度为 N 个字符的宽度
android:maxEms	设置 TextView 的宽度最长为 N 个字符的宽度,与 ems 同时使用时将覆盖 ems 选项
android:minEms	设置 TextView 的宽度最短为 N 个字符的宽度,与 ems 同时使用时将覆盖 ems 选项
android:maxLength	限制显示的文本长度,超出部分不显示
android:lines	设置文本的行数,设置两行就显示两行,即使第二行没有数据
android:maxLines	设置文本的最大显示行数,与 width 或者 layout_width 结合使用,超出部分自动换行,超出行数将不显示
android:minLines	设置文本的最小行数,与 lines 类似
android:linksClickable	设置链接是否可以单击,即使设置了 autoLink
android:lineSpacingExtra	设置行间距

续表

XML 属性	功 能 描 述
android:lineSpacingMultiplier	设置行间距的倍数,例如 1.2
android:numeric	如果被设置,该控件将有一个数字输入法。该属性对 TextView 无用,设置后唯一效果是有单击效果,对输入控件如 EditView 才有实际意义
android:password	以小点"."显示文本
android:phoneNumber	设置为电话号码的输入方式
android:scrollHorizontally	设置文本超出 TextView 的宽度的情况下,是否出现横向滚动条
android:selectAllOnFocus	如果文本是可选的,则使其获取焦点,而不是将光标移动至文本的开始位置或者末尾位置。在 TextView 中设置后无效果
android:shadowColor	指定文本阴影的颜色,需要与 shadowRadius 一起使用
android:shadowDx	设置阴影横向坐标开始位置
android:shadowDy	设置阴影纵向坐标开始位置
android:shadowRadius	设置阴影的半径。设置为 0.1 就变成字体的颜色,一般设置为 3.0 的效果较好
android:singleLine	设置单行显示。如果和 layout_width 一起使用,当文本不能全部显示时,后面用 … 来表示,例如 android:text=" test _ singleLine" android:singleLine="true" android:layout_width="20dp"将只显示 t…。如果不设置 singleLine 或者设置为 false,文本将自动换行
android:text	设置显示文本
android:textAppearance	设置文字外观。如"? android:attr/textAppearanceLargeInverse",此处引用的是系统自带的一个外观,"?"表示系统是否有这种外观,否则使用默认的外观。取值情况如下:textAppearanceButton、textAppearanceInverse、textAppearanceLarge、textAppearanceLargeInverse、textAppearanceMedium、textAppearanceMediumInverse、textAppearanceSmall、textAppearanceSmallInverse
android:textColor	设置文本颜色
android:textColorHighlight	被选中文字的底色,默认为蓝色
android:textColorHint	设置提示信息文字的颜色,默认为灰色。与 hint 一起使用
android:textColorLink	文字链接的颜色
android:textScaleX	设置文字缩放,默认为 1.0f。可设置为 0.5f、1.0f、1.5f、2.0f
android:textSize	设置文字大小,推荐度量单位 sp,如 15sp
android:textStyle	设置字形,该属性有 3 个常量值:bold(粗体) 0,italic(斜体) 1,bolditalic(又粗又斜) 2。设置时可以使用一个或多个常量值,多个常量值之间用"|"隔开
android:height	设置文本区域的高度,支持度量单位为 px(像素)、dp、sp、in、mm
android:maxHeight	设置文本区域的最大高度
android:minHeight	设置文本区域的最小高度
android:width	设置文本区域的宽度,支持度量单位:px(像素)、dp、sp、in、mm

表 3-16 中介绍了 TextView 最常用的属性,其他不常用的属性此处没有介绍,读者可在实际使用时再具体查询。

TextView 提供了大量的 XML 属性,通过这些属性开发人员可以控制 TextView 中文本的行为,下面通过简单示例讲解 TextView 的基本用法,代码如下所示。

【代码3-18】 textview_demo.xml

```xml
<LinearLayout xmlns:android="http://schemas.android.com/apk/res/android" …>
    <!-- 设置字号为20sp -->
    <TextView
        android:layout_width="match_parent"
        android:layout_height="wrap_content"
        android:text="TextView演示"
        android:textSize="20sp" />
    <!-- 设置中间省略,所有字母大写,字号为20sp,内容一行 -->
    <TextView
        android:layout_width="match_parent"
        android:layout_height="wrap_content"
        android:singleLine="true"
        android:ellipsize="middle"
        android:text="TextView演示 TextView演示 TextView演示 TextView
            演示 TextView演示 TextView演示 TextView演示"
        android:textAllCaps="true"
        android:textSize="20sp" />
    <!-- 邮件、电话添加链接 -->
    <TextView
        android:layout_width="match_parent"
        android:layout_height="wrap_content"
        android:singleLine="true"
        android:text="邮件: qst@163.com 电话: 053212345678"
        android:autoLink="email|phone"
        android:textSize="20sp" />
    <!-- 测试密码框 -->
    <TextView
        android:layout_width="match_parent"
        android:layout_height="wrap_content"
        android:text="TextView演示"
        android:password="true"
        android:textSize="20sp" />
</LinearLayout>
```

代码解释如下：第一个 TextView 通过设置 android:textSize="20sp" 指定了字号为 20sp。其中,sp(scaled pixels)为比例像素,主要用于处理字体的大小,可以根据用户的字体大小首选项进行缩放；第二个 TextView 设置了 android:ellipsize="middle" 属性,当文本内容大余文本框宽度时,从中间省略文本,通过设定 android:textAllCaps="true" 属性,将该文本框中的所有字母都以大写形式进行显示；第三个 TextView 设置了 android:autoLink="email|phone" 属性,该文本框会自动为文本框内的 Email、电话号码添加超链接；第四个 TextView 设置了 android:password="true" 属性,指定了该文本框会用点来显示所有字符。

运行 TextViewDemoActivity,效果如图 3-19 所示。

图 3-19 TextView 效果演示

3.4.3 EditText 编辑框

EditText 是 TextView 的子类，继承了 TextView 的 XML 属性和方法，EditText 和 TextView 的最大区别是：EditText 可以接收用户的输入。EditText 作为用户与系统之间的文本输入接口，用于接收用户输入的数据并传给系统，从而使系统获取所需要的数据。EditText 组件最重要是 inputType 属性，该属性用于指定在 EditText 输入值时所启动的虚拟键盘风格，例如经常需要虚拟键盘只提供字符或数字。在开发过程中，常用的 inputType 属性值如表 3-17 所示。

表 3-17 inputType 属性值

属 性 值	功 能 描 述	属 性 值	功 能 描 述
text	普通文本，默认	textLongMessage	长信息
textCapCharacters	字母大写	textPassword	密码
textCapWords	每个单词的首字母大写	number	数字
textAutoCorrect	自动完成	numberSigned	带符号数字格式
textMultiLine	多行输入	numberDecimal	带小数点的浮点格式
textNoSuggestions	不提示	phone	拨号键盘
textUri	网址	datetime	时间日期
textEmailAddress	电子邮件地址	date	日期键盘
textEmailSubject	邮件主题	time	时间键盘
textShortMessage	短信		

下面通过简单示例演示 inputType 属性的使用，代码如下所示。

【代码 3-19】 edittext_demo.xml

```
<LinearLayout xmlns:android = "http://schemas.android.com/apk/res/android" …>
    <!-- 多行效果 -->
    <EditText
        android:layout_width = "match_parent"
        android:layout_height = "wrap_content"
        android:inputType = "textMultiLine"
        android:hint = "多行效果"
        android:textSize = "20sp" />
    <!-- 拨号键盘 -->
    <EditText
        android:layout_marginTop = "15dp"
        android:layout_width = "match_parent"
        android:layout_height = "wrap_content"
        android:inputType = "phone"
        android:textSize = "20sp" />
    <!-- 密码类型 -->
    <EditText
        android:layout_marginTop = "15dp"
        android:layout_width = "match_parent"
        android:layout_height = "wrap_content"
        android:inputType = "textPassword"
        android:hint = "输入密码"
        android:textSize = "20sp" />
</LinearLayout>
```

代码解释如下：上述代码中，三个 EditText 都通过 android:hint 属性指定了文本框的提示信息；当用户输入内容之前，文本框内默认显示指定的提示信息，当用户输入信息时，提示信息被清除，第一个 EditText 通过设置 android:inputType="textMultiLine"属性来指定该文本框允许多行输入；第二个 EditText 通过设置 android:inputType="phone"属性来指定该文本框只能接收数值的输入；第三个 EditText 通过设置 android:inputType="textPassword"属性来指定该文本框是一个密码框。

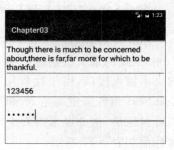

图 3-20　EditText 效果演示

通过 Activity 运行上述代码，效果如图 3-20 所示。

3.4.4　Button 按钮

Button 继承于 TextView，主要用于在 UI 界面上生成一个按钮，当用户单击按钮时，会触发一个 OnClick 事件。按钮的使用相对比较容易，通过 android:backgroud 属性为按钮指定背景颜色或背景图片，使用各式各样的背景图片可以实现各种不规则形状的按钮。

Button 类通过继承父类的方法来实现对按钮组件的操作，表 3-18 列举了 Button 类的常用方法。

表 3-18　Button 类的方法

方　　法	功　能　描　述
onKeyDown()	当用户按键时，该方法被调用
onKeyUp()	当用户按键弹起后，该方法被调用
onKeyLongPress()	当用户保持按键时，该方法被调用
onKeyMultiple()	当用户多次按键时，该方法被调用
invalidateDrawable()	用于刷新 Drawable 对象
onPreDraw()	用于设置视图显示，例如在视图显示之前调整滚动轴的边界
setOnKeyListener()	用于设置按键监听器
setOnClickListener()	用于设置单击监听器

下述代码以 setOnClickListener()为例，通过一个模拟登录操作来演示 Button、TextView 和 EditView 的使用。界面布局的代码如下所示。

【代码 3-20】　login.xml

```xml
<LinearLayout xmlns:android="http://schemas.android.com/apk/res/android" …>
    <!-- 标题 ① -->
    <TextView
        android:layout_width = "wrap_content"
        android:layout_height = "wrap_content"
        android:layout_gravity = "center"
        android:text = "用户登录"
        android:textSize = "35sp" />
    <!-- 用户名 ② -->
    <LinearLayout
        android:layout_width = "match_parent"
```

```xml
        android:layout_height = "wrap_content"
        android:layout_marginTop = "10dp" >
    <TextView
        android:layout_width = "wrap_content"
        android:layout_height = "wrap_content"
        android:text = "用户名: " />
    <EditText
        android:id = "@ + id/userNameTxt"
        android:layout_width = "match_parent"
        android:layout_height = "wrap_content" />
</LinearLayout>
<!-- 密码 ③ -->
<LinearLayout
    android:layout_width = "match_parent"
    android:layout_height = "wrap_content" >
    <TextView
        android:layout_width = "wrap_content"
        android:layout_height = "wrap_content"
        android:inputType = "textPassword"
        android:text = "密    码: " />
    <EditText
        android:id = "@ + id/passwordTxt"
        android:layout_width = "match_parent"
        android:layout_height = "wrap_content" />
</LinearLayout>
<!-- 登录按钮 ④ -->
<Button
    android:id = "@ + id/loginBtn"
    android:layout_width = "match_parent"
    android:layout_height = "wrap_content"
    android:text = "登录" />
<!-- 成功或失败提示 ⑤ -->
<TextView
    android:id = "@ + id/tipsTxt"
    android:layout_width = "wrap_content"
    android:layout_height = "wrap_content"
    android:layout_gravity = "center"
    android:layout_marginTop = "5dp"
    android:text = "显示成功或失败"
    android:visibility = "gone" />
</LinearLayout>
```

代码解释如下：标号①处显示的是当前界面的标题，并将 TextView 的字号设为 35sp，相对比较醒目；标号②处通过 LinearLayout 将 TextView 和 EditText 组合起来，用于接收用户名的输入；标号③处通过 LinearLayout 将 TextView 和 EditText 组合起来，用于接收密码的输入，其中 EditText 的 inputType 设置为 textPassword，表示该文本框当密码框使用；标号④定义了一个登录按钮，在 Activity 中用于实现登录的业务逻辑处理；标号⑤定义了一个 TextView，用于显示用户登录成功或失败后的提示，例如用户名不存在、密码错误等。通过设置 android:visibility = "gone"，默认隐藏该提示。

接下来在相应的 Activity 中实现登录的业务逻辑，此处仅实现一个简单的登录验证，并不进行真正的登录跳转,代码如下所示。

【代码3-21】 LoginActivity.java

```java
public class LoginActivity extends AppCompatActivity {
    EditText userNameTxt;              //用户名 ①
    EditText passwordTxt;              //密码
    Button loginBtn;                   //登录按钮
    TextView tipsTv;                   //提示
    @Override
    protected void onCreate(Bundle savedInstanceState) {
        super.onCreate(savedInstanceState);
        setContentView(R.layout.login);
        //初始化各个组件 ②
        userNameTxt = (EditText)findViewById(R.id.userNameTxt);
        passwordTxt = (EditText) findViewById(R.id.passwordTxt);
        tipsTv = (TextView) findViewById(R.id.tipsTxt);
        loginBtn = (Button)findViewById(R.id.loginBtn);
        //实现单击 Button 的业务逻辑 ③
        loginBtn.setOnClickListener(new View.OnClickListener() {
            @Override
            public void onClick(View v) {
                String userName = userNameTxt.getText().toString();      //获取用户名
                String password = passwordTxt.getText().toString();      //获取密码
                //判断用户名
                if(!"admin".equals(userName)) {
                    tipsTv.setText("用户名不存在!");
                    tipsTv.setVisibility(View.VISIBLE);
                    return;}
                if(!"1".equals(password)) {
                    tipsTv.setText("密码不正确!");
                    tipsTv.setVisibility(View.VISIBLE);
                    return; }
                if("admin".equals(userName)&&"1".equals(password)) {
                    tipsTv.setText("登录成功!");
                    tipsTv.setVisibility(View.VISIBLE);}}});}
}
```

代码解释如下:标号①处定义了一个 EditText 类型的属性变量 userNameTxt,用于获取界面传递过来的对象,其他属性的定义功能类似,此处不再赘述。标号②处对标号①处所定义的各个属性变量进行初始化,通过对属性变量的赋值,使其能够进行后续的业务逻辑操作。标号③处实现了登录按钮 loginBtn 的业务逻辑,逻辑相对比较简单,如果用户名无效,则显示"用户名不存在";如果密码不对,则显示"密码不正确!";如果用户输入的用户名和密码都正确,则显示"登录成功!"。

注意

此处仅演示 Button、TextView 和 EditText 的使用,并没用牵扯到更复杂的应用逻辑。在实际开发中,用户登录时需要查询后台数据库来验证用户名和密码是否正确,请参见本书贯穿任务中的登录模块。

运行上述代码,当用户输入错误时进行相应的错误提示,例如用户名输入 admi,密码任意输入 1,显示的界面如图 3-21 所示。当用户输入正确的用户名和密码时,显示的界面如图 3-22 所示。

图 3-21　用户输入错误时的情况　　　图 3-22　用户输入正确时的情况

读者可以进一步完善该示例,例如增加用户名和密码的空校验等功能。

3.4.5　单选按钮和单选按钮组

在一组按钮中有且仅有一个按钮能够被选中,当选择按钮组中某个按钮时会取消其他按钮的选中状态。上述效果需要 RadioButton 和 RadioGroup 配合使用才能实现。RadioGroup 是单选按钮组,是一个允许容纳多个 RadioButton 的容器。在没有 RadioGroup 的情况下,RadioButton 可以分别被选中;当多个 RadioButton 同在一个 RadioGroup 按钮组中时,RadioButton 只允许选择其中之一。RadioButton 和 RadioGroup 的关系如下:RadioButton 表示单个圆形单选框,而 RadioGroup 是一个可以容纳多个 RadioButton 的容器;同一个 RadioGroup 中,只能有一个 RadioButton 被选中;不同的 RadioGroup 中,RadioButton 互不影响,即如果组 A 中有一个被选中,组 B 中依然可以有一个被选中;通常情况下,一个 RadioGroup 中至少有两个 RadioButton;大部分应用场景下,建议一个 RadioGroup 中的 RadioButton 默认会有一个被选中,并将其放在 RadioGroup 中的起始位置。

RadioGroup 类是 LinearLayout 的子类,其常用的设置和控制单选按钮组的方法如表 3-19 所示。

表 3-19　RadioButton 相关方法

方　　法	功　能　描　述
getCheckedRadioButtonId()	获取被选中按钮的 ID
clearCheck()	清除选中状态
check(int id)	通过参数 ID 来设置该选项为选中状态;如果传入 −1 则清除单选按钮组的勾选状态,相当于调用 clearCheck()操作
setOnCheckedChangeListener（RadioGroup. OnCheckedChangeListener listener)	在一个单选按钮组中,当该单选按钮勾选状态发生改变时所要调用的回调函数。当 RadioButton 的 checked 属性为 true 时,check(id)方法不会触发 onCheckedChanged 事件
addView(View child, int index, ViewGroup. LayoutParams params)	使用指定的布局参数添加一个子视图。其中:child 为所要添加的子视图;index 为将要添加子视图的位置;params 为所要添加的子视图的布局参数
getText()	用于获取单选框的值

此外，通过 OnCheckedChangeListener 监听器可对单选按钮的状态切换进行监听并处理。

下面通过一个简单的实例演示 RadioButton 和 RadioGroup 的使用，界面布局的代码如下所示。

【代码 3-22】 radiobutton_demo.xml

```xml
<LinearLayout xmlns:android = "http://schemas.android.com/apk/res/android" …>
    <!-- 显示选择的内容 ① -->
    <TextView
        android:id = "@+id/chooseTxt"
        android:layout_width = "match_parent"
        android:layout_height = "wrap_content"
        android:text = "我选择的是…?"
        android:textSize = "30sp" />
    <!-- 单选按钮组 ② -->
    <RadioGroup
        android:id = "@+id/radioGroup"
        android:layout_width = "match_parent"
        android:layout_height = "match_parent" >
        <RadioButton
            android:id = "@+id/radioButton1"
            android:layout_width = "wrap_content"
            android:layout_height = "match_parent"
            android:text = "按钮 1" />
        <RadioButton
            android:id = "@+id/radioButton2"
            android:layout_width = "wrap_content"
            android:layout_height = "match_parent"
            android:text = "按钮 2" />
    </RadioGroup>
    <!-- 清除所有选中状态 ③ -->
    <Button
        android:id = "@+id/radio_clearBtn"
        android:layout_width = "match_parent"
        android:layout_height = "match_parent"
        android:text = "清除选中" />
    <!-- 往按钮组中添加新的单选按钮 ④ -->
    <Button
        android:id = "@+id/radio_addBtn"
        android:layout_width = "match_parent"
        android:layout_height = "match_parent"
        android:text = "添加子项" />
</LinearLayout>
```

代码解释如下：标号①处用于显示当前选中按钮的标题；标号②处定义了一个单选按钮组，并为该按钮组添加了两个单选按钮；标号③处定义了一个"清除"按钮，用于清除按钮组中所有单选按钮的选中状态；标号④处定义了一个"添加子项"按钮，用于向按钮组中添加新的互斥的单选按钮。

接下来在对应的 Activity 中演示按钮组的使用，实现清除按钮组中所有按钮的选中状态，以及向按钮组中添加新的单选按钮的功能，代码如下所示。

【代码 3-23】 RadioButtonActivity.java

```java
public class RadioButtonActivity extends AppCompatActivity {
    private TextView chooseTxt;              //显示选择的单选按钮文本 ①
    private RadioGroup radioGroup;           //按钮组
    //多个单选按钮
    private RadioButton radioButton1;
    private RadioButton radioButton2;
    private Button radioClearBtn;            //清除按钮
    private Button radioAddBtn;              //新增按钮
    @Override
    protected void onCreate(Bundle savedInstanceState) {
        super.onCreate(savedInstanceState);
        setContentView(R.layout.radiobutton_demo);
        //初始化按钮 ②
        chooseTxt = (TextView)findViewById(R.id.chooseTxt);
        radioGroup = (RadioGroup)findViewById(R.id.radioGroup);
        radioButton1 = (RadioButton)findViewById(R.id.radioButton1);
        radioButton2 = (RadioButton)findViewById(R.id.radioButton2);
        radioGroup.setOnCheckedChangeListener(onCheckedChangeListener);
        //清除选中状态
        radioClearBtn = (Button)findViewById(R.id.radio_clearBtn);
        radioClearBtn.setOnClickListener(onClickListener);
        //增加子选项
        radioAddBtn = (Button)findViewById(R.id.radio_addBtn);
        radioAddBtn.setOnClickListener(onClickListener);}
    /**定义按钮组的监听事件 ③*/
    private OnCheckedChangeListener onCheckedChangeListener
        = new OnCheckedChangeListener() {
        @Override
        public void onCheckedChanged(RadioGroup group, int checkedId) {
        int id = group.getCheckedRadioButtonId();
            switch (group.getCheckedRadioButtonId()) {   //获取当前选中的按钮的 Id
            case R.id.radioButton1:
                chooseTxt.setText("我选择的是:" + radioButton1.getText());
                break;
            case R.id.radioButton2:
                chooseTxt.setText("我选择的是:" + radioButton2.getText());
                break;
            default:
                chooseTxt.setText("我选择的是:新增");
                break;}} };
    //定义清除状态按钮和新增按钮的单击事件 ④
    private OnClickListener onClickListener = new OnClickListener() {
        @Override
        public void onClick(View view) {
            switch (view.getId()) {
            case R.id.radio_clearBtn:
                radioGroup.check(-1);                    //清除选项
                chooseTxt.setText("我选择的是...?");
                break;
            case R.id.radio_addBtn:
                //新增子选项
```

```
                    RadioButton  newRadio = new RadioButton(RadioButtonActivity.this);
                    newRadio.setLayoutParams(new LayoutParams(
                            LayoutParams.MATCH_PARENT, LayoutParams.MATCH_PARENT));
                    newRadio.setText("新增");
                    radioGroup.addView(newRadio,radioGroup.getChildCount());
                    break;
                default:
                    break;}}}
}
```

代码解释如下：标号①处定义了一个 TextView 类型的属性变量 chooseTxt，用于获取当前被选中按钮的文本，其他属性的定义请见注释，此处不再赘述；标号②处对标号①处所定义的各个属性变量进行初始化，通过对属性变量的赋值，使其可以进行后续的业务逻辑操作；标号③处定义了一个按钮组监听器对象，用于获取当前在按钮组中选中的单选按钮对象，并将文本显示在 chooseTxt 对象上；标号④处定义一个普通按钮监听器对象，用于实现 radioClearBtn 和 radioAddBtn 的业务逻辑功能，当用户单击 radioClearBtn 按钮时，按钮组中被选中的单选按钮状态被清空；当用户单击 radioAddBtn 时，系统会在 radioGroup 对象中增加一个单选按钮对象。

运行上述代码，效果如图 3-23 所示。当用户单击"添加子项"，并选中新增的选项后，效果如图 3-24 所示。

可以通过 android:drawableRight 属性使单选按钮在文本的右边显示，对布局文件进行修改，代码如下所示。

【代码 3-24】 radiobutton_demo.xml

```xml
<LinearLayout xmlns:android = "http://schemas.android.com/apk/res/android" …>
    <!-- 显示选择的内容   -->
    <TextView
        android:id = "@ + id/chooseTxt"
        android:layout_width = "match_parent"
        android:layout_height = "wrap_content"
        android:text = "我选择的是…?"
        android:textSize = "30sp" />
    <!-- 单选按钮组  -->
    <RadioGroup
        android:id = "@ + id/radioGroup"
        android:layout_width = "match_parent"
        android:layout_height = "match_parent" >
        <RadioButton
            android:id = "@ + id/radioButton1"
            android:layout_width = "wrap_content"
            android:layout_height = "match_parent"
            android:button = "@null"
            android:drawableRight = "@android:drawable/btn_radio"
            android:text = "按钮 1" />
…代码省略
```

运行修改后的代码，效果如图 3-25 所示。

图 3-23　选中的图示　　　图 3-24　添加子项　　　图 3-25　改变 RadioButton 样式

> **注意**
>
> 关于按钮默认样式的自定义与相关理论知识，请参考 2.6 节。

3.4.6　CheckBox 复选框

CheckBox 复选框是一种具有双状态的按钮，具有选中或者未选中两种状态。在布局文件中定义复选按钮时，对每一个按钮注册 OnCheckedChangeListener 事件监听，然后在 onCheckedChanged() 事件处理方法中根据 isChecked 参数来判断选项是否被选中。

CheckBox 和 RadioButton 的主要区别如下：

（1）RadioButton 单选按钮被选中后，再次单击时无法改变其状态，而 CheckBox 复选框被选中后，可以通过单击来改变其状态。

（2）在 RadioButton 单选按钮组中，只允许选中一个；而在 CheckBox 复选框中，允许同时选中多个。

（3）在大部分 UI 框架中 RadioButton 默认都以圆形表示，CheckBox 默认都以矩形表示。

下述代码通过一个简单的示例演示 CheckBox 的用法，以"体育爱好"的多选为例，人们的"体育爱好"可能有足球、篮球等，而人的性别选择有所不同，性别只能选择"男"或"女"，且两者互斥。示例的界面布局代码如下所示。

【代码 3-25】　checkbox_demo.xml

```xml
<LinearLayout xmlns:android = "http://schemas.android.com/apk/res/android" …>
    <!-- 基本显示 ① -->
    <TextView
        android:layout_width = "match_parent"
        android:layout_height = "wrap_content"
        android:text = "@string/title"
        android:textSize = "20sp"
        android:textStyle = "bold"
        android:textColor = "#FFFFFF"
        />
    <!-- 足球 ② -->
    <CheckBox
        android:id = "@+id/checkbox1"
        android:layout_width = "wrap_content"
```

```xml
        android:layout_height = "wrap_content"
        android:text = "@string/football"
        android:textSize = "16sp"
        />
    <!-- 篮球 ③ -->
    <CheckBox
        android:id = "@+id/checkbox2"
        android:layout_width = "wrap_content"
        android:layout_height = "wrap_content"
        android:text = "@string/basketball"
        android:textSize = "16sp"
        />
    <!-- 排球 ④ -->
    <CheckBox
        android:id = "@+id/checkbox3"
        android:layout_width = "wrap_content"
        android:layout_height = "wrap_content"
        android:text = "@string/volleyball"
        android:textSize = "16sp"
        />
</LinearLayout>
```

代码解释如下：标号①处的 TextView 用于显示用户的标题；标号②处定义的是"足球"复选框；标号③处定义的是"篮球"复选框；标号④处定义的是"排球"复选框。

上述代码中，复选框的文本部分使用了字符串资源，例如"足球"的文本是引用的 strings.xml 文件中的字符串，其中 strings.xml 中的字符串定义如下所示。

【代码 3-26】 strings.xml

```xml
<?xml version = "1.0" encoding = "utf-8"?>
<resources>
    <string name = "title">你喜欢的运动是：</string>
    <string name = "app_name">复选框测试</string>
    <string name = "football">足球</string>
    <string name = "basketball">篮球</string>
    <string name = "volleyball">排球</string>
</resources>
```

通常在开发过程中使用 strings.xml 文件的目的如下。

（1）为了国际化。Android 建议将屏幕中显示的文字定义在 strings.xml 中，如果今后需要进行国际化时仅需要修改 string.xml 文件即可。例如，原本开发的应用是面向国内用户的，在屏幕上使用中文，当需要将应用国际化时，用户希望屏幕上所显示的内容是英文，此时如果没有把文字信息定义在 string.xml 中，就需要修改程序的内容来实现。但如果把所有屏幕上出现的文字信息都集中存放在 string.xml 文件中，在需要国际化时只需修改 string.xml 中所定义的字符串资源即可，android 操作系统会根据用户手机的语言环境和国家来自动选择相应的 string.xml 文件，实现起来更加方便。

（2）为了减少应用的体积，降低数据的冗余。例如在应用中要使用"我们一直在努力"这段文字 1000 次，如果不将"我们一直在努力"定义在 string.xml 文件中，而是在每次使用时直接使用该字符串，这样会浪费大量的空间，并且维护起来较为麻烦。

> 关于各种类型资源文件的讲解会在后面的章节中进行,此处不再赘述。

下面在相应的 Activity 中演示复选框的使用,当用户选择不同的"爱好"时,在屏幕上显示用户的选择结果,代码如下所示。

【代码 3-27】 CheckBoxDemoActivity.java

```java
public class CheckBoxDemoActivity extends AppCompatActivity {
    //声明复选框 ①
    private CheckBox footballChx;
    private CheckBox basketballChx;
    private CheckBox volleyballChx;
    @Override
    public void onCreate(Bundle savedInstanceState) {
        super.onCreate(savedInstanceState);
        setContentView(R.layout.checkbox_demo);
        //通过 findViewById 获得 CheckBox 对象 ②
        footballChx = (CheckBox) findViewById(R.id.footballChx);
        basketballChx = (CheckBox) findViewById(R.id.basketballChx);
        volleyballChx = (CheckBox) findViewById(R.id.volleyballChx);
        //注册事件监听器 ③
        footballChx.setOnCheckedChangeListener(listener);
        basketballChx.setOnCheckedChangeListener(listener);
        volleyballChx.setOnCheckedChangeListener(listener);}
    //响应事件 ④
    private OnCheckedChangeListener listener = new OnCheckedChangeListener() {
        @Override
        public void onCheckedChanged(CompoundButton buttonView,
             boolean isChecked) {
                switch (buttonView.getId()) {
            case R.id.footballChx:
                //选择足球
                if (isChecked) {
                    //Toast 的使用 ⑤
                    Toast.makeText(CheckBoxDemoActivity.this, "你喜欢足球",
                        Toast.LENGTH_LONG).show();}
                break;
            case R.id.basketballChx:
                //选择篮球
                if (isChecked) {
                    Toast.makeText(CheckBoxDemoActivity.this, "你喜欢篮球",
                        Toast.LENGTH_LONG).show();}
            case R.id.volleyballChx:
                //选择排球
                if (isChecked) {
                    Toast.makeText(CheckBoxDemoActivity.this, "你喜欢排球",
                        Toast.LENGTH_LONG).show();}
            default:
                break;}}};
}
```

代码解释如下：标号①处定义了 3 个 CheckBox 复选框，供用户进行选择；标号②处对标号①处所定义的各个属性变量初始化，通过对属性变量的赋值，使其可以进行后续的业务逻辑操作；标号③处分别为 3 个 CheckBox 对象设置监听器，用于监听各自的选中或取消事件；标号④处定义了一个监听器对象，用于监听并实现 3 个 CheckBox 的业务逻辑功能，当用户单击不同的 CheckBox 时，屏幕上会通过 Toast 对象显示相应的文本信息。

图 3-26　爱好的选择

运行上述 Activity，并选择"篮球"时，界面效果如图 3-26 所示。

> Toast 是 Android 中用来显示提示信息的一种机制，与 Dialog 不同的是：Toast 提示没有焦点，且时间有限，在一定的时间后会自动消失。Toast 使用简单，主要用于向用户显示提示消息，在后续章节会有详细的介绍。

3.4.7　开关控件

ToggleButton、Switch、CheckBox 和 RadioButton 组件均继承自 android.widget.CompoundButton，都是选择类型的按钮，因此这些组件的用法非常相似。CompoundButton 按钮共有两种状态：选中（checked）和未选中（unchecked）状态，而 Switch 控件是 Android 4.0 后出现的控件。

ToggleButton 所支持的 XML 属性和方法如表 3-20 所示。

表 3-20　ToggleButton 的 XML 属性和相关方法

XML 属性	对应方法	功能描述
android:checked	setChecked(boolean)	设置该按钮是否被选中
android:textOff	setTextOff(CharSequence)	设置按钮的状态关闭时所显示的文本
android:textOn	setTextOn(CharSequence)	设置按钮的状态打开时所显示的文本

下述代码通过一个简单的示例演示 ToggleButton 的用法。当 ToggleButton 处于选中的状态时，文本显示"已开启"；当 ToggleButton 处于未选中状态时，文本显示"已关闭"。首先实现界面布局，代码如下所示。

【代码 3-28】　togglebutton_demo.xml

```
<LinearLayout xmlns:android = "http://schemas.android.com/apk/res/android"
    android:layout_width = "match_parent"
    android:layout_height = "match_parent"
    android:orientation = "horizontal" >
    <!-- 文本展示 ① -->
    <TextView
        android:id = "@ + id/tvSound"
        android:layout_width = "wrap_content"
        android:layout_height = "wrap_content"
        android:text = "已开启"
        android:textColor = "@android:color/black"
```

```xml
                android:textSize = "14.0sp" />
        <!-- 定义 ToggleButton 对象 ② -->
        <ToggleButton
                android:id = "@ + id/tglSound"
                android:layout_width = "wrap_content"
                android:layout_height = "wrap_content"
                android:layout_marginLeft = "10dp"
                android:checked = "true"
                android:text = ""
                android:textOff = "Off"
                android:textOn = "On"/>
</LinearLayout>
```

代码解释如下：标号①处的 TextView 用于显示 ToggleButton 按钮 On/Off 时的标题；标号②处定义了一个 ToggleButton，用于测试按钮的开启或关闭。

然后，在 Activity 中展示 ToggleButton 的使用：当用户选择了 On/Off 后，在屏幕上显示用户的选择。对应的 Activity 代码如下所示。

【代码 3-29】 ToggleButtonDemoActivity.java

```java
public class ToggleButtonDemoActivity extends AppCompatActivity {
    //声明 xml 中定义的组件 ①
    private ToggleButton mToggleButton;
    private TextView tvSound;
    @Override
    public void onCreate(Bundle savedInstanceState) {
        super.onCreate(savedInstanceState);
        setContentView(R.layout.toggleswitch_demo);
        initView();}
    //初始化控件方法 ②
    private void initView() {
        //获取到控件
        mToggleButton = (ToggleButton) findViewById(R.id.tglSound);
        tvSound = (TextView) findViewById(R.id.tvSound);
        //注册监听器,添加监听事件 ③
        mToggleButton.setOnCheckedChangeListener(new OnCheckedChangeListener() {
            @Override
            public void onCheckedChanged(CompoundButton buttonView,
                    boolean isChecked) {
                if (isChecked) {
                    tvSound.setText("已开启");
                } else {
                    tvSound.setText("已关闭");}}});}
}
```

代码解释如下：标号①处分别声明了 ToggleButton 类型和 TextView 类型的属性变量；标号②处对标号①处所定义的各个属性变量进行初始化，通过对属性变量的赋值，使其可以进行后续的业务逻辑操作；标号③处对 ToggleButton 对象注册监听器，用来监听按钮的开启或关闭事件。

运行上述代码后，当用户单击 On 后，显示效果如图 3-27 所示，当用户切换按钮变为 On 状态时，显示的文本变为"已开启"。从图 3-27 中可以看出，默认的 ToggleButton 比较难看，在实际开发中可以通过设置按钮的背景图片来实现较为美观的按钮。本示例中提供了用于切换 On/Off 的较为美观的图片，通过对其设置按钮的背景图片来实现较为美观的效果，修改后的代码如下所示。

【代码 3-30】 togglebutton_demo.xml

```xml
<LinearLayout xmlns:android="http://schemas.android.com/apk/res/android" …>
    <!-- 文本展示 ① -->
    <TextView
        android:id="@+id/tvSound"
        android:layout_width="wrap_content"
        android:layout_height="wrap_content"
        android:text="已开启"
        android:textColor="@android:color/black"
        android:textSize="14.0sp" />
    <!-- 定义 ToggleButton 对象   ② -->
    <ToggleButton
        android:id="@+id/tglSound"
        android:layout_width="wrap_content"
        android:layout_height="wrap_content"
        android:layout_marginLeft="10dp"
        android:background="@drawable/selector_btn_toggle"
        android:checked="true"
        android:text=""
        android:textOff=""
        android:textOn="" />
</LinearLayout>
```

代码解释如下：标号①处的 TextView 用于显示 On/Off 后的标题。标号②处定义了一个 ToggleButton 按钮，用于显示开启或关闭，此时需要将属性 android:textOff 和 android:textOn 设置为空，否则将会覆盖在图片上。然后将 android:backgroud 的属性值设置为 selector_btn_toggle，该值对应的并不是一幅图片，而是一个资源文件 selector_btn_toggle.xml。

在资源目录 res/drawable 下，创建 selector_btn_toggle.xml 资源文件，代码如下所示。

【代码 3-31】 selector_btn_toggle.xml

```xml
<?xml version="1.0" encoding="utf-8"?>
<selector xmlns:android="http://schemas.android.com/apk/res/android">
    <item android:state_checked="true" android:drawable="@drawable/btn_open" />
    <item android:drawable="@drawable/btn_close" />
</selector>
```

在上述资源文件中，指定了 ToggleButton 在默认状态下的背景图片和选中状态下的图片背景。运行更改后的代码，效果如图 3-28 所示。

图 3-27　选中状态

图 3-28　ToggleButton 背景优化

通过效果可以看出,美化后的 ToggleButton 更注重用户的体验。

Switch 的使用方式与 ToggleButton 类似,Switch 所支持的 XML 属性和方法如表 3-21 所示。

表 3-21 Switch 的 XML 属性及对应方法

XML 属性	对 应 方 法	功 能 描 述
android:checked	setChecked(boolean)	设置当前按钮的状态,选中或未选中
android:textOff	setTextOff(CharSequence)	设置按钮关闭状态所显示的文本
android:textOn	setTextOn(CharSequence)	设置按钮打开状态所显示的文本
android:switchMinWidth	setSwitchMinWidth(int)	设置开关的最小宽度
android:textStyle	setSwitchTypeface(Typeface,int)	设置开关的文本风格
android:typeface	setSwitchTypeface(Typeface)	设置开关的文本的字体风格
android:switchPadding	setSwitchPadding(int)	设置开关与标题文本之间的空白
android:thumb	setThumbResource(int)	使用自定义的 Drawable 来绘制开关的开关按钮
android:track	setTrackResource(int)	使用自定义的 Drawable 来绘制开关的开关轨道

下述代码通过一个示例演示 Switch 的使用。通过改变 Switch 状态来实现界面布局中的 LinearLayout 布局的方向在水平和垂直布局之间切换,布局文件代码如下所示。

【代码 3-32】 switch_demo.xml

```xml
<LinearLayout xmlns:android="http://schemas.android.com/apk/res/android" …>
    <!-- 定义一个 Switch 组件 ① -->
    <Switch
        android:id="@+id/switcher"
        android:layout_width="wrap_content"
        android:layout_height="wrap_content"
        android:checked="true"
        android:textOff="横向排列"
        android:textOn="纵向排列"
        android:showText="ture"
        android:thumb="@drawable/check" />
    <!-- 定义一个可以动态改变方向的线性布局 ② -->
    <LinearLayout
        android:id="@+id/test"
        android:layout_width="match_parent"
        android:layout_height="match_parent"
        android:orientation="vertical" >
        <TextView
            android:layout_width="wrap_content"
            android:layout_height="wrap_content"
            android:text="测试文本 1" />
        <TextView
            android:layout_width="wrap_content"
            android:layout_height="wrap_content"
            android:text="测试文本 2" />
        <TextView
            android:layout_width="wrap_content"
            android:layout_height="wrap_content"
            android:text="测试文本 3" />
    </LinearLayout>
</LinearLayout>
```

代码解释如下：标号①处定义了一个 Switch 组件，并将 android:thumb 的属性设置为 @drawable/check；标号②处定义了一个可以动态改变方向的线性布局，其中包含 3 个文本框用于显示效果。

下面在 Activity 中演示 Switch 的使用：当用户选择了"横向排列/纵向排列"时，界面布局发生相应的变化。Activity 代码的实现如下所示。

【代码 3-33】 SwitchDemoActivity.java

```java
public class SwitchDemoActivity extends AppCompatActivity {
    //定义变量 ①
    Switch switcher;
    @Override
    public void onCreate(Bundle savedInstanceState) {
        super.onCreate(savedInstanceState);
        setContentView(R.layout.switch_demo);
        //初始化组件②
        switcher = (Switch) findViewById(R.id.switcher);
        final LinearLayout test = (LinearLayout) findViewById(R.id.test);
        OnCheckedChangeListener listener = new OnCheckedChangeListener() {
            @Override
            public void onCheckedChanged(CompoundButton button,
                    boolean isChecked) {
                if (isChecked) {
                    //设置 LinearLayout 垂直布局
                    test.setOrientation(LinearLayout.VERTICAL);
                    switcher.setChecked(true);
                } else {
                    //设置 LinearLayout 水平布局
                    test.setOrientation(LinearLayout.HORIZONTAL);
                    switcher.setChecked(false);}}};
        //为 switch 组件添加事件监听器 ③
        switcher.setOnCheckedChangeListener(listener);}
}
```

代码解释如下：标号①处分别声明了一个 Switch 类型的属性变量；标号②处用于初始化标号①处所声明的属性变量，通过对属性变量的赋值，使其可以进行后续的业务逻辑操作；标号③处对 Switch 对象设置监听器，用于监听按钮的开启或关闭事件，当事件发生时，判断按钮的"开启/关闭"状态，并切换界面的布局。

运行上述代码后，界面效果如图 3-29 所示。当用户再次单击"纵向排列"按钮时，系统会自动切换到"横向排列"状态，效果如图 3-30 所示。

图 3-29 Switch 实现纵向布局

图 3-30 Switch 实现横向布局

3.4.8 图片视图(ImageView)

ImageView 继承自 View 组件,主要用于显示图像资源(例如图片等),ImageView 还可以定义所显示的尺寸等。此外,ImageView 还派生了 ImageButton、ZoomButton 等组件。ImageView 所支持的 XML 属性和方法如表 3-22 所示。

表 3-22　ImageView 的 XML 属性及方法

XML 属性	对应方法	功能描述
android:adjustViewBounds	setAdjustViewBounds (boolean)	是否保持宽高比。需要与 maxWidth、MaxHeight 一起使用,单独使用没有效果
android:cropToPadding	setCropToPadding (boolean)	截取指定区域是否使用空白代替。单独设置无效果,需要与 scrollY 一起使用
android:maxHeight	setMaxHeight(int)	设置 View 的最大高度,单独使用无效,需要与 setAdjustViewBounds()方法一起使用。如果想设置图片固定大小,又想保持图片宽高比,需要如下设置:设置 setAdjustViewBounds 为 true;设置 maxWidth 和 MaxHeight 属性;设置 layout_width 和 layout_height 均为 wrap_content
android:maxWidth	setMaxWidth(int)	设置 View 的最大宽度。用法同 android:maxHeight
android:src	setImageResource(int)	设置 ImageView 所显示的 Drawable 对象
android:scaleType	setScaleType(ImageView. ScaleType)	设置所显示的图片如何缩放或移动以适应 ImageView 的大小

ImageView 的 android:scaleType 属性可以指定如下属性值。
- matrix:用矩阵来绘图。
- fitXY:拉伸图片(不按比例)以填充 View 的宽高。
- fitStart:按比例拉伸图片,图片拉伸后的高度为 View 的高度,且显示在 View 的左边。
- fitCenter:按比例拉伸图片,图片拉伸后的高度为 View 的高度,且显示在 View 的中间。
- fitEnd:按比例拉伸图片,图片拉伸后的高度为 View 的高度,且显示在 View 的右边。
- center:按原图大小显示图片,当图片宽高大于 View 的宽高时,截图图片中间部分显示。
- centerCrop:按比例放大原图,直至等于 View 某边的宽高。
- centerInside:当原图宽高等于 View 的宽高时,按原图大小居中显示;反之将原图缩放至 View 的宽高居中显示。

此外,为了控制 ImageView 所显示的图片,该组件提供了如下方法。
- setImageBitmap(Bitmap):使用 Bitmap 位图来设置 ImageView 所显示的图片。
- setImageDrawable(Drawable):使用 Drawable 对象来设置 ImageView 所显示的图片。
- setImageResource(int):使用图片资源 ID 来设置 ImageView 所显示的图片。
- setImagURI(Uri):使用图片的 URI 来设置 ImageView 所显示的图片。

下面通过一个简单示例演示 ImageView 的使用。通过单击"下一页/上一页"按钮，实现国旗的切换，对应的布局文件代码如下所示。

【代码 3-34】 imageview_demo.xml

```xml
<LinearLayout xmlns:android = "http://schemas.android.com/apk/res/android" …>
    <!-- 国旗及文字 ① -->
    <LinearLayout
        android:layout_width = "wrap_content"
        android:layout_height = "wrap_content"
        android:layout_gravity = "center"
        android:orientation = "vertical" >
        <ImageView
            android:id = "@ + id/guoqiImageView"
            android:layout_width = "200dp"
            android:layout_height = "wrap_content"
            android:layout_gravity = "center"
            android:layout_marginTop = "30dp"
            android:background = "@android:color/white"
            android:scaleType = "fitCenter"
            android:src = "@drawable/china" />
        <TextView
            android:id = "@ + id/guoqiTxt"
            android:layout_width = "wrap_content"
            android:layout_height = "wrap_content"
            android:layout_gravity = "center"
            android:text = "中国" />
    </LinearLayout>
    <!-- 分页 ② -->
    <LinearLayout
        android:layout_width = "match_parent"
        android:layout_height = "wrap_content"
        android:gravity = "center"
        android:orientation = "horizontal" >
        <ImageButton
            android:id = "@ + id/backImageBtn"
            android:layout_width = "50dp"
            android:layout_height = "50dp"
            android:layout_gravity = "center"
            android:src = "@drawable/back" />        <ImageButton
            android:id = "@ + id/forwardImageBtn"
            android:layout_marginLeft = "20dp"
            android:layout_width = "50dp"
            android:layout_height = "50dp"
            android:layout_gravity = "center"
            android:src = "@drawable/forward" />
    </LinearLayout>
</LinearLayout>
```

代码解释如下：标号①处定义了一个 ImageView 组件，用来显示国旗的图片，通过设置 android:scaleType = "fitCenter" 属性，使用拉伸后图片的高度作为 View 的高度，且在 View 的中间显示，同时还定义了一个 TextView 组件用来显示国别；标号②处定义了两个 ImageButton 组件，分别用于显示"上一页"和"下一页"的国旗和国别。

下面在 Activity 中演示图片分页效果：当用户分别单击"上一页/下一页"按钮时，在屏幕上会显示相应的国旗和国别。对应的 Activity 代码实现如下所示。

【代码 3-35】 ImageViewDemoActivity.java

```java
public class ImageViewDemoActivity extends AppCompatActivity {
    //定义变量 ①
    //国旗对应的 ImageView
    ImageView flagImageView;
    TextView flagTxt;
    ImageButton backImageBtn;                        //上一页
    ImageButton forwardImageBtn;                     //下一页
    //国旗数组 中国 德国 英国 ②
    int[]flag = {R.drawable.china,R.drawable.germany,R.drawable.britain};
    String[]flagNames = {"中国","德国","英国"};
    //当前页 默认第一页
    int currentPage = 0;
    @Override
    public void onCreate(Bundle savedInstanceState) {
        super.onCreate(savedInstanceState);
        setContentView(R.layout.imageview_demo);
        //初始化组件 ③
        flagImageView = (ImageView) findViewById(R.id.flagImageView);
        guoqiTxt = (TextView)findViewById(R.id.flagTxt);          //国旗名称
        //上一页、下一页
        backImageBtn = (ImageButton)findViewById(R.id.backImageBtn);
        forwardImageBtn = (ImageButton)findViewById(R.id.forwardImageBtn);
        //注册监听器
        backImageBtn.setOnClickListener(onClickListener);
        forwardImageBtn.setOnClickListener(onClickListener);}
    //定义单击事件监听器 ④
    private OnClickListener onClickListener = new OnClickListener() {
        @Override
        public void onClick(View v) {
            switch (v.getId()) {
            case R.id.backImageBtn:
                if(currentPage == 0){
                    Toast.makeText(ImageViewDemoActivity.this,
                            "第一页,前面没有了", Toast.LENGTH_SHORT).show();
                    return;   }
                //上翻一页
                currentPage--;
                //设置国旗图片
                flagImageView.setImageResource(flag[currentPage]);
                //设置国旗名字
                flagTxt.setText(flagNames[currentPage]);
                break;
            case R.id.forwardImageBtn:
                if(currentPage == (flag.length-1)){
                    Toast.makeText(ImageViewDemoActivity.this,
                            "最后一页,后面没有了", Toast.LENGTH_SHORT).show();
                    return;   }
                currentPage++;                                        //下翻一页
                flagImageView.setImageResource(flag[currentPage]);    //设置国旗图片
```

```
                flagTxt.setText(flagNames[currentPage]);          //设置国旗名字
            break;
        default:
            break;}}};
}
```

代码解释如下：标号①处分别声明了 ImageView 类型和 ImageButton 类型的属性变量；标号②处定义了两个数组并进行初始化，分别表示国旗图片资源和国旗名称；标号③处用于初始化标号①处所声明的属性变量，并对 backImageBtn 和 forwardImageBtn 对象设置监听器，来监听各自的单击事件；标号④处定义了一个 OnClickListener 类型的监听器，用于处理按钮单击事件，当单击按钮时根据所单击的按钮不同来显示不同的国旗和国别。

运行上述代码后，默认的界面效果如图 3-31 所示。当用户单击>按钮（下一页）后，界面显示效果如图 3-32 所示。

图 3-31 切换国旗 1

图 3-32 切换国旗 2

> **注意**
>
> 在实际开发中，如果要实现页面的切换功能，通常使用 ViewPager 类；该类是 Android Support Liberary 中自带的一个附加包的一个类，用来实现屏幕间的切换。

3.5 Dialog 对话框

Dialog 对话框对于应用程序而言是一个必不可少的组件，在 Android 中，对话框对于一些重要的提示信息或者需要用户额外的交互是很有帮助的。所有的对话框都直接或间接继承自 Dialog 类，其中 AlterDialog 直接继承自 Dialog 类，而其他的几个类则继承自 AlterDialog 类。

对话框就是一个小窗口，并不会填满整个屏幕，通常以模态显示，需要用户采取行动才能进行后续操作。Android 提供了丰富的对话框支持，其中常用的对话框有以下 4 种。

- **AlertDialog 提示对话框**：是一种使用广泛且功能丰富的对话框。AlertDialog 不仅可以包含 0～3 个响应按钮，还可以包含一个单选框、复选框或列表。警告对话框通常用于创建交互式界面，是最常用的对话框类型之一。

- **ProgressDialog 进度条对话框**：只是对进度条进行了简单的封装，用于显示一个进度环或进度条，由于 ProgressDialog 是 AlertDialog 的扩展，所以也支持按钮选项。
- **DatePickerDialog 日期对话框**：用于用户选择日期的对话框。
- **TimePickerDialog 时间对话框**：用于用户选择时间的对话框。

3.5.1 AlertDialog 提示对话框

AlertDialog 继承自 Dialog 类，可以包含一个标题、一个内容消息或者一个选择列表以及 0~3 个按钮。在创建 AlterDialog 时推荐使用 AlterDialog 的 Builder 内部类来创建。首先使用 Builder 对象来设置 AlterDialog 的各种属性，然后通过 Builder.create()方法生成一个 AlterDialog 对象。如果只是显示一个 AlterDialog 对话框，则可以直接使用 Builder.show()方法返回一个 AlterDialog 对象并显示。

当仅仅提示一段信息时，可以直接使用 AlterDialog 的属性设置提示信息，相关方法如表 3-23 所示。

表 3-23 AlterDialog 的相关方法

方法	功能描述
void create()	根据设置的属性，创建一个 AlterDialog
void show()	根据设置的属性，显示已创建的 AlterDialog
AlterDialog.Builder setTitle()	设置标题
AlterDialog.Builder setIcon()	设置标题的图标
AlterDialog.Builder setMessage()	设置标题的内容
AlterDialog.Builder setCancelable()	设置是否模态。一般设置为 false，表示采用模态形式，要求用户必须采取行动才能继续进行剩下的操作
AlterDialog setPositiveButton()	为对话框添加 Yes 按钮
AlterDialog setNegativeButton()	为对话框添加 No 按钮

AlterDialog.Builder 的大部分设置属性的方法返回此 AlterDialog.Builder 对象，所以可以使用链式方式编写代码，这样更方便。

当一个对话框调用 show()方法将其展示到屏幕上时，如果需要消除该对话框，可以使用 DialogInterface 接口所提供的 cancel()方法来取消对话框或者 dismiss()方法来消除对话框。cancel()和 dismiss()方法的作用是一样的，但推荐使用 dismiss()方法。Dialog 和 AlterDialog 都实现了 DialogInterface 接口，因此所有的对话框都可以使用这两个方法来消除对话框。对话框使用的场景较多，主要分为如下几种形式。

1. 普通对话框

下面通过一个简单的示例，演示使用 AlterDialog 如何提示信息，布局文件代码如下所示。

【代码 3-36】 dialog_demo.xml

```xml
<LinearLayout xmlns:android = "http://schemas.android.com/apk/res/android" …>
    <!-- 普通对话框   ① -->
    <Button
        android:id = "@+id/normalBtn"
        android:layout_width = "match_parent"
        android:layout_height = "wrap_content"
        android:text = "普通对话框" />
</LinearLayout>
```

上述代码中,标号①处定义了一个 Button 组件,当用户单击时,弹出一个普通对话框。

接下来在对应的 Activity 中实现按钮事件的业务逻辑:当用户单击"普通对话框"时,在屏幕上显示对话框;当用户单击对话框中的"确认"按钮时,将退出应用程序;当用户单击"取消"按钮时,程序返回到主界面。代码实现如下。

【代码 3-37】 DialogDemoActivity.java

```java
public class DialogDemoActivity extends AppCompatActivity {
    //普通对话框 ①
    Button normalBtn;
    @Override
    protected void onCreate(Bundle savedInstanceState) {
        super.onCreate(savedInstanceState);
        setContentView(R.layout.dialog_demo);
        normalBtn = (Button) findViewById(R.id.normalBtn);      //初始化组件 ②
        normalBtn.setOnClickListener(onClickListener);    }     //设置监听器对象
    //定义单击事件监听器 ③
    private OnClickListener onClickListener = new OnClickListener() {
        @Override
        public void onClick(View v) {
            switch (v.getId()) {
                case R.id.normalBtn: {
                    //普通对话框 ④
                    AlertDialog.Builder builder = new AlertDialog.Builder(
                            DialogDemoActivity.this);
                    builder.setMessage("确认退出吗?")
                            .setTitle("提示");
                    //单击确认按钮后触发事件
                    builder.setPositiveButton("确认",
                            new DialogInterface.OnClickListener() {
                                @Override
                                public void onClick(DialogInterface dialog,
                                        int which) {
                                    dialog.dismiss();
                                    DialogDemoActivity.this.finish();}    });
                    //单击取消后触发事件
                    builder.setNegativeButton("取消",
                            new DialogInterface.OnClickListener() {
                                @Override
                                public void onClick(DialogInterface dialog,
                                        int which) {
                                    dialog.dismiss();}});
```

```
                    builder.create().show();
                    break;}
            default:
                    break;}}};
}
```

代码解释如下：标号①处声明了 Button 类型的属性变量，当用户单击该按钮时，弹出普通对话框。标号②处对标号①处所声明的属性变量进行初始化，通过对属性变量的赋值，使其可以进行后续的业务逻辑操作。标号③处创建了一个监听器，用来监听用户单击按钮时所触发的事件。标号④处实现了 OnClickListener 事件处理的业务逻辑：当事件触发时，弹出一个带有"确认"和"取消"的对话框；单击"确认"按钮退出当前应用，单击"退出"按钮返回程序的主界面。

运行上述代码后，在屏幕上单击"默认对话框"，则打开提示对话框，如图 3-33 所示。

图 3-33　普通对话框

2．内容型对话框

在上述实例的基础上演示内容型对话框的使用。在 dialog_demo.xml 文件中添加一个"内容型对话框"按钮，当用户单击该按钮时，弹出一个含有三个按钮的对话框。以观众对电影的喜好为例实现上述功能。首先在布局文件中添加"内容型对话框"按钮，代码如下所示。

【代码 3-38】　dialog_demo.xml

```xml
<LinearLayout xmlns:android="http://schemas.android.com/apk/res/android" …>
    <!-- 普通对话框      …省略 -->
    <!-- 内容型对话框      -->
    <Button
        android:id="@+id/contentBtn"
        android:layout_width="match_parent"
        android:layout_height="wrap_content"
        android:text="内容型对话框" />
</LinearLayout>
```

然后，在 DialogDemoActivity.java 中按照"普通对话框"的编写步骤，添加"内容型对话框"对应的代码，代码如下所示。

【代码 3-39】　DialogDemoActivity.java

```java
…省略
public class DialogDemoActivity extends AppCompatActivity {
    Button normalBtn;           //普通对话框
    Button contentBtn;          //内容型对话框
    @Override
    protected void onCreate(Bundle savedInstanceState) {
        super.onCreate(savedInstanceState);
        setContentView(R.layout.dialog_demo);
        //初始化组件
        normalBtn = (Button) findViewById(R.id.normalBtn);
        //设置监听器对象
```

```java
            normalBtn.setOnClickListener(onClickListener);
            contentBtn = (Button)findViewById(R.id.contentBtn);
            contentBtn.setOnClickListener(onClickListener);}
    private OnClickListener onClickListener = new OnClickListener() {
        @Override
        public void onClick(View v) {
            switch (v.getId()) {
            case R.id.normalBtn:
                …//省略      }
            case R.id.contentBtn:{
                //处理内容型的对话框
                AlertDialog.Builder builder
                        = new AlertDialog.Builder(DialogDemoActivity.this);
                builder.setIcon(android.R.drawable.btn_star)
                    .setTitle("喜欢度调查").setMessage("你喜欢成龙的电影吗?")
                    .setPositiveButton("很喜欢",
                        new DialogInterface.OnClickListener() {
                        @Override
                        public void onClick(DialogInterface dialog, int which) {
                            Toast.makeText(DialogDemoActivity.this,
                                "我很喜欢他的电影.",Toast.LENGTH_LONG).show();}});
                //不喜欢
                builder.setNegativeButton("不喜欢",
                    new DialogInterface.OnClickListener() {
                        @Override
                        public void onClick(DialogInterface dialog, int which) {
                            Toast.makeText(DialogDemoActivity.this,
                                "我不喜欢他的电影.", Toast.LENGTH_LONG).show();}});
                builder.setNeutralButton("一般",
                    new DialogInterface.OnClickListener() {
                        @Override
                        public void onClick(DialogInterface dialog, int which) {
                            Toast.makeText(DialogDemoActivity.this,
                                "一般吧,谈不上喜欢.", Toast.LENGTH_LONG).show();}});
                builder.show();}      //显示对话框
            default:
                break;}}};
}
```

运行上述代码后,在屏幕上单击"内容型对话框"按钮,弹出一个内容型对话框,效果如图 3-34 所示。

图 3-34　内容型对话框

第3章 UI编程基础

> **注意**
>
> 除了上述两种类型的对话框之外,开发人员还可以在对话框中实现一组单选框、多选框或列表项等多种形式,限于篇幅,此处不再赘述。

3.5.2 ProgressDialog 进度对话框

在用户使用 App 的过程中,有些操作需要提示用户等待,例如在执行耗时较多的操作时,可以使用进度对话框来显示一个进度信息来提示用户等待,此时可以使用 ProgressDialog 组件来实现。

ProgressDialog 有两种显示方式:①滚动的环状图标,是一个包含标题和提示内容的等待对话框;②带刻度的进度条,和常规进度条的用法一致。

上述两种方式的显示样式可以通过 ProgressDialog.setProgressStyle()方法进行设置,该方法的参数取值情况为:STYLE_HORIZONTAL,刻度滚动;STYLE_SPINNER,图标滚动,为默认选项。

其中,图标滚动可以使用两种方式来实现:一种是使用构造方法创建 ProgressDialog 对象,再设置对象的属性;另外一种是直接使用 ProgressDialog.show()静态方法返回一个 ProgressDialog 对象,然后调用 show()方法来显示。

下面通过一个简单示例演示 ProgressDialog 的使用。当用户单击"滚动等待对话框"按钮时,弹出滚动等待类型的对话框;当用户单击"进度条对话框"按钮时,弹出进度条类型的对话框,对应的布局文件代码如下所示。

【代码 3-40】 progress_demo.xml

```xml
<LinearLayout xmlns:android = "http://schemas.android.com/apk/res/android" ...>
    <!-- 滚动等待对话框    -->
    <Button
        android:id = "@ + id/progressCircleBtn"
        android:layout_width = "match_parent"
        android:layout_height = "wrap_content"
        android:text = "滚动等待对话框" />
    <!-- 进度条对话框    -->
    <Button
        android:id = "@ + id/progressBarBtn"
        android:layout_width = "match_parent"
        android:layout_height = "wrap_content"
        android:text = "进度条对话框" />
</LinearLayout>
```

上述代码中定义了两个按钮,分别用于实现弹出"滚动等待对话框"和"进度条对话框"。

下面在相应的 Activity 中实现以下功能:当用户单击"滚动等待对话框"按钮时,屏幕上显示滚动等待对话框;当用户单击"进度条对话框"按钮时,屏幕上会显示进度条对话框。代码实现如下。

【代码 3-41】 ProgressDemoActivity.java

```java
public class ProgressDemoActivity extends AppCompatActivity {
    Button progressCircleBtn;           //滚动等待对话框
    Button progressBarBtn;              //进度条对话框
    //存储进度条当前值,初始为 0
    int count = 0;
    @Override
    protected void onCreate(Bundle savedInstanceState) {
        super.onCreate(savedInstanceState);
        setContentView(R.layout.progress_demo);
        //初始化组件
        progressCircleBtn = (Button) findViewById(R.id.progressCircleBtn);
        //设置监听器对象
        progressCircleBtn.setOnClickListener(onClickListener);
        progressBarBtn = (Button) findViewById(R.id.progressBarBtn);
        progressBarBtn.setOnClickListener(onClickListener);}
    private OnClickListener onClickListener = new OnClickListener() {
        @Override
        public void onClick(View v) {
            switch (v.getId()) {
            case R.id.progressCircleBtn: {
                //滚动等待对话框
                final ProgressDialog progressDialog = new ProgressDialog(
                        ProgressDemoActivity.this);
                progressDialog.setIcon(R.drawable.ic_launcher);
                progressDialog.setTitle("等待");
                progressDialog.setMessage("正在加载...");
                progressDialog.show();
                new Thread(new Runnable() {
                    @Override
                    public void run() {
                        try {
                            Thread.sleep(5000);
                        } catch (Exception e) {
                            e.printStackTrace();
                        } finally {
                            progressDialog.dismiss();}}
                }).start();}
            case R.id.progressBarBtn: {
                //滚动等待对话框
                final ProgressDialog progressDialog = new ProgressDialog(
                        ProgressDemoActivity.this);//得到一个对象
                //设置为矩形进度条
                progressDialog.setProgressStyle(ProgressDialog.STYLE_HORIZONTAL);
                progressDialog.setTitle("提示");
                progressDialog.setMessage("数据加载中,请稍后...");
                progressDialog.setIcon(R.drawable.ic_launcher);
                //设置进度条是否为不明确
                progressDialog.setIndeterminate(false);
                progressDialog.setCancelable(true);
                progressDialog.setMax(200);              //设置进度条的最大值
                progressDialog.setProgress(0);           //设置当前默认进度为 0
                progressDialog.setSecondaryProgress(1000);   //设置第二条进度值为 100
```

```
        progressDialog.show();  //显示进度条
        new Thread() {
            public void run() {
                while (count <= 200) {
                    progressDialog.setProgress(count++);
                    try {
                        Thread.sleep(100);   //暂停 0.1s
                    } catch (Exception e) {
                    }}}
        }.start();}
        default:
            break;}}};
}
```

运行上述代码,在屏幕上单击"滚动等待对话框",效果如图 3-35 所示。当用户单击"进度条对话框"按钮后,效果如图 3-36 所示。

图 3-35 滚动等待对话框

图 3-36 进度条对话框

3.6 贯穿任务实现

本章任务完成礼品、攻略的展示,包括列表界面和详情界面。

3.6.1 实现【任务 3-1】

本任务编写完成主界面 Activity。

1. 主界面 MainActivity

主界面中需要完成下列功能:

(1) 显示"礼物推荐"、"礼品中心"和"送礼攻略"三个模块的入口,每个模块都整屏显示,通过左右切换的方式在三个模块间转换;

(2) 礼物推荐界面上部自动轮播多个推荐的礼物,下部分为左右两部分,也分别显示推荐的礼物;

(3) 主界面右下角显示一个扇形的菜单,单击可以展开和收缩,单击此菜单可进入"应用设置"、"用户日历"、"购物车"和"个人中心"。

编写主界面布局文件 main.xml,代码如下所示。

【任务 3-1】 main.xml

```xml
<RelativeLayout xmlns:android="http://schemas.android.com/apk/res/android" ...
    tools:context=".MainActivity">
    <RelativeLayout
        android:id="@+id/titleLayout"
        android:layout_width="match_parent"
        android:layout_height="50dp"
        android:background="@color/background"
        android:gravity="center_vertical|left"
        android:paddingBottom="7dp"
        android:paddingLeft="7dp"
        android:paddingTop="7dp">
        <ImageView
            android:id="@+id/logoImageView"
            android:layout_width="70dp"
            android:layout_height="90dp"
            android:src="@drawable/logo1" />
        <TextView
            android:id="@+id/titleTextView"
            style="@style/text1"
            android:layout_width="match_parent"
            android:layout_height="match_parent"
            android:layout_toRightOf="@id/logoImageView"
            android:text="礼物推荐"
            android:textColor="@color/selected"
            android:textSize="26sp" />
    </RelativeLayout>
    <android.support.v4.view.ViewPager
        android:id="@+id/viewPager"
        android:layout_width="match_parent"
        android:layout_height="match_parent"
        android:layout_below="@id/titleLayout" />
    <include layout="@layout/corner_menu" />
</RelativeLayout>
```

在主界面布局文件 main.xml 中,使用 RelativeLayout 进行布局:上部 ID 为 titleLayout 的 RelativeLayout 中包含一个 logo 图标和当前切换到的模块的标题。中部 ID 为 ViewPager 的视图是一个 android.support.v4.ViewPager 视图,负责显示多个模块的入口。ViewPager 是非常常用的一种视图组件,多用于在多个视图之间切换显示,其内置了对滑动手势的支持,使用非常方便。在 MainActivity 中将通过代码方式指定 ViewPager 需要显示的视图。最后通过 include 元素将右下角的扇形菜单引入。

扇形菜单的布局 corner_menu.xml 代码如下所示。

【任务 3-1】 corner_menu.xml

```xml
<RelativeLayout xmlns:android="http://schemas.android.com/apk/res/android"
    xmlns:tools="http://schemas.android.com/tools"
    android:layout_width="140dp"
    android:layout_height="140dp"
    android:layout_alignParentBottom="true"
    android:layout_alignParentRight="true">
```

```xml
<RelativeLayout
    android:id = "@ + id/cornerMenuOpenRelativeLayout"
    android:layout_width = "match_parent"
    android:layout_height = "match_parent"
    android:visibility = "gone" >
    <ImageView
        android:layout_width = "match_parent"
        android:layout_height = "match_parent"
        android:layout_alignParentBottom = "true"
        android:layout_alignParentRight = "true"
        android:scaleType = "fitXY"
        android:src = "@drawable/corner_menu_open" />
    <ImageView
        android:id = "@ + id/settingImageView"
        android:layout_width = "35dp"
        android:layout_height = "35dp"
        android:layout_marginLeft = "100dp"
        android:layout_marginTop = "10dp"
        android:scaleType = "fitXY"
        android:src = "@drawable/icon_setting" />
    <ImageView
        android:id = "@ + id/calendarImageView"
        android:layout_width = "30dp"
        android:layout_height = "30dp"
        android:layout_marginLeft = "63dp"
        android:layout_marginTop = "33dp"
        android:scaleType = "fitXY"
        android:src = "@drawable/icon_calendar" />
    <ImageView
        android:id = "@ + id/shoppingBagImageView"
        android:layout_width = "35dp"
        android:layout_height = "35dp"
        android:layout_marginLeft = "27dp"
        android:layout_marginTop = "60dp"
        android:scaleType = "fitXY"
        android:src = "@drawable/icon_shopping_bag" />
    <ImageView
        android:id = "@ + id/personalImageView"
        android:layout_width = "33dp"
        android:layout_height = "33dp"
        android:layout_marginLeft = "10dp"
        android:layout_marginTop = "102dp"
        android:scaleType = "fitXY"
        android:src = "@drawable/icon_personal" />
</RelativeLayout>
<ImageView
    android:id = "@ + id/showCornerMenuImageView"
    android:layout_width = "91dp"
    android:layout_height = "95dp"
    android:layout_alignParentBottom = "true"
    android:layout_alignParentRight = "true"
    android:scaleType = "fitXY"
    android:src = "@drawable/corner_menu_close"
    android:visibility = "visible" />
</RelativeLayout>
```

布局文件编写完成后，修改 MainActivity 代码，如下所示。

【任务 3-1】 MainActivity.java

```java
//主界面 public class MainActivity extends Activity {
    //标题
    private TextView titleTextView;
    //ViewPager
    private ViewPager viewPager;
    //右下角扇形菜单
    private RelativeLayout cornerMenuOpenRelativeLayout;
    private ImageView settingImageView;              //设置
    private ImageView calendarImageView;             //日历
    private ImageView personalImageView;             //个人中心
    private ImageView shoppingBagImageView;          //购物袋
    private ImageView showCornerMenuImageView;       //展开/收缩
    @Override
    protected void onCreate(Bundle savedInstanceState) {
        super.onCreate(savedInstanceState);
        setContentView(R.layout.main);
        titleTextView = (TextView) findViewById(R.id.titleTextView);
        viewPager = (ViewPager) findViewById(R.id.viewPager);
        cornerMenuOpenRelativeLayout
                = (RelativeLayout) findViewById(R.id.cornerMenuOpenRelativeLayout);
        settingImageView = (ImageView) findViewById(R.id.settingImageView);
        calendarImageView = (ImageView) findViewById(R.id.calendarImageView);
        personalImageView = (ImageView) findViewById(R.id.personalImageView);
        shoppingBagImageView = (ImageView) findViewById(R.id.shoppingBagImageView);
        showCornerMenuImageView = (ImageView) findViewById(R.id.showCornerMenuImageView);
        titleTextView.setText("礼物推荐");
        init();
    }
    /** 只是检查是否保存了 mobileToken,没有到服务器验证 */
    boolean checkLogin() {
        App app = (App) getApplication();
        if (app.user == null || app.user.mobileToken == null) {
            Dialogs.showSimpleDialog(this, "登录后才能继续操作,您现在登录吗?", true,
                    new OnOkListener() {
                        @Override
                        public void onOk() {
                            Intent intent = new Intent(getApplicationContext(),
                                    LoginActivity.class);
                            startActivity(intent);}});
            return false;}
        return true;}
    private void init() {
        viewPager.setOnPageChangeListener(new OnPageChangeListener() {
            @Override
            public void onPageSelected(int index) {
                switch (index) {
                    case 0:
                        titleTextView.setText("礼物推荐");
                        break;
                    case 1:
                        titleTextView.setText("礼品中心");
```

```java
                    break;
                case 2:
                    titleTextView.setText("礼品攻略");
                    break;}}
            @Override
            public void onPageScrolled(int arg0, float arg1, int arg2) {
            }
            @Override
            public void onPageScrollStateChanged(int arg0) {
            }});
    viewPager.setAdapter(new PagerAdapter() {
        @Override
        public int getCount() {
            return 3; }
        @Override
        public boolean isViewFromObject(View v, Object obj) {
            return v == obj; }
        @Override
        public void destroyItem(View container, int position, Object object) {
            ((ViewPager) container).removeView((View) object);}
        @Override
        public Object instantiateItem(View container, int position) {
            final Context context = getApplicationContext();
            View v = null;
            if (position == 0) {
                v = getLayoutInflater().inflate(R.layout.index_recommend,
                        viewPager, false);
                ViewFlipper vf = (ViewFlipper) v
                        .findViewById(R.id.topViewFlipper);
                vf.setInAnimation(context, R.anim.in_from_right);
                vf.setOutAnimation(context, R.anim.out_to_left);
                ViewFlipper.LayoutParams params = new ViewFlipper.LayoutParams(
                        ViewFlipper.LayoutParams.MATCH_PARENT,
                        ViewFlipper.LayoutParams.MATCH_PARENT);
                ImageView iv1 = new ImageView(context);
                iv1.setScaleType(ScaleType.CENTER_CROP);
                iv1.setImageResource(R.drawable.index_recommend_top1);
                vf.addView(iv1, params);
                ImageView iv2 = new ImageView(context);
                iv2.setScaleType(ScaleType.CENTER_CROP);
                iv2.setImageResource(R.drawable.index_recommend_top2);
                vf.addView(iv2, params);
                ImageView iv3 = new ImageView(context);
                iv3.setScaleType(ScaleType.CENTER_CROP);
                iv3.setImageResource(R.drawable.index_recommend_top3);
                vf.addView(iv3, params);
                ImageView bottomLeft = (ImageView) v
                        .findViewById(R.id.bottomLeftImageView);
                bottomLeft
                        .setImageResource(R.drawable.index_recommend_leftbottom);
                ImageView bottomRight = (ImageView) v
                        .findViewById(R.id.bottomRightImageView);
                bottomRight
```

```java
                .setImageResource(R.drawable.index_recommend_rightbottom);
        } else if (position == 1) {
            v = new ImageView(context);
            ImageView iv = (ImageView) v;
            iv.setScaleType(ScaleType.CENTER_CROP);
            iv.setImageResource(R.drawable.index_giftcenter);
        } else if (position == 2) {
            v = new ImageView(context);
            ImageView iv = (ImageView) v;
            iv.setScaleType(ScaleType.CENTER_CROP);
            iv.setImageResource(R.drawable.index_strategy);}
        final ViewGroup.LayoutParams p = new ViewGroup.LayoutParams(
                ViewGroup.LayoutParams.MATCH_PARENT,
                ViewGroup.LayoutParams.MATCH_PARENT);
        ((ViewPager) container).addView(v, p);
        return v;} });
settingImageView.setOnClickListener(new OnClickListener() {
    @Override
    public void onClick(View v) {
        Intent intent = new Intent(getApplicationContext(),
                SettingActivity.class);
        startActivity(intent);} });
calendarImageView.setOnClickListener(new OnClickListener() {
    @Override
    public void onClick(View v) {
        //TODO 后续章节添加功能
    } });
shoppingBagImageView.setOnClickListener(new OnClickListener() {
    @Override
    public void onClick(View v) {
        //TODO 后续章节添加功能
    }});
personalImageView.setOnClickListener(new OnClickListener() {
    @Override
    public void onClick(View v) {
        if (!checkLogin())
            return;
        Intent intent = new Intent(getApplicationContext(),
                PersonalActivity.class);
        startActivity(intent);}});
cornerMenuOpenRelativeLayout.setOnClickListener(new OnClickListener() {
    @Override
    public void onClick(View v) {
        cornerMenuOpenRelativeLayout.setVisibility(View.GONE);
        showCornerMenuImageView.setVisibility(View.VISIBLE);}});
showCornerMenuImageView.setOnClickListener(new OnClickListener() {
    @Override
    public void onClick(View v) {
        cornerMenuOpenRelativeLayout.setVisibility(View.VISIBLE);
        showCornerMenuImageView.setVisibility(View.GONE);}});}
}
```

上述代码中,将布局文件 main.xml 和 cornet_menu.xml 中的视图元素声明为变量,并对各个控件添加了 OnClickListener 监听器,单击时展开、收缩扇形菜单,或者打开对应的 Activity。MainActivity 中的 checkLogin() 方法用于检查 Application 中保存的 user 对象是否有登录标识,如果没有则提示登录。

2. ViewPager

ViewPager 控件是最常用的复杂控件之一。在新建项目时,Android Studio 会自动导入 android-support-v4.jar,其中包含了 ViewPager 类。ViewPager 用于切换显示多个页面视图,其中提供了多个方法用于控制子视图的显示,并提供了子视图切换时的监听机制。在 MainActivity 代码中,通过 setOnPageChangeListener() 方法为 viewPager 设置页面切换的监听器,在切换页面时修改标题 TextView 的文字。

需要注意,ViewPager 中需要切换的子视图是通过设置适配器的方式进行指定的,例如 MainActivity 代码中,通过 setAdapter() 方法为 viewPager 设置了适配器 PagerAdapter。

当创建 PagerAdapter 时,需要重写以下 4 个方法。

(1) public int getCount():ViewPager 通过此方法决定返回页面的数量。

(2) public Object instantiateItem(View container, int position):ViewPager 通过此方法创建一个指定位置的页面视图。其中参数 container 代表 ViewPager,参数 position 表示第几页。返回值是 Object 类型,代表该方法调用完毕后所创建的页面视图对应的 key。

(3) public void destroyItem(View container, int position, Object object):ViewPager 通过此方法移除页面视图。其中参数 container 代表 ViewPager,参数 position 表示第几页,参数 object 表示 instantiateItem() 方法创建页面视图时所返回的 key 对象。

(4) public boolean isViewFromObject(View v, Object obj):ViewPager 通过此方法判断某个页面视图与 instantiateItem() 方法创建页面视图时所返回的 key 对象之间的对应关系。

在 MainActivity 中,ViewPager 所使用的 PagerAdapter 对象重写了上述 4 个方法。

getCount() 方法返回 3,表示 ViewPager 共有 3 个页面(分别是"礼物推荐"、"礼品中心"和"送礼攻略")。

在 instantiateItem() 方法中,根据页面位置的不同,分别创建了不同的视图:第 1 页通过 LayoutInflater 创建了布局文件 index_recommend 对应的视图;第 2 页构造了 ImageView 对象,并指定显示的内容;第 3 页与第 2 页类似,构造了 ImageView 对象,并指定显示的内容。

在 instantiateItem() 方法的尾部,将根据页码构造视图并添加到 ViewPager 中,最后将该视图对象作为页面的 key 返回。其中涉及的 index_recommend 布局文件代码如下所示。

【任务 3-1】 index_recommend.xml

```
<LinearLayout xmlns:android = "http://schemas.android.com/apk/res/android" …>
    <ViewFlipper
        android:id = "@ + id/topViewFlipper"
        android:layout_width = "match_parent"
        android:layout_height = "0dp"
```

```xml
        android:layout_weight = "1"
        android:autoStart = "true"
        android:flipInterval = "4000" >
    </ViewFlipper >
    < LinearLayout
        android:layout_width = "match_parent"
        android:layout_height = "0dp"
        android:layout_weight = "1.00"
        android:orientation = "horizontal" >
        < ImageView
            android:id = "@ + id/bottomLeftImageView"
            android:layout_width = "0dp"
            android:layout_height = "match_parent"
            android:layout_weight = "1"
            android:scaleType = "centerCrop" >
        </ImageView >
        < ImageView
            android:id = "@ + id/bottomRightImageView"
            android:layout_width = "0dp"
            android:layout_height = "match_parent"
            android:layout_weight = "1"
            android:scaleType = "centerCrop" >
        </ImageView >
    </LinearLayout >
</LinearLayout >
```

上述代码中,布局整体分为上下两部分：上部是一个 ViewFlipper 控件,用于自动轮播；下部是左右两个 ImageView 控件,用于显示推荐礼品的图片。在 instantiateItem()方法中,向 ViewFlipper 控件中添加了 3 个 ImageView 并指定显示的内容,也对下部的两个 ImageView 指定了显示的内容。

在 destroyItem()方法中,由于 instantiateItem()方法将页面视图作为 key 返回,所以 destroyItem()方法的 object 参数实际上就是页面视图,可以直接从 ViewPager 中移除。

isViewFromObject()方法与 destroyItem()相似,也是因为 instantiateItem()方法将页面视图作为 key 返回,所以 isViewFromObject()的 object 参数实际上就是页面视图,因此将 object 参数与 View 参数直接通过＝＝比较来判断是否对应。

MainActivity 编写完毕后,运行项目,主页显示如图 3-37 所示。

图 3-37　主页

3.6.2 实现【任务 3-2】

需求的 UI 设计中,各个界面都包含上部的标题栏和右下角的扇形菜单,并且各个标题栏的功能图标相似。另外,在各个 Activity 中还会包含很多需要重用的功能,例如检查登录状态、显示统一的消息对话框等。因此,要为所有 Activity 设定一个统一的父类,将上述功能实现后供各个子 Activity 调用。下面首先在 com.qst.giftems.activity 包中编写 Activity 父类 BaseActivity,然后编写各个功能 Activity。

1. 定义 Activity 的父类 BaseActivity

【任务 3-2】 BaseActivity.java

```java
/** Activity 父类,包含标题栏和右下角操作区 */
public abstract class BaseActivity extends Activity {
    //顶部标题栏
    protected TextView titleTextView;
    protected View backView;                     //返回
    protected View shareView;                    //分享
    protected View collectView;                  //收藏
    protected View recycleView;                  //删除
    protected View helpView;                     //帮助
    protected View quitView;                     //退出
    //右下角扇形菜单
    protected RelativeLayout cornerMenuOpenRelativeLayout;
    protected ImageView settingImageView;        //设置
    protected ImageView calendarImageView;       //日历
    protected ImageView personalImageView;       //个人中心
    protected ImageView shoppingBagImageView;    //购物袋
    protected ImageView showCornerMenuImageView; //展开/收缩
    //等待 HTTP 响应的 dialog
    protected SimpleProgressDialog progressDialog;
    protected int layoutId;                      //布局资源 ID
    protected String title;                      //标题
    protected BaseActivity(int layoutId, String title) {
        this.layoutId = layoutId;
        this.title = title;}
    @Override
    protected void onCreate(Bundle savedInstanceState) {
        super.onCreate(savedInstanceState);
        setContentView(layoutId);
        titleTextView = (TextView) findViewById(R.id.titleTextView);
        backView = findViewById(R.id.backView);
        shareView = findViewById(R.id.shareView);
        collectView = findViewById(R.id.collectView);
        recycleView = findViewById(R.id.recycleView);
        helpView = findViewById(R.id.helpView);
        quitView = findViewById(R.id.quitView);
        cornerMenuOpenRelativeLayout
            = (RelativeLayout) findViewById(R.id.cornerMenuOpenRelativeLayout);
        settingImageView = (ImageView) findViewById(R.id.settingImageView);
```

```java
calendarImageView = (ImageView) findViewById(R.id.calendarImageView);
personalImageView = (ImageView) findViewById(R.id.personalImageView);
shoppingBagImageView = (ImageView) findViewById(R.id.shoppingBagImageView);
showCornerMenuImageView = (ImageView) findViewById(R.id.showCornerMenuImageView);
progressDialog = new SimpleProgressDialog(this);
progressDialog.setCancelable(false);
titleTextView.setText(title);
backView.setOnClickListener(new OnClickListener() {
    @Override
    public void onClick(View v) {
        finish();} });
quitView.setOnClickListener(new OnClickListener() {
    @Override
    public void onClick(View v) {
        Dialogs.showSimpleDialog(BaseActivity.this, "您确定退出登录吗?", true,
            new Dialogs.OnOkListener() {
                @Override
                public void onOk() {
                    App app = (App) getApplication();
                    app.user = null;
                    showToast("已退出登录");
                    quitView.setVisibility(View.GONE);
                    Intent intent = new Intent(
                        getApplicationContext(),
                        MainActivity.class);
                    startActivity(intent);}});} });
settingImageView.setOnClickListener(new OnClickListener() {
    @Override
    public void onClick(View v) {
        if (BaseActivity.this.getClass() == SettingActivity.class)
            return;
        Intent intent = new Intent(getApplicationContext(),
            SettingActivity.class);
        startActivity(intent);} });
calendarImageView.setOnClickListener(new OnClickListener() {
    @Override
    public void onClick(View v) {
        //TODO 后续章节添加功能
    } });
shoppingBagImageView.setOnClickListener(new OnClickListener() {
    @Override
    public void onClick(View v) {
        //TODO 后续章节添加功能
    }});
personalImageView.setOnClickListener(new OnClickListener() {
    @Override
    public void onClick(View v) {
        if (BaseActivity.this.getClass() == PersonalActivity.class)
            return;
        if (!checkLogin())
            return;
        Intent intent = new Intent(getApplicationContext(),
            PersonalActivity.class);
```

```
                    startActivity(intent);}});
            cornerMenuOpenRelativeLayout.setOnClickListener(new OnClickListener() {
                @Override
                public void onClick(View v) {
                    cornerMenuOpenRelativeLayout.setVisibility(View.GONE);
                    showCornerMenuImageView.setVisibility(View.VISIBLE);}});
            showCornerMenuImageView.setOnClickListener(new OnClickListener() {
                @Override
                public void onClick(View v) {
                    cornerMenuOpenRelativeLayout.setVisibility(View.VISIBLE);
                    showCornerMenuImageView.setVisibility(View.GONE);}}); }
    public void showMessage(String message) {
        Dialogs.showSimpleDialog(this, message, false, null);}
    public void showToast(String message) {
        Toast.makeText(getApplicationContext(), message, Toast.LENGTH_SHORT)
                .show();}
    public void showNetError() {
        Toast.makeText(getApplicationContext(), "无法连接到服务器,请检查网络后再试.",
                Toast.LENGTH_LONG).show();}
    public void showWaitDialog() {
        progressDialog.show();}
    public void dismissWaitDialog() {
        progressDialog.dismiss(); }
    /** 只是检查是否保存了mobileToken,没有到服务器验证 */
    public boolean checkLogin() {
        App app = (App) getApplication();
        if (app.user == null || app.user.mobileToken == null) {
            Dialogs.showSimpleDialog(this, "登录后才能继续操作,您现在登录吗?", true,
                    new OnOkListener() {
                        @Override
                        public void onOk() {
                            Intent intent = new Intent(getApplicationContext(),
                                    LoginActivity.class);
                            startActivity(intent);}});
            return false;}
        return true;}
}
```

在上述代码中,实现了以下功能:定义了一个继承 Activity 的抽象类 BaseActivity,作为项目中其他 Activity 的父类;在 BaseActivity 类中声明了顶部标题栏中的标题 TextView 和所有可能用到的图标元素;在 BaseActivity 类中声明了右下角扇形菜单中的图标元素;在 BaseActivity 构造方法中,需要传入布局文件的 ID 和标题;在返回图标的单击事件中,调用 Activity 的 finish()方法结束当前 Activity;在退出图标的单击事件中,弹出确认对话框,用户确认退出后,修改 Application 中的 user 对象为 null,提示已退出并返回到主界面;在扇形菜单的各个功能图标的单击事件中,打开对应的 Activity。

2. 顶部标题栏

各个 Activity 中包含相似的顶部标题栏,顶部标题栏部分包含标题 TextView 和所有 Activity 中可能用到的图标,其布局代码如下所示。

【任务 3-2】 title_bar.xml

```xml
<RelativeLayout xmlns:android="http://schemas.android.com/apk/res/android"
    android:layout_height="50dp"
    android:background="@color/title_bottom" >
    <TextView
        android:id="@+id/titleTextView"
        style="@style/text1"
        android:layout_width="match_parent"
        android:layout_height="match_parent"
        android:textColor="#FFFFFF"
        android:textSize="20sp" />
    <View
        android:id="@+id/backView"
        android:layout_width="40dp"
        android:layout_height="40dp"
        android:layout_centerVertical="true"
        android:layout_marginLeft="10dp"
        android:background="@drawable/back" />
    <LinearLayout
        android:layout_width="wrap_content"
        android:layout_height="40dp"
        android:layout_alignParentRight="true"
        android:layout_centerVertical="true"
        android:layout_marginRight="10dp" >
        <View
            android:id="@+id/shareView"
            android:layout_width="40dp"
            android:layout_height="40dp"
            android:layout_marginLeft="5dp"
            android:background="@drawable/share1"
            android:visibility="gone" />
        … //此处省略 collectView、recycleView、helpView 和 quitView 的 4 个<View>
    </LinearLayout>
</RelativeLayout>
```

在所有继承 BaseActivity 的 Activity 所对应的布局文件中,都需要使用<include>标签引入扇形菜单和顶部标题栏的布局文件。

3. 统一的信息提示对话框

BaseActivity 中用到了一个统一的消息对话框,其布局代码如下所示。

【任务 3-2】 dialog_simple.xml

```xml
<LinearLayout xmlns:android="http://schemas.android.com/apk/res/android"
    android:layout_width="wrap_content"
    android:layout_height="wrap_content"
    android:background="@drawable/background_border"
    android:orientation="vertical" >
    <TextView
        android:id="@+id/messageTextView"
        style="@style/text1"
```

```xml
        android:layout_width = "match_parent"
        android:layout_height = "wrap_content"
        android:layout_marginTop = "10dp"
        android:padding = "20dp"
        android:singleLine = "false" />
    <LinearLayout
        android:layout_width = "match_parent"
        android:layout_height = "wrap_content"
        android:background = "@color/title_bottom"
        android:gravity = "center"
        android:orientation = "horizontal"
        android:padding = "5dp" >
        <Button
            android:id = "@ + id/okButton"
            style = "@style/button"
            android:layout_width = "80dp"
            android:layout_height = "40dp"
            android:layout_marginLeft = "20dp"
            android:layout_marginRight = "20dp"
            android:text = "确定" />
        <Button
            android:id = "@ + id/cancelButton"
            style = "@style/button"
            android:layout_width = "80dp"
            android:layout_height = "40dp"
            android:layout_marginLeft = "20dp"
            android:layout_marginRight = "20dp"
            android:text = "取消" />
    </LinearLayout>
    <View
        android:layout_width = "wrap_content"
        android:layout_height = "10dp" />
</LinearLayout>
```

上述代码用于布局一个简单对话框,其中包含一个显示信息的 TextView 和"确定"、"取消"两个按钮。

下述代码在 com.qst.giftems 包中创建了一个对话框类,提供了显示指定格式的对话框方法,并为相关按钮提供了相应的事件处理,代码如下所示。

【任务 3-2】 Dialogs.java

```java
public class Dialogs {
    /* * * 显示一个信息提示对话框
     * @param context
     *               Context
     * @param message
     *               信息
     * @param cancelable
     *               是否可取消
     * @param onOkListener
     *               "确定"按钮单击事件 */
    public static void showSimpleDialog(Context context, String message,
            boolean cancelable, final OnOkListener onOkListener) {
```

```java
        View view = LayoutInflater.from(context).inflate(
                R.layout.dialog_simple, null);
        AlertDialog.Builder builder = new AlertDialog.Builder(context);
        builder.setView(view);
        final AlertDialog dialog = builder.create();
        TextView messageTextView = (TextView) view
                .findViewById(R.id.messageTextView);
        Button okButton = (Button) view.findViewById(R.id.okButton);
        Button cancelButton = (Button) view.findViewById(R.id.cancelButton);
        messageTextView.setText(message);
        okButton.setOnClickListener(new OnClickListener() {
            @Override
            public void onClick(View v) {
                if (onOkListener != null)
                    onOkListener.onOk();
                dialog.dismiss();}});
        dialog.setCancelable(cancelable);
        if (cancelable) {
            cancelButton.setOnClickListener(new OnClickListener() {
                @Override
                public void onClick(View v) {
                    dialog.cancel();}});
        } else {
            cancelButton.setVisibility(View.GONE);}
        dialog.show();}
    public interface OnOkListener {
        void onOk();}
}
```

上述代码中,showSimpleDialog()方法用于显示信息提示对话框,可以传入信息文本、是否可取消、单击确定按钮需要执行的操作。编写完毕后,在需要提示信息的位置,直接调用 Dialogs.showSimpleDialog()方法即可,效果如图 3-38 所示。

图 3-38　信息提示对话框

3.6.3　实现【任务 3-3】

本任务编写"GIFT-EMS 礼记"的辅助功能对应的 Activity,这些 Activity 中涉及的业务处理相对比较简单,主要包括"登录""注册用户""个人中心""修改密码""我的资料""应用设置""应用更新"和"关于"。

1. 登录 LoginActivity

礼记 App 的首页是不需要登录即可查看的,但是购买礼物等主要功能则必须登录才可使用,登录 Activity 对应的布局代码如下所示。

【任务 3-3】　login.xml

```xml
<RelativeLayout xmlns:android = "http://schemas.android.com/apk/res/android"
    xmlns:tools = "http://schemas.android.com/tools"
```

```xml
    android:layout_width = "match_parent"
    android:layout_height = "match_parent"
    android:background = "@drawable/background_body"
    android:focusable = "true"
    android:focusableInTouchMode = "true" >
    <include
        android:id = "@ + id/titleBarRelativeLayout"
        layout = "@layout/title_bar" />
    <RelativeLayout
        android:id = "@ + id/bottomRelativeLayout"
        android:layout_width = "match_parent"
        android:layout_height = "80dp"
        android:layout_alignParentBottom = "true"
        android:background = "@color/title_bottom"
        android:gravity = "center" >
        <Button
            android:id = "@ + id/loginButton"
            style = "@style/button"
            android:layout_width = "100dp"
            android:layout_height = "40dp"
            android:text = "登录" />
        <Button
            android:id = "@ + id/registerButton"
            style = "@style/button"
            android:layout_width = "100dp"
            android:layout_height = "40dp"
            android:layout_marginLeft = "120dp"
            android:text = "注册" />
    </RelativeLayout>
    <TextView
        android:id = "@ + id/userNameTextView"
        style = "@style/text1"
        android:layout_width = "wrap_content"
        android:layout_height = "wrap_content"
        android:layout_below = "@id/titleBarRelativeLayout"
        android:layout_marginLeft = "20dp"
        android:layout_marginTop = "40dp"
        android:text = "账号:" />
    <EditText
        android:id = "@ + id/userNameEditText"
        style = "@style/text1"
        android:layout_width = "match_parent"
        android:layout_height = "35dp"
        android:layout_alignBaseline = "@id/userNameTextView"
        android:layout_marginRight = "20dp"
        android:layout_toRightOf = "@id/userNameTextView"
        android:background = "@drawable/background_edit"
        android:hint = "请输入用户名或绑定的手机号码"
        android:maxLength = "12" />
    <TextView
        android:id = "@ + id/passwordTextView"
        style = "@style/text1"
        android:layout_width = "wrap_content"
```

```xml
        android:layout_height = "wrap_content"
        android:layout_alignLeft = "@id/userNameTextView"
        android:layout_below = "@id/userNameTextView"
        android:layout_marginTop = "40dp"
        android:text = "密码: " />
    <EditText
        android:id = "@+id/passwordEditText"
        style = "@style/text1"
        android:layout_width = "match_parent"
        android:layout_height = "35dp"
        android:layout_alignBaseline = "@id/passwordTextView"
        android:layout_marginRight = "20dp"
        android:layout_toRightOf = "@id/passwordTextView"
        android:background = "@drawable/background_edit"
        android:inputType = "textPassword"
        android:maxLength = "12" />
    <TextView
        android:id = "@+id/errorTextView"
        style = "@style/text3"
        android:layout_width = "wrap_content"
        android:layout_height = "wrap_content"
        android:layout_alignParentRight = "true"
        android:layout_below = "@id/passwordTextView"
        android:layout_marginRight = "20dp"
        android:layout_marginTop = "10dp"
        android:text = "账号或密码错误,请重新输入"
        android:visibility = "invisible" />
    <include
        android:id = "@+id/cornerMenuRelativeLayout"
        layout = "@layout/corner_menu" />
</RelativeLayout>
```

上述布局中包含"用户名""密码"两个输入框和"登录""注册"两个按钮,还包含一个 TextView 用于显示错误提示信息。

下述代码中,在 com. qst. giftems. activity 包中定义了 LoginActivity 类并继承 BaseActivity,其中实现用户输入的校验,代码如下所示。

【任务 3-3】 LoginActivity. java

```java
public class LoginActivity extends BaseActivity {
    EditText userNameEditText;
    EditText passwordEditText;
    TextView errorTextView;
    Button loginButton;
    Button registerButton;
    public LoginActivity() {
        super(R.layout.login, "登录");
    }
    @Override
    protected void onCreate(Bundle savedInstanceState) {
        super.onCreate(savedInstanceState);
        userNameEditText = (EditText) findViewById(R.id.userNameEditText);
        passwordEditText = (EditText) findViewById(R.id.passwordEditText);
```

```
        errorTextView = (TextView) findViewById(R.id.errorTextView);
        loginButton = (Button) findViewById(R.id.loginButton);
        registerButton = (Button) findViewById(R.id.registerButton);
        loginButton.setOnClickListener(new OnClickListener() {
            @Override
            public void onClick(View v) {
                View view = getWindow().peekDecorView();
                if (view != null) {
                    InputMethodManager inputmanger
                        = (InputMethodManager) getSystemService(Context.INPUT_METHOD_SERVICE);
                    inputmanger.hideSoftInputFromWindow(view.getWindowToken(),0);}
                String userName = userNameEditText.getText().toString();
                String password = passwordEditText.getText().toString();
                if (userName.length() == 0 || password.length() == 0) {
                    errorTextView.setVisibility(View.VISIBLE);
                    return;}
                errorTextView.setVisibility(View.INVISIBLE);
                //TODO 后续章节中添加连接服务器验证登录的功能
        }});
        registerButton.setOnClickListener(new OnClickListener() {
            @Override
            public void onClick(View v) {
                Intent intent = new Intent(getApplicationContext(),
                    RegisterActivity.class);
                startActivity(intent);}});}
}
```

上述代码中，在"登录"按钮的单击事件中，检查了用户名和密码的输入；在"注册"按钮的单击事件中，打开了用于注册用户的 RegisterActivity。运行应用程序时，在主界面单击右下角扇形菜单中的"个人中心"图标，会提示需要登录才能继续操作，然后单击"确定"按钮，进入登录界面，如图 3-39 所示。

图 3-39 登录

2. 注册 RegisterActivity

编写注册用户的 Activity 对应的布局文件 register.xml,该布局文件与登录布局类似,需要包含注册用户时需要录入的用户名、密码、手机号等信息,由于篇幅所限,此处不再给出代码。

编写相应的 RegisterActivity,代码如下所示。

【任务 3-3】 RegisterActivity.java

```java
public class RegisterActivity extends BaseActivity {
    EditText userNameEditText;
    TextView userNameErrorTextView;
    EditText passwordEditText;
    EditText password2EditText;
    TextView passwordErrorTextView;
    EditText mobileEditText;
    TextView mobileErrorTextView;
    EditText validateCodeEditText;
    Button sendValidateCodeButton;
    Button okButton;
    public RegisterActivity() {
        super(R.layout.register, "注册账号");}
    @Override
    protected void onCreate(Bundle savedInstanceState) {
        super.onCreate(savedInstanceState);
        userNameEditText = (EditText) findViewById(R.id.userNameEditText);
        userNameErrorTextView
                = (TextView) findViewById(R.id.userNameErrorTextView);
        passwordEditText = (EditText) findViewById(R.id.passwordEditText);
        password2EditText = (EditText) findViewById(R.id.password2EditText);
        passwordErrorTextView
                = (TextView) findViewById(R.id.passwordErrorTextView);
        mobileEditText = (EditText) findViewById(R.id.mobileEditText);
        mobileErrorTextView = (TextView) findViewById(R.id.mobileErrorTextView);
        validateCodeEditText
                = (EditText) findViewById(R.id.validateCodeEditText);
        sendValidateCodeButton
                = (Button) findViewById(R.id.sendValidateCodeButton);
        okButton = (Button) findViewById(R.id.okButton);
        userNameEditText.setOnFocusChangeListener(new OnFocusChangeListener() {
            @Override
            public void onFocusChange(View v, boolean hasFocus) {
                if (!hasFocus) {
                    String userName = userNameEditText.getText().toString();
                    if (userName.length() == 0) {
                        userNameErrorTextView.setText("请输入用户名");
                        return; }
                    if (userName.length() < 3) {
                        userNameErrorTextView.setText("用户名长度至少 3 位");
                        return;}}}});
        OnFocusChangeListener ofcl = new OnFocusChangeListener() {
            @Override
```

```java
            public void onFocusChange(View v, boolean hasFocus) {
                if (!hasFocus) {
                    String password = ((EditText) v).getText().toString();
                    if (password.length() < 6) {
                        passwordErrorTextView.setText("密码长度至少6位");
                    } else {
                        passwordErrorTextView.setText("");}}}};
passwordEditText.setOnFocusChangeListener(ofcl);
password2EditText.setOnFocusChangeListener(ofcl);
mobileEditText.setOnFocusChangeListener(new OnFocusChangeListener() {
    @Override
    public void onFocusChange(View v, boolean hasFocus) {
        if (!hasFocus) {
            String mobile = mobileEditText.getText().toString();
            if (!StringUtils.isMobile(mobile)) {
                mobileErrorTextView.setText("请输入正确的手机号码");
                return;}}}});
sendValidateCodeButton.setOnClickListener(new OnClickListener() {
    @Override
    public void onClick(View v) {
        String mobile = mobileEditText.getText().toString();
        if (!StringUtils.isMobile(mobile)) {
            showMessage("请输入正确的手机号码");
            return;}}});
okButton.setOnClickListener(new OnClickListener() {
    @Override
    public void onClick(View v) {
        String userName = userNameEditText.getText().toString();
        if (userName.length() == 0) {
            showMessage("请输入用户名");
            return; }
        if (userName.length() < 3) {
            showMessage("用户名长度至少3位");
            return; }
        String password = passwordEditText.getText().toString();
        String password2 = password2EditText.getText().toString();
        if (!password.equals(password2)) {
            showMessage("密码两次输入不一致");
            return; }
        if (password.length() < 6) {
            showMessage("密码长度至少6位");
            return; }
        String mobile = mobileEditText.getText().toString();
        if (!StringUtils.isMobile(mobile)) {
            showMessage("请输入正确的手机号码");
            return; }
        String validateCode = validateCodeEditText.getText().toString();
        if (validateCode.length() == 0) {
            showMessage("请输入您收到的验证码");
            return; }
        //TODO 后续章节添加向服务器发送注册数据的功能
    }});}
}
```

上述代码中,为用户名、密码等编辑框注册了 OnFocusChangeListener;当编辑框失去焦点时验证输入的内容是否符合要求,单击"注册"按钮时会对所有编辑框整体验证一遍。运行应用程序时,进入"登录"界面,单击"注册"按钮后可进入"注册账号"界面,如图 3-40 所示。

图 3-40 注册用户界面

3. 个人中心 PersonalActivity

编写个人中心 Activity 对应的布局文件,代码如下所示。

【任务 3-3】 personal.xml

```xml
<RelativeLayout xmlns:android = "http://schemas.android.com/apk/res/android" …>
    <include
        android:id = "@ + id/titleBarRelativeLayout"
        layout = "@layout/title_bar" />
    <LinearLayout
        android:layout_width = "match_parent"
        android:layout_height = "wrap_content"
        android:layout_below = "@id/titleBarRelativeLayout"
        android:orientation = "vertical"
        android:padding = "10dp" >
        <Button
            android:id = "@ + id/myInfoButton"
            style = "@style/button"
            android:layout_width = "match_parent"
            android:layout_height = "40dp"
            android:text = "我的资料" />
            …//省略"修改密码""我的积分""我的订单""我的收藏夹""登记收礼人"和"扫一扫"
六个<Button>代码
```

```xml
<View
    android:layout_width = "match_parent"
    android:layout_height = "0.5dp"
    android:layout_marginTop = "20dp"
    android:background = "@color/title_bottom" />
<TextView
    style = "@style/text3"
    android:layout_width = "match_parent"
    android:layout_height = "wrap_content"
    android:layout_marginTop = "20dp"
    android:text = "温馨提示:如果您想索要发票,请联系客服"
    android:textSize = "12sp" />
<TextView
    style = "@style/text1"
    android:layout_width = "match_parent"
    android:layout_height = "wrap_content"
    android:layout_marginTop = "10dp"
    android:text = "客服电话: 88888888" />
</LinearLayout >
< include
    android:id = "@ + id/cornerMenuRelativeLayout"
    layout = "@layout/corner_menu" />
</RelativeLayout >
```

在 com.qst.giftems.activity 包中编写个人中心 PersonalActivity, 代码如下所示。

【任务 3-3】 PersonalActivity.java

```java
public class PersonalActivity extends BaseActivity {
    Button myInfoButton,              //我的资料
           changePasswordButton,      //修改密码
           userAddressButton,         //登记收礼人
           orderButton,               //我的订单
           pointButton,               //我的积分
           scanButton,                //扫一扫
           favorityButton;            //我的收藏夹
    public PersonalActivity() {
        super(R.layout.personal, "个人中心");}
    @Override
    protected void onCreate(Bundle savedInstanceState) {
        super.onCreate(savedInstanceState);
        myInfoButton = (Button) findViewById(R.id.myInfoButton);
        changePasswordButton = (Button) findViewById(R.id.changePasswordButton);
        userAddressButton = (Button) findViewById(R.id.userAddressButton);
        orderButton = (Button) findViewById(R.id.orderButton);
        pointButton = (Button) findViewById(R.id.pointButton);
        favorityButton = (Button) findViewById(R.id.favorityButton);
        scanButton = (Button) findViewById(R.id.scanButton);
        App app = (App) getApplication();
        if (app.user == null || StringUtils.isEmpty(app.user.mobileToken))
            quitView.setVisibility(View.GONE);
        else
            quitView.setVisibility(View.VISIBLE);
```

```java
myInfoButton.setOnClickListener(new OnClickListener() {
    @Override
    public void onClick(View v) {
        if (!checkLogin())
            return;
        Intent intent = new Intent(getApplicationContext(),
                MyActivity.class);
        startActivity(intent);}});
changePasswordButton.setOnClickListener(new OnClickListener() {
    @Override
    public void onClick(View v) {
        if (!checkLogin())
            return;
        Intent intent = new Intent(getApplicationContext(),
                ChangePasswordActivity.class);
        startActivity(intent);} });
userAddressButton.setOnClickListener(new OnClickListener() {
    @Override
    public void onClick(View v) {
        if (!checkLogin())
            return;
        //TODO 后续章节实现此功能
    }});
orderButton.setOnClickListener(new OnClickListener() {
    @Override
    public void onClick(View v) {
        if (!checkLogin())
            return;
        //TODO 后续章节实现此功能
    }});
pointButton.setOnClickListener(new OnClickListener() {
    @Override
    public void onClick(View v) {
        if (!checkLogin())
            return;
        //TODO 后续章节实现此功能
    }});
favorityButton.setOnClickListener(new OnClickListener() {
    @Override
    public void onClick(View v) {
        if (!checkLogin())
            return;
        //TODO 后续章节实现此功能
    }});
scanButton.setOnClickListener(new OnClickListener() {
    @Override
    public void onClick(View v) {
        if (!checkLogin())
            return;
        //TODO 后续章节实现此功能
    });}
}
```

"个人中心"界面的布局文件和 Activity 代码都比较简单,此处不再详细解释。

进入"个人中心"界面时会验证用户是否已登录,目前还未实现完整的登录功能,因此修改 App 类,模拟一个登录用户,代码如下所示。

【任务 3-3】 App.java

```java
public class App extends Application {
    /** 当前登录用户 */
    public User user;
    @Override
    public void onCreate() {
        //模拟一个登录用户.后续章节完成登录功能后修改此处
        user = new User();
        user.id = 1;
        user.name = "test";
        user.mobileToken = "ABCDEFGH";}
}
```

运行应用程序,单击菜单中的个人中心图标,进入"个人中心"界面,如图 3-41 所示。

图 3-41 "个人中心"界面

"我的资料""修改密码""应用设置""应用更新""关于"等 Activity 的界面结构和代码都非常简单,此处不再将代码一一列出,这些 Activity 运行后的界面如图 3-42 所示。

注意

> 由于此应用程序的大部分业务功能都需要与服务器端进行数据交互,而本书目前尚未介绍相关的技术,因此上述的 Activity 都没有完整实现功能,这些将留待后续章节逐步完善。

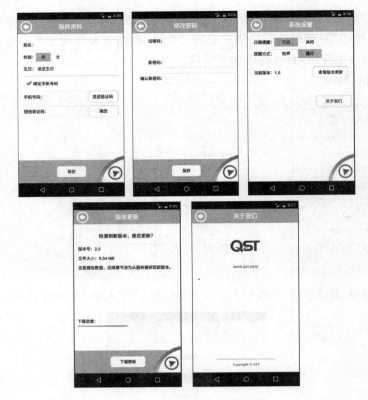

图 3-42　个人中心的相关操作

本章总结

小结

- Android 应用的绝大部分 UI 组件都放在 android.widget 包及其子包中，Android 应用程序的所有 UI 组件都继承了 View 类。
- Android 中的界面元素主要由以下几个部分构成：视图、视图容器、Fragment、Activity 和布局管理器。
- Android 的所有 UI 组件都是建立在 View、ViewGroup 基础之上的，Android 采用了"组合器"模式来设计 View 和 ViewGroup，其中 ViewGroup 是 View 的子类。
- 布局管理器可以根据运行平台来调整组件的大小，程序员的工作只是为容器选择合适的布局管理器即可。
- Android 提供了多种布局，常用的布局有以下几种：LinearLayout（线性布局）、RelativeLayout（相对布局）、TableLayout（表格布局）和 AbsoluteLayout（绝对布局）。
- Android 提供了两种方式的事件处理：基于回调的事件处理和基于监听的事件处理。
- Android 系统中引用了 Java 的事件处理机制，包括事件、事件源和事件监听器三个事件模型。
- Android 的事件处理机制是一种委派式事件处理方式，该处理方式类似于人类社会

- 的分工协作。这种委派式的处理方式将事件源和事件监听器分离,从而提供更好的程序模型,有利于提高程序的可维护性和代码的健壮性。
- 对于基于回调的事件处理模型而言,事件源和事件监听器是统一的,当用户在 GUI 组件上触发某个事件时,组件自身的方法将会负责处理该事件。
- 对 Widget 组件进行 UI 设计时,既可以采用 XML 布局方式也可以采用编码方式来实现,其中 XML 布局方式更加简单易用,被广泛使用。

Q&A

问题:列举 View 类的主要子类。

回答:Android 应用的绝大部分 UI 组件都放在 android.widget 包及其子包中,Android 应用程序的所有 UI 组件都继承了 View 类,例如:TextView、EditText、Button、Checkbox、RadioGroup、Spinner 等。

章节练习

习题

1. 下列_____可做 EditText 编辑框的提示信息。
   A. android:inputType　　　　　　B. android:text
   C. android:digits　　　　　　　　D. android:hint
2. 关于 widget 组件属性的写法,下面正确的是_____(多选)。
   A. android:id="@+id/tv_username"　　B. android:layout_width="100px"
   C. android:src="@drawable/icon"　　　D. android:id="@id/tabhost"
3. 下面_____不是 Android SDK 中的 ViewGroup(视图容器)。
   A. LinearLayout　　B. ListView　　C. GridView　　D. Button
4. Android 提供了_____和_____两种事件处理方式。
5. Android 中的所有 UI 组件都是建立在_____基础之上。
6. 使用_____来设置 AlterDialog 的各种属性。
7. 简述 Android 中常用的几种布局方式。

上机

1. 训练目标:Android 页面布局使用。

培养能力	熟练使用 Android 页面布局			
掌握程度	★★★★★	难度	中	
代码行数	400	实施方式	重复编码	
结束条件	熟练使用 Android 页面布局并完成计算器页面的编写			
参考训练内容 利用线性布局或相对布局实现一个加、减、乘、除及数字 1~9 的完整布局				

2. 训练目标：熟练使用 Android 组件的事件。

培养能力	熟练使用 Android 组件的事件		
掌握程度	★★★★★	难度	高
代码行数	800	实施方式	重复编码
结束条件	熟练使用 Android 组件的事件，并完善计算器		
参考训练内容 对计算器页面的按钮进行事件注册，并编写相应的处理事件来完成整个计算器的编写			

第4章

UI进阶

本章任务是完成"GIFT-EMS 礼记"项目的礼品和送礼攻略的展示功能:
- 【任务 4-1】 礼品和送礼攻略的列表界面。
- 【任务 4-2】 礼品展示界面。
- 【任务 4-3】 攻略展示界面。
- 【任务 4-4】 完成收礼人列表界面。
- 【任务 4-5】 完成收礼人编辑界面。
- 【任务 4-6】 完成我的收藏界面。

知 识 点	Listen(听)	Know(懂)	Do(做)	Revise(复习)	Master(精通)
Fragment	★	★	★	★	
Menu 和 Toolbar	★	★	★	★	★
高级控件	★	★	★	★	★

4.1 Fragment

Android 从 3.0 开始引入 Fragment(碎片),允许将 Activity 拆分成多个完全独立封装的可重用的组件,每个组件拥有自己的生命周期和 UI 布局。使用 Fragment 为不同型号、尺寸、分辨率的设备提供统一的UI设计方案，Fragment的最大优点是让开发者更加灵活地根据屏幕大小(包括小屏幕的手机、大屏幕的平板电脑)来创建相应的UI界面。

以新闻列表为例，当进行小屏幕手机开发时，开发者通常需要编写两个Activity，分别是ActivityA和ActivityB，其中ActivityA用于显示所有的新闻列表，列表内容为新闻的标题，ActivityB用于显示新闻的详细信息。当用户单击某个新闻标题时，由ActivityA启动ActivityB，并显示该标题所对应的新闻内容，两个Activity界面如图4-1所示。

当进行平板电脑开发时，使用FragmentA来显示标题列表，使用FragmentB来显示新闻的详细内容，将这两个Fragment在同一个Activity中并排显示。当单击FragmentA中的新闻标题时，通过FragmentB来显示该标题对应的新闻内容，显示效果如图4-2所示。每个Fragment都有自己的生命周期和相应的响应事件，通过切换Fragment同样可以实现显示切换的效果。

图4-1 手机上显示新闻列表　　　　图4-2 平板上显示新闻列表及内容

每个Fragment都是独立的模块，并与其所绑定的Activity紧密地联系在一起，Fragment通常会被封装成可重用的模块。对于一个界面允许有多个UI模块的设备(如平板电脑等)，Fragment拥有更好的适应性和动态构建UI的能力，在Activity中可以动态地添加、删除或更换Fragment。

由于Fragment具有独立的布局，能够进行事件响应，且具有自身的生命周期和行为，所以开发人员还可以在多个Activity中共用一个Fragment实例，即当程序运行在大屏设备时启动一个包含多个Fragment的Activity，当程序运行在小屏设备时启动一个包含少量Fragment的Activity。同样以新闻列表为例，对程序进行如下设置：当检测到程序运行在大屏设备时，启动ActivityA，并将标题列表和新闻内容所对应的两个Fragment都放在ActivityA中；当检测到程序运行在小屏设备时，依然启动ActivityA，但此时ActivityA中只包含一个标题列表Fragment，当用户单击某个新闻标题时，ActivityA将启动ActivityB，再通过ActivityB加载新闻内容所对应的Fragment。

4.1.1 使用Fragment

创建Fragment的过程与Activity类似，自定义的Fragment必须继承Fragment类或其子类。Fragment的继承体系如图4-3所示。

与Activity类似，同样需要实现Fragment中的回调方法，如onCreate()、onCreateView()、

图 4-3　Fragment 的继承体系

onStart()和 onResume()等方法。

通常在创建 Fragment 时,需要实现以下三个方法。

- **onCreate**():系统在创建 Fragment 对象时调用此方法,用于初始化相关的组件,例如一些在暂停或者停止时依然需要保留的组件。
- **onCreateView**():系统在第一次绘制 Fragment 对应的 UI 时调用此方法,该方法将返回一个 View,如果 Fragment 未提供 UI 则返回 null。当 Fragment 继承自 ListFragment 时,onCreateView()方法默认返回一个 ListView。
- **onPause**():当用户离开 Fragment 时首先调用此方法;当用户无须返回时,可以通过该方法来保存相应的数据。

Fragment 不能独立运行,必须嵌入在 Activity 中使用,因此 Fragment 的生命周期与其所在的 Activity 密切相关。将 Fragment 加载到 Activity 中主要有以下两种方式:①把 Fragment 添加到 Activity 的布局文件中;②在 Activity 的代码中动态添加 Fragment。

在上述两种方式中,第一种方式虽然简单但灵活性不够。如果把 Fragment 添加到 Activity 的布局文件中,就会使得 Fragment 及其视图与 Activity 的视图绑定在一起,在 Activity 的生命周期中,无法灵活地切换 Fragment 视图。因此在实际开发过程中,多采用第二种方式。相对而言,第二种方式要比第一种方式复杂,但也是唯一可以在运行时控制 Fragment 的方式,可以动态地添加、删除或替换 Fragment 实例。

1. 创建 Fragment

下述代码演示了 Fragment 的基本用法。屏幕分为左右两部分,通过单击屏幕左侧的按钮,可以在右侧动态地显示 Fragment。其布局文件代码如下所示。

【代码 4-1】　fragment_main.xml

```
<LinearLayout xmlns:android = "http://schemas.android.com/apk/res/android" …
```

```xml
    android:orientation = "horizontal">
    <LinearLayout
        android:id = "@+id/left"
        android:layout_width = "0dp"
        android:layout_height = "match_parent"
        android:layout_weight = "1"
        android:background = "#FFFFFF"
        android:orientation = "vertical" >
        <Button
            android:id = "@+id/displayBtn"
            android:layout_width = "wrap_content"
            android:layout_height = "wrap_content"
            android:text = "显示" />
    </LinearLayout>
    <LinearLayout
        android:id = "@+id/right"
        android:layout_width = "0dp"
        android:layout_height = "match_parent"
        android:layout_weight = "3"
        android:background = "#D3D3D3"
        android:orientation = "vertical" >
    </LinearLayout>
</LinearLayout>
```

上述代码较为简单，定义了一个 ID 为 left 的 LinearLayout 和一个 ID 名为 right 的 LinearLayout，将整个布局分为左右两部分：左边占 1/4，右边占 3/4。其中，ID 为 right 的 LinearLayout 是一个包含 Fragment 的容器。

接下来创建 FragmentDemoActivity，用于显示上面的布局效果，代码如下所示。

【代码 4-2】 FragmentDemoActivity.java

```java
public class FragmentDemoActivity extends AppCompatActivity {
    //展示内容 Button
    Button displayBtn;
    @Override
    protected void onCreate(Bundle savedInstanceState) {
        super.onCreate(savedInstanceState);
        setContentView(R.layout.fragment_main);
        displayBtn = (Button) findViewById(R.id.displayBtn);
        displayBtn.setOnClickListener(new OnClickListener() {
            @Override
            public void onClick(View v) {
                //TODO
            }});}
}
```

上述代码中，声明并初始化名为 displayBtn 的 Button 组件，并为其添加了 OnClickListener 事件监听器，此处仅作演示，事件处理部分暂未实现。运行上述代码，界面效果如图 4-4 所示。

图 4-4　fragment_main.xml 界面效果

当用户单击"显示"按钮时,需要在屏幕右侧动态地显示内容,此处通过动态添加 Fragment 来实现。在工程中创建一个名为 fragment_right.xml 的文件,用于显示右侧的内容,代码如下所示。

【代码 4-3】　fragment_right.xml

```
<LinearLayout xmlns:android = "http://schemas.android.com/apk/res/android" …>
    <TextView
        android:id = "@ + id/textView1"
        android:layout_width = "wrap_content"
        android:layout_height = "wrap_content"
        android:singleLine = "false"
        android:text = "新闻内容新闻内容新闻内容新闻内容新闻内容新闻内容新闻内容"/>
    <Button
        android:id = "@ + id/frgBtn"
        android:layout_width = "wrap_content"
        android:layout_height = "wrap_content"
        android:text = "show" />
</LinearLayout>
```

上述代码较为简单,定义了两个组件:①TextView 组件用于显示普通的文本;②Button 按钮用于演示 Fragment 中的事件处理机制。

下述代码定义一个 Fragment 类,代码实现如下所示。

【代码 4-4】　RightFragment.java

```
public class RightFragment extends Fragment {
    @Override
    public void onCreate(Bundle savedInstanceState) {
        super.onCreate(savedInstanceState);}
    //重写 onCreateView()方法 ①
    public View onCreateView(LayoutInflater inflater, ViewGroup container,
        Bundle savedInstanceState) {
        //获取 view 对象 ②
        View view = inflater.inflate(R.layout.fragment_right, null);
        //从 view 容器中获取组件③
        Button button = (Button) view.findViewById(R.id.frgBtn);
```

```
    button.setOnClickListener(new OnClickListener() {
        @Override
        public void onClick(View v) {
            Toast.makeText(getActivity(),
              "我是Fragment", Toast.LENGTH_SHORT).show();}});
    return view;}
@Override
public void onPause() {
    super.onPause();}
}
```

上述代码需要注意以下几点：标号①处重写了onCreateView()方法，该方法返回的View对象将作为该Fragment显示的View组件，当Fragment绘制界面组件时将会回调该方法；标号②处通过LayoutInflater对象的inflate()方法加载fragment_right.xml布局文件，并返回对应的View容器对象，其他组件对象都是从该View对象中获取的；标号③处从View容器中获取Button对象，并为该对象添加OnClickListener事件监听器，然后实现相应的事件处理功能。

修改代码4-2中displayBtn按钮的事件处理方法，当用户单击"显示"按钮时，右侧动态显示Fragment对象对应的布局，所修改的代码如下所示。

```
...//省略
displayBtn.setOnClickListener(new OnClickListener() {
    @Override
    public void onClick(View v) {
        //步骤1：获得一个FragmentTransaction的实例
        FragmentManager fragmentManager = getFragmentManager();
        FragmentTransaction transaction = fragmentManager
             .beginTransaction();
        //步骤2：用add()方法加上Fragment的对象rightFragment
        RightFragment rightFragment = new RightFragment();
        transaction.add(R.id.right, rightFragment);
        //步骤3：调用commit()方法使得FragmentTransaction实例的改变生效
        transaction.commit();}
});
...//省略
```

再次运行FragmentDemoActivity并单击"显示"按钮时，显示右侧的Fragment；单击show按钮时，弹出"我是Fragment"提示信息，界面效果如图4-5所示。

2. 管理Fragment

通过FragmentManager实现Fragment对象的管理。在Activity中可以通过getFragmentManager()方法来获取FragmentManager对象。

FragmentManager能够完成以下几方面的操作：

（1）通过findFragmentById()或findFragmentByTag()

图4-5 动态显示Fragment对象

方法来获取 Activity 中已存在的 Fragment 对象；

（2）通过 popBackStack() 方法将 Fragment 从 Activity 的后退栈中弹出（模拟用户按下 Back 按键）；

（3）通过 addOnBackStackChangedListener() 方法来注册一个侦听器以监视后退栈的变化。

当需要添加、删除或替换 Fragment 对象时，需要借助于 FragmentTransaction 对象来实现，FragmentTransaction 用于实现 Activity 对 Fragment 的操作，例如添加或删除 Fragment 操作。

Fragment 的最大特点是根据用户的输入，可以灵活地对 Fragment 进行添加、删除、替换以及其他操作。开发人员可以把每个事务保存在 Activity 的后退栈中，使得用户能够在 Fragment 之间进行导航（与在 Activity 之间导航相同）。

针对一组 Fragment 的变化称为一个事务，事务通过 FragmentTransaction 来执行，而 FragmentTransaction 对象需要通过 FragmentManager 来获取，示例代码如下所示。

【示例】 获取 FragmentTransaction 对象

```
FragmentManager fragmentManager = getFragmentManager();
FragmentTransaction fragmentTransaction = fragmentManager.beginTransaction();
```

事务是指在同一时刻执行的一组动作，要么一起成功，要么同时失败。事务可以用 add()、remove()、replace() 等方法构成，最后使用 commit() 方法来提交事务。

在调用 commit() 之前，可以使用 addToBackStack() 方法把事务添加到一个后退栈中，这个后退栈属于所对应的 Activity。当用户按下返回键时，就可以返回到 Fragment 执行事务之前的状态。

> 注意
>
> FragmentTransaction 被称作 Fragment 事务，与数据库事务类似，Fragment 事务代表了 Activity 对 Fragment 执行的多个改变操作。

下述代码演示了如何用一个 Fragment 代替另一个 Fragment，并在后退栈中保存被代替的 Fragment 的状态。

【示例】 使用 FragmentTransaction

```
Fragment newFragment = new ExampleFragment();    //创建一个新的 Fragment 对象
FragmentTransaction transaction               //通过 FragmentManager 获取 Fragment 事务对象
        = getFragmentManager().beginTransaction();
//通过 replace() 方法把 fragment_container 替换成新的 Fragment 对象
transaction.replace(R.id.fragment_container,newFragment);
transaction.addToBackStack(null);             //添加到回退栈
transaction.commit();                         //提交事务
```

上述代码中，ExampleFragment 类是一个自定义的 Fragment 子类。通过 replace() 方法使用 newFragment 代替组件 R.id.fragment_container 所指向的 ViewGroup 中包含的 Fragment 对象。然后调用 addToBackStack() 方法，将被代替的 Fragment 放入回退栈。当

用户按 Back 键时,回退到事务提交之前的状态,即界面重新展示原来的 Fragment 对象。如果向事务中添加了多个动作,例如多次调用了 add()、remove()方法之后又调用了 addToBackStack()方法,那么在 commit()之前调用的所有方法都被作为一个事务。当用户按返回键时,所有的动作都会回滚。

事务中动作的执行顺序可以随意,但需要注意以下两点:

(1) 程序的最后必须调用 commit()方法。

(2) 如果程序中添加了多个 Fragment 对象,则显示的顺序跟添加顺序一致(即后添加的覆盖之前的)。如果在执行的事务中有删除 Fragment 对象的动作,而且没有调用 addToBackStack()方法,那么当事务提交时被删除的 Fragment 就会被销毁。反之,那些 Fragment 就不会被销毁,而是处于停止状态,当用户返回时,这些 Fragment 将会被恢复。

注意

> 调用 commit()后,事务并不会马上提交,而是会在 Activity 的 UI 线程(主线程)中等待,直到线程能执行的时候才执行,不过可以在 UI 线程中调用 executePendingTransactions()方法来立即执行事务。但一般不需这样做,除非有其他线程在等待事务的执行。

3. 与 Activity 通信

Fragment 的实现是独立于 Activity,可以用于多个 Activity 中;而每个 Activity 允许包含同一个 Fragment 类的多个实例。在 Fragment 中,通过调用 getActivity()方法可以获得其所在的 Activity 实例,然后使用 findViewById()方法查找 Activity 中的组件,示例代码如下所示。

【示例】 **Fragment 获取其所在的 Activity 中的组件**

```
View listView = getActivity().findViewById(R.id.list);
```

在 Activity 中还可以通过 FragmentManager 的 findFragmentById()等方法来查找其所包含的 Frament 实例,示例代码如下所示。

【示例】 **Activity 获取指定 Frament 实例**

```
ExampleFragment fragment = (ExampleFragment)getFragmentManager()
        .findFragmentById(R.id.example_fragment);
```

有时需要 Fragment 与 Activity 共享事件,通常做法是在 Fragment 中定义一个回调接口,然后在 Activity 中实现该回调接口。

下面以新闻列表为例,在 Activity 中包含两个 Fragment:FragmentA 用于显示新闻标题,FragmentB 用于显示标题对应的内容。在 FragmentA 中,用户单击某个标题时通知 Activity,然后 Activity 再通知 FragmentB,此时 FragmentB 就会显示该标题所对应的新闻内容。在 FragmentA 中定义 OnNewsSelectedListener 接口,代码如下所示。

【示例】 在 Fragment 中定义回调接口

```
public static class FragmentA extends ListFragment {
...//省略
    //Activity必须实现下面的接口
    public interface OnNewsSelectedListener{
        //传递当前被选中的标题的 ID
            public void onNewsSelected(long id); }
...//省略
}
```

然后在 Activity 中实现 OnNewsSelectedListener 接口，并重写 onNewsSelected()方法来通知 FragmentB。当 Fragment 添加到 Activity 中时，会调用 Fragment 的 onAttach()方法，在该方法中检查 Activity 是否实现了 OnNewsSelectedListener 接口，并对传入的 Activity 实例进行类型转换，代码如下所示。

【示例】 使用 onAttach()方法检查 Activity 是否实现了回调接口

```
public static class FragmentA extends ListFragment {
    OnNewsSelectedListener mListener;
    ...//省略
    @Override
    public void onAttach(Activity activity){
        super.onAttach(activity);
        try{
            mListener = (OnNewsSelectedListener)activity;
        }catch(ClassCastException e){
            throw new ClassCastException(activity.toString()
                + "必须继承接口 OnNewsSelectedListener");}}
    ...//省略
}
```

上述代码中，如果 Activity 没有实现该接口，则 FragmentB 会抛出 ClassCastException 异常。mListener 成员变量用于保存 OnNewsSelectedListener 的实例，FragmentA 通过调用 mListener 的方法实现与 Activity 共享事件。由于 FragmentA 继承自 ListFragment 类，所以每次选中列表项时，就会调用 FragmentA 的 onListItemClick()方法，在 onListItemClick()方法中调用 onNewsSelected()方法实现与 Activity 的共享事件，示例代码如下所示。

【示例】 Fragment 与 Activity 共享事件

```
public static class FragmentA extends ListFragment {
    OnNewsSelectedListener mListener;
    ...//省略
    @Override
        public void onListItemClick(ListView l,View v,int position,long id){
            mListener.onNewsSelected(id); }
    ...//省略
}
```

上述代码中，onListItemClick()方法中的参数 id 是列表中被选项的 ID，Fragment 通过该 ID 实现从程序的某个存储单元中取得标题的内容。

> **注意**
>
> 在数据传递时,也可以直接把数据从 FragmentA 传递给 FragmentB,不过该方式降低了 Fragment 的可重用的能力。现在的处理方式只需要把发生的事件告诉宿主,由宿主决定如何处置,以使 Fragment 的重用性更好。

4.1.2 Fragment 的生命周期

Fragment 的生命周期与 Activity 的生命周期类似,也具有以下几个状态。
- **活动状态**:当前 Fragment 位于前台时,用户可见并且可以获取焦点;
- **暂停状态**:其他 Activity 位于前台,该 Fragment 仍然可见,但不能获取焦点;
- **停止状态**:该 Fragment 不可见,失去焦点;
- **销毁状态**:该 Fragment 被完全删除或该 Fragment 所在的 Activity 结束。

Fragment 的生命周期及相关回调方法如图 4-6 所示。

图 4-6 Fragment 的生命周期

Fragment 生命周期中的方法说明如表 4-1 所示。

表 4-1 Fragment 生命周期中的方法

序号	方 法	功 能 描 述
1	onAttach()	当一个 Fragment 对象关联到一个 Activity 时被调用
2	onCreate()	初始化创建 Fragment 对象时被调用
3	onCreateView()	当 Activity 获得 Fragment 的布局时调用此方法,Fragment 在其中创建自己的界面
4	onActivityCreated()	当 Activity 对象完成自己的 onCreate() 方法时调用
5	onStart()	Fragment 对象在 UI 界面可见时调用
6	onResume()	Fragment 对象的 UI 可以与用户交互时调用
7	onPause()	Fragment 对象可见,但不可交互,由 Activity 对象转为 onPause 状态时调用
8	onStop()	有组件完全遮挡,或者宿主 Activity 对象转为 onStop 状态时调用
9	onDestroyView()	Fragment 对象清理 View 资源时调用,即移除 Fragment 中的视图
10	onDestroy()	Fragment 对象完成对象清理 View 资源时调用
11	onDetach()	当 Fragment 被从 Activity 中删掉时被调用

上述方法中,当一个 Fragment 被创建的时候执行方法 1~4;当 Fragment 创建完毕并呈现到前台时,执行方法 5 和 6;当该 Fragment 从可见状态转换为不可见状态时,执行方法 7 和 8;当该 Fragment 被销毁(或者持有该 Fragment 的 Activity 被销毁)时,执行方法 9 和 11;此外在 3 和 5 过程中,可以使用 Bundle 对象保存一个 Fragment 的对象。

无论是在布局文件中包含 Fragment,还是在 Activity 动态添加 Fragment,Fragment 必须依存于 Activity,因此 Activity 的生命周期会直接影响到 Fragment 的生命周期。Fragment 和 Activity 生命周期对比如图 4-7 所示。

Activity 直接影响其所包含的 Fragment 的生命周期,所以对 Activity 生命周期中的某个方法调用时,也会产生对 Fragment 相应的方法调用。例如,当 Activity 的 onPause() 方法被调用时,其中包含的所有 Fragment 的 onPause() 方法都将被调用。

在生命周期中,Fragment 的回调方法要比 Activity 多,多出的方法主要用于与 Activity 的交互,例如 onAttach()、onCreateView()、onActivityCreated()、onDestroyView() 和 onDetach() 方法。

只有当 Activity 进入运行状态时(即 running 状态),才允许添加或删除 Fragment。因此,只有当 Activity 处于 resumed 状态时,Fragment 的生命周期才能独立运转,其他阶段依赖于 Activity 的生命周期。

为了使读者更好地理解 Fragment 的生命周期,下面分别介绍静态方式和动态方式。

1. 静态方式

所谓静态方式,是指将 Fragment 组件在布局文件中进行布局。Fragment 的生命周期会随其所在的 Activity 的生命周期而发生变化,生命周期方法的调用过程如下。

(1) 当首次展示布局页面时,其生命周期方法调用的顺序是:onAttach()→onCreate()→onCreateView()→onActivityCreated()→onStart()→onResume()。

(2) 当关闭手机屏幕或者手机屏幕变暗时,其生命周期方法调用的顺序是:onPause()→

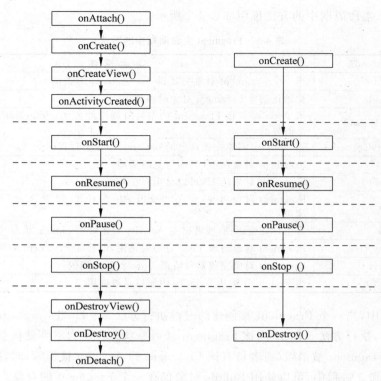

图 4-7 Activity 与 Fragment 生命周期对比

onStop()。

(3) 当对手机屏幕解锁或者手机屏幕变亮时,其生命周期方法调用的顺序是:onStart()→onResume()。

(4) 当对 Fragment 所在屏幕按返回键时,其生命周期方法调用的顺序是:onPause()→onStop()→onDestroyView()→onDestroy()→onDetach()。

2. 动态方式

当使用 FragmentManager 动态地管理 Fragment 并且涉及 addToBackStack 时,其生命周期的展现显得有些复杂。

动态方式主要通过重写 Fragment 生命周期的方法,然后在 Activity 代码中动态使用 Fragment。例如,定义两个 Fragment 分别为 FragmentA 和 FragmentB,在其生命周期的各个方法中打印(Log 输出方式)相关信息来验证方法的调用顺序,定义 FragmentA 的代码如下所示。

【代码 4-5】 FragmentA.java

```
public class FragmentA extends Fragment {
    private static final String TAG = FragmentA.class.getSimpleName();
    @Override
    public void onAttach(Activity activity) {
        super.onAttach(activity);
        Log.i(TAG, "onAttach");}
```

第4章 UI进阶

```java
        @Override
        public void onCreate(Bundle savedInstanceState) {
            super.onCreate(savedInstanceState);
            Log.i(TAG, "onCreate");}
        @Override
        public View onCreateView(LayoutInflater inflater, ViewGroup container,
                Bundle savedInstanceState) {
            Log.i(TAG, "onCreateView");
            return inflater.inflate(R.layout.fragment_a, null, false);}
        @Override
        public void onViewCreated(View view, Bundle savedInstanceState) {
            Log.i(TAG, "onViewCreated");
            super.onViewCreated(view, savedInstanceState);}
        @Override
        public void onDestroy() {
            Log.i(TAG, "onDestroy");
            super.onDestroy();}
        @Override
        public void onDetach() {
            Log.i(TAG, "onDetach");
            super.onDetach();}
        @Override
        public void onDestroyView() {
            Log.i(TAG, "onDestroyView");
            super.onDestroyView();}
        @Override
        public void onStart() {
            Log.i(TAG, "onStart");
            super.onStart();}
        @Override
        public void onStop() {
            Log.i(TAG, "onStop");
            super.onStop();}
        @Override
        public void onResume() {
            Log.i(TAG, "onResume");
            super.onResume();}
        @Override
        public void onPause() {
            Log.i(TAG, "onPause");
            super.onPause();}
        @Override
        public void onActivityCreated(Bundle savedInstanceState) {
            Log.i(TAG, "onActivityCreated");
            super.onActivityCreated(savedInstanceState);}
}
```

定义 FragmentB 的代码如下所示。

【代码 4-6】 FragmentB.java

```java
public class FragmentB extends Fragment {
    private static final String TAG = FragmentB.class.getSimpleName();
```

```java
@Override
public void onAttach(Activity activity) {
    super.onAttach(activity);
    Log.i(TAG, "onAttach");}
@Override
public void onCreate(Bundle savedInstanceState) {
    super.onCreate(savedInstanceState);
    Log.i(TAG, "onCreate");}
@Override
public View onCreateView(LayoutInflater inflater, ViewGroup container,
        Bundle savedInstanceState) {
    Log.i(TAG, "onCreateView");
    return inflater.inflate(R.layout.fragment_b, null, false);}
@Override
public void onViewCreated(View view, Bundle savedInstanceState) {
    Log.i(TAG, "onViewCreated");
    super.onViewCreated(view, savedInstanceState);}
@Override
public void onDestroy() {
    Log.i(TAG, "onDestroy");
    super.onDestroy();}
@Override
public void onDetach() {
    Log.i(TAG, "onDetach");
    super.onDetach();}
@Override
public void onDestroyView() {
    Log.i(TAG, "onDestroyView");
    super.onDestroyView();}
@Override
public void onStart() {
    Log.i(TAG, "onStart");
    super.onStart();}
@Override
public void onStop() {
    Log.i(TAG, "onStop");
    super.onStop();}
@Override
public void onResume() {
    Log.i(TAG, "onResume");
    super.onResume();}
@Override
public void onPause() {
    Log.i(TAG, "onPause");
    super.onPause();}
@Override
public void onActivityCreated(Bundle savedInstanceState) {
    Log.i(TAG, "onActivityCreated");
    super.onActivityCreated(savedInstanceState);}
}
```

在 Activity 中调用 FragmentA 和 FragmentB, 代码如下所示。

【代码 4-7】 FragmentLifecircleActivity.java

```java
public class FragmentLifecircleActivity extendsAppCompatActivity
        implements View.OnClickListener {
    //声明 Fragment 管理器 ①
    private FragmentManager fragmentManager;
    //声明变量
    private Button fragABtn;
    private Button fragBBtn;
    //Fragments
    private FragmentA fragmentA;
    private FragmentB fragmentB;
    //Fragment 名称列表
    private String[] fragNames = {"FragmentA","FragmentB"};
    @Override
    protected void onCreate(Bundle savedInstanceState) {
        super.onCreate(savedInstanceState);
        setContentView(R.layout.activity_frg_life);
        //初始化 Fragment 管理器 ②
        fragmentManager = getFragmentManager();
        //初始化组件
        fragABtn = (Button) findViewById(R.id.fragABtn);
        fragBBtn = (Button) findViewById(R.id.fragBBtn);
        //设置事件监听器 ③
        fragABtn.setOnClickListener(this);
        fragBBtn.setOnClickListener(this);}
    //单击事件监听 ④
    @Override
    public void onClick(View v) {
        FragmentTransaction fragmentTransaction
                = fragmentManager.beginTransaction();
        switch (v.getId()) {
            case R.id.fragABtn:
                if (fragmentA == null) {
                    fragmentA = new FragmentA();
                    fragmentTransaction.replace(R.id.frag_container,
                        fragmentA, fragNames[0]);
                    //把 FragmentA 对象添加到后退栈中
                    //fragmentTransaction.addToBackStack(fragNames[0]);
                } else {
                    Fragment fragment
                        = fragmentManager.findFragmentByTag(fragNames[0]);
                    //替换 Fragment
                    fragmentTransaction.replace(R.id.frag_container,
                        fragment, fragNames[0]);}
                break;
            case R.id.fragBBtn:
                if (fragmentB == null) {
                    fragmentB = new FragmentB();
                    fragmentTransaction.replace(R.id.frag_container,
                        fragmentB, fragNames[1]);
                    //把 FragmentB 对象添加到后退栈中
```

```
                            //fragmentTransaction.addToBackStack(fragNames[1]);
                    } else {
                        Fragment fragment
                             = fragmentManager.findFragmentByTag(fragNames[1]);
                        //替换 Fragment
                        fragmentTransaction.replace(R.id.frag_container,
                            fragment, fragNames[1]);}
                    break;
                default:
                    break;}
            fragmentTransaction.commit();}
}
```

上述代码需要说明以下几点：标号①处分别声明了 FragmentManager、Button 类型的变量属性；标号②处对标号①处所声明的属性变量进行初始化，使其可以进行后续的业务逻辑操作；标号③处对 fragABtn、fragBBtn 对象注册监听器，来监听用户单击时触发的事件；标号④处重写了 OnClickListener 监听器中的 onClick()方法，用于处理单击事件发生时根据按钮的 ID 来决定相应的处理逻辑功能。

运行 FragmentLifecircleActivity 代码时，产生的效果如图 4-8 所示。

图 4-8 选中的图示

当第一次单击"显示 FragA"按钮时，实际执行的代码如下。

```
...//省略
fragmentA = new FragmentA();
fragmentTransaction.replace(R.id.frag_container, fragmentA, fragNames[0]);
...//省略
fragmentTransaction.commit();
```

在 LogCat 控制台中打印的日志如下所示。

```
12-07 03:30:06.542: I/FragmentA(2215): onAttach
12-07 03:30:06.542: I/FragmentA(2215): onCreate
12-07 03:30:06.542: I/FragmentA(2215): onCreateView
12-07 03:30:06.542: I/FragmentA(2215): onViewCreated
12-07 03:30:06.542: I/FragmentA(2215): onActivityCreated
12-07 03:30:06.542: I/FragmentA(2215): onStart
12-07 03:30:06.543: I/FragmentA(2215): onResume
```

由上述打印日志可知，FragmentA 的生命周期和在布局文件中静态设置的表现完全一致，此处不再赘述。当继续单击"显示 FragB"按钮时，在 LogCat 控制台打印的日志如下所示。

```
12-07 03:45:17.240: I/FragmentA(2507): onPause
12-07 03:45:17.240: I/FragmentA(2507): onStop
12-07 03:45:17.240: I/FragmentA(2507): onDestroyView
12-07 03:45:17.240: I/FragmentA(2507): onDestroy
12-07 03:45:17.240: I/FragmentA(2507): onDetach
```

```
12-07 03:45:17.240: I/FragmentB(2507): onAttach
12-07 03:45:17.240: I/FragmentB(2507): onCreate
12-07 03:45:17.240: I/FragmentB(2507): onCreateView
12-07 03:45:17.241: I/FragmentB(2507): onViewCreated
12-07 03:45:17.241: I/FragmentB(2507): onActivityCreated
12-07 03:45:17.241: I/FragmentB(2507): onStart
12-07 03:45:17.241: I/FragmentB(2507): onResume
```

由上述打印结果可知，FragmentA 从运行状态到销毁状态所调用方法的顺序为：onPause()→onStop()→onDestroyView()→onDestroy()→onDetach()。此时，FragmentA 已经由 FragmentManager 进行销毁，取而代之的是 FragmentB 对象。如果此时按 Back 键，FragmentB 所调用的方法与 FragmentA 调用的顺序一样。在添加 Fragment 的过程中如果没有调用 addToBackStack()方法进行保存，那么使用 FragmentManager 更换 Fragment 时，是不会保存 Fragment 的状态的。

如果取消 FragmentLifecircleActivity 中 addToBackStack()部分的代码注释，如下所示。

```
//把 FragmentA 对象添加到后退栈中
fragmentTransaction.addToBackStack(fragNames[0]);
    …//省略部分代码
//把 FragmentB 对象添加到后退栈中
fragmentTransaction.addToBackStack(fragNames[1]);
```

重新运行 FragmentLifecircleActivity，然后单击"显示 FragA"按钮，在 Logcat 控制台打印的日志如下。

```
12-07 04:16:29.333: I/FragmentA(2751): onAttach
12-07 04:16:29.333: I/FragmentA(2751): onCreate
12-07 04:16:29.333: I/FragmentA(2751): onCreateView
12-07 04:16:29.335: I/FragmentA(2751): onViewCreated
12-07 04:16:29.335: I/FragmentA(2751): onActivityCreated
12-07 04:16:29.335: I/FragmentA(2751): onStart
12-07 04:16:29.335: I/FragmentA(2751): onResume
```

由上述日志可以得知，此时 FragmentA 所调用的方法与没有添加 addToBackStack()方法时没有任何区别。

然后继续单击"显示 FragB"按钮，在 Logcat 控制台打印的日志如下。

```
12-07 04:18:43.927: I/FragmentA(2751): onPause
12-07 04:18:43.927: I/FragmentA(2751): onStop
12-07 04:18:43.927: I/FragmentA(2751): onDestroyView
12-07 04:18:43.927: I/FragmentB(2751): onAttach
12-07 04:18:43.927: I/FragmentB(2751): onCreate
12-07 04:18:43.927: I/FragmentB(2751): onCreateView
12-07 04:18:43.928: I/FragmentB(2751): onViewCreated
12-07 04:18:43.929: I/FragmentB(2751): onActivityCreated
12-07 04:18:43.929: I/FragmentB(2751): onStart
12-07 04:18:43.929: I/FragmentB(2751): onResume
```

由上述日志可以得知,FragmentA 生命周期方法只是调用了 onDestroyView(),而没有调用 onDestroy()和 onDetach()方法,即 FragmentA 界面虽然被销毁,但 FragmentManager 并没有完全销毁 FragmentA,FragmentA 仍然存在并保存在 FragmentManager 中。

继续单击"显示 FragA"按钮,使用 FragmentA 来替换当前显示的 FragmentB,此时实际上执行的代码如下。

```
Fragment fragment = fragmentManager.findFragmentByTag(fragNames[0]);
//替换 Fragment
fragmentTransaction.replace(R.id.frag_container,fragment, fragNames[0]);
```

此时 Logcat 控制台打印的日志为

```
12-07 04:27:48.822: I/FragmentA(2859): onCreateView
12-07 04:27:48.823: I/FragmentA(2859): onViewCreated
12-07 04:27:48.823: I/FragmentA(2859): onActivityCreated
12-07 04:27:48.823: I/FragmentA(2859): onStart
12-07 04:27:48.823: I/FragmentA(2859): onResume
12-07 04:27:48.822: I/FragmentB(2859): onPause
12-07 04:27:48.822: I/FragmentB(2859): onStop
12-07 04:27:48.822: I/FragmentB(2859): onDestroyView
```

由上述日志可以得知,使用 FragmentA 替换 FragmentB 的方法调用顺序与使用 FragmentB 替换 FragmentA 时的调用顺序一致,其作用只是销毁视图,但依然保留了 Fragment 的状态。此时 FragmentA 直接调用 onCreateView()方法重新创建视图,并使用上次被替换时的 Fragment 状态。

Fragment 的生命周期对于初学者有些难度,希望读者通过实际 Log 观察的方式对本节有关生命周期的案例进行运行并认真比对,深刻地理解 Fragment 生命周期的原理和机制,这对后期 Fragment 的进一步使用会有很大的帮助。

4.2 Menu 和 Toolbar

Menu(菜单)和 ToolBar(活动条)都是在 Android 应用开发过程中必不可少的元素。Menu 在桌面应用中使用十分广泛,几乎所有的桌面应用都有菜单。而由于受到手机屏幕大小的制约,菜单在手机应用中的使用减少了很多,但为了增强用户的体验,仍然在手机应用中提供菜单功能。ToolBar 在 Android 5.0 以上版本中提供,ToolBar 位于传统标题栏的位置,即显示在屏幕的顶部,用于显示应用的图标和 Activity 标题。下面结合具体案例对 Menu 和 ToolBar 进行详细的介绍。

4.2.1 Menu 菜单

Android 中提供的菜单有如下几种。

- 选项菜单（Option Menu）：是最常规的菜单，通过单击 Android 设备的菜单栏来启动。
- 子菜单：单击子菜单会弹出悬浮窗口来显示子菜单项，子菜单不支持嵌套，即子菜单中只能包含菜单项而不能再包含其他子菜单。
- 上下文菜单：长按视图控件时所弹出的菜单，在 Windows 中单击右键弹出的菜单也是上下文菜单。
- 图标菜单：带 icon 的菜单项。
- 扩展菜单：选项菜单最多只能显示 6 个菜单项，当超过 6 个时第 6 个菜单项会被系统替换为一个"更多"的子菜单，显示不出的菜单项都作为"更多"菜单的子菜单项。

注意

子菜单项、上下文菜单项、扩展菜单项均无法显示图标。

在 Android 中，android.view.Menu 接口代表一个菜单，用来管理各种菜单项。在开发过程中，一般不需要自己创建菜单，因为每个 Activity 默认都自带了一个菜单，只要为菜单添加菜单项及相关事件处理即可。MenuItem 类代表菜单中的菜单项，SubMenu 代表子菜单，两者均位于 android.view 包中。Menu、MenuItem 和 SubMenu 三者的关系如图 4-9 所示。

图 4-9　Menu、SubMenu 和 MenuItem

每个 Activity 都包含一个菜单，在菜单中又可以包含多个菜单项和子菜单。由于子菜单实现了 Menu 接口，所以子菜单本身也是菜单，其中可以包含多个菜单项。通常系统创建菜单的方法主要有以下两种：①onCreateOptionsMenu()，创建选项菜单；②onCreateContextMenu()，创建上下文菜单。

OnCreateOptionsMenu()和 OnOptionsMenuSelected()方法是 Activity 中提供的两个回调方法，分别用于创建菜单项和响应菜单项的单击事件。

下面介绍如何创建菜单项、菜单项分组及菜单事件的处理方法。

1. Options Menu 选项菜单

前面介绍过 Android 的 Activity 中已经封装了 Menu 对象，并提供了 onCreateOptionsMenu()回调方法供开发人员对菜单进行初始化，该方法只会在选项菜单第一次显示时被调用。如果需要动态改变选项菜单的内容，可以使用 onPrepareOptionsMenu()方法来实现。初始化菜单内容的代码如下所示。

【代码4-8】 MenuDemoActivity.java

```java
public class MenuDemoActivity extends AppCompatActivity{
    protected void onCreate(Bundle savedInstanceState) {
        super.onCreate(savedInstanceState);}
    public boolean onCreateOptionsMenu(Menu menu) {
        super.onCreateOptionsMenu(menu);      //调用父类方法来加入系统菜单
        //添加菜单项
        menu.add("菜单项1");
        menu.add("菜单项2");
        menu.add("菜单项3");
        //如果希望显示菜单,请返回true
        return true;}
}
```

上述代码重写了 Activity 的 onCreateOptionsMenu()方法;在该方法中获得系统提供的 Menu 对象,然后通过 Menu 对象的 add()方法向菜单中添加菜单项。运行上述代码并单击右侧图标 时,效果如图4-10所示。

图4-10 菜单项效果图

添加菜单项时,除了使用 add(CharSequence title)方法,还可以使用以下两种方法。

- **add(int resId)**:使用资源文件中的文本来设置菜单项的内容,例如 add(R.string.menu1),其中 R.string.menu1 对应的是在 res/string.xml 中定义的文本。
- **add(int groupId,int itemId,int order,CharSequence title)**:该方法的参数 groupId 表示组号,开发人员可以给菜单项进行分组,以便快速地操作同一组菜单。参数 itemId 为菜单项指定唯一的 ID 号,该项用户可以自己指定,也可以让系统来自动分配,在响应菜单时通过 ID 号来判断被单击的菜单;参数 order 表示菜单项显示顺序的编号,编号小的显示在前面;参数 title 用于设置菜单项的内容。

下面使用多个参数的 add()方法实现菜单项的添加,代码如下所示。

【代码4-9】 MenuDemoActivity.java

```java
...//省略代码
    public boolean onCreateOptionsMenu(Menu menu) {
        super.onCreateOptionsMenu(menu);
        //添加4个菜单项,分成2组
        int group1 = 1;
        int gourp2 = 2;
        menu.add(group1, 1, 1, "菜单项1");
        menu.add(group1, 2, 2, "菜单项2");
        menu.add(gourp2, 3, 3, "菜单项3");
        menu.add(gourp2, 4, 4, "菜单项4");
        //显示菜单
        return true;
    }
```

运行效果与图4-10完全相同。对菜单项分组之后,使用 Menu 接口中提供的方法对菜单按组进行操作,该常用的方法如下:removeGroup(int group)用于删除一组菜单;

第4章 UI进阶

setGroupVisible(int group, boolean visible)用于设置一组菜单是否可见;
setGroupEnabled(int group, boolean enabled)用于设置一组菜单是否可单击;
setGroupCheckable(int group,boolean checkable,boolean exclusive)用于设置一组菜单的勾选情况。

注意

> 限于篇幅,此处不再演示对组进行的操作,读者可以自行验证。

2.响应菜单项

Android为菜单提供了onOptionsItemSelected和onMenuItemSelected两种响应方式。

1)onOptionsItemSelected()方法

通过重写Activity类的onOptionsItemSelected()方法来响应菜单项事件,此种方式也是最常用的菜单响应方式;当菜单项被单击时,Android会自动调用该方法,并传入当前所单击的菜单项,其核心代码如下所示。

【代码4-10】 MenuDemoActivity.java

```java
...//省略
@Override
    public boolean onCreateOptionsMenu(Menu menu) {
        super.onCreateOptionsMenu(menu);
        //添加4个菜单项,分成2组
        int group1 = 1;
        int gourp2 = 2;
        menu.add(group1, 1, 1, "菜单项1");
        menu.add(group1, 2, 2, "菜单项2");
        menu.add(gourp2, 3, 3, "菜单项3");
        menu.add(gourp2, 4, 4, "菜单项4");
        //显示菜单
        return true;}
@Override
    public boolean onOptionsItemSelected(MenuItem item) {
        switch (item.getItemId()) {
          case 1:
              Toast.makeText(this, "菜单项1", Toast.LENGTH_SHORT).show();
              break;
          case 2:
              Toast.makeText(this, "菜单项2", Toast.LENGTH_SHORT).show();
              break;
          case 3:
              Toast.makeText(this, "菜单项3", Toast.LENGTH_SHORT).show();
              break;
          case 4:
              Toast.makeText(this, "菜单项4", Toast.LENGTH_SHORT).show();
              break;}
        return super.onOptionsItemSelected(item);
}
```

上述代码通过重写onOptionsItemSelected()方法来响应菜单事件,为方便代码演示,此处将菜单项 ID 硬编码在程序中。运行上述代码,并单击"菜单项 1"时,效果如图 4-11 所示。

2)onMenuItemSelected()方法

通过重写 Activity 类的 onMenuItemSelected()方法也可以实现菜单项的响应。当菜单项被单击时,Android 会自动调用该方法,并传入当前所单击的菜单项。

onMenuItemSelected 与 onOptionsItemSelected 菜单事件响应的区别如下。

- **onMenuItemSelected()**:当选择选项菜单或上下文菜单时都会触发该事件处理方法;
- **onOptionsItemSelected()**:该方法只在选项菜单被选中时才会被触发。

图 4-11 单击菜单项效果图

onMenuItemSelected()的使用方式与 onOptionsItemSelected()类似,下面以 onMenuItemSelected()为例,代码如下所示。

```
...//省略
    @Override
    public boolean onMenuItemSelected(int featureId, MenuItem item) {
        switch (item.getItemId()) {
            case 1:
                Toast.makeText(this, "菜单项 11", Toast.LENGTH_SHORT).show();
                break;
            case 2:
                Toast.makeText(this, "菜单项 22", Toast.LENGTH_SHORT).show();
                break;
            case 3:
                Toast.makeText(this, "菜单项 33", Toast.LENGTH_SHORT).show();
                break;
            case 4:
                Toast.makeText(this, "菜单项 44", Toast.LENGTH_SHORT).show();
                break;}
        return super.onMenuItemSelected(featureId, item);}
```

注意

> 如果 Activity 中同时重写 onMenuItemSelected()和 onOptionsItemSelected()方法,当单击同一个菜单项时,将先调用 onMenuItemSelected()方法,然后调用 onOptionsItemSelected()方法,限于篇幅,此处不再进行演示,读者可以自行验证。

3. SubMenu 子菜单

子菜单是一种组织式菜单项,被大量地运用在 Windows 和其他操作系统的 GUI 设计中。Android 同样支持子菜单,开发人员可以通过 addSubMenu()方法来创建子菜单。创建子菜单的步骤如下。

(1)重写 Activity 类的 onCreateOptionsMenu()方法,调用 Menu 的 addSubMenu()方法来添加子菜单。

(2) 调用 SubMenu 的 add()方法为子菜单添加菜单项。

(3) 重写 Activity 类的 onOptionsItemSelected()方法,以响应子菜单的单击事件。

下述代码演示创建子菜单的过程。

【代码 4-11】 SubMenuDemoActivity.java

```java
public class SubMenuDemoActivity extends AppCompatActivity {
    @Override
    protected void onCreate(Bundle savedInstanceState) {
        super.onCreate(savedInstanceState);
        //setContentView(R.layout.activity_main);}
    //初始化菜单
    @Override
    public boolean onCreateOptionsMenu(Menu menu) {
        //添加子菜单
        SubMenu subMenu = menu.addSubMenu(0, 2, Menu.NONE, "基础操作");
        //为子菜单添加菜单项
        //重命名菜单项
        MenuItem renameItem = subMenu.add(2, 201, 1, "重命名");
        renameItem.setIcon(android.R.drawable.ic_menu_edit);
        //分享菜单项
        MenuItem shareItem = subMenu.add(2, 202, 2, "分享");
        shareItem.setIcon(android.R.drawable.ic_menu_share);
        //删除菜单项
        MenuItem delItem = subMenu.add(2, 203, 3, "删除");
        delItem.setIcon(android.R.drawable.ic_menu_delete);
        return true;}
    //根据菜单执行相应内容
    @Override
    public boolean onOptionsItemSelected(MenuItem item) {
        switch (item.getItemId()) {
            case 201:
                Toast.makeText(getApplicationContext(), "重命名...",
                        Toast.LENGTH_SHORT).show();
                Break;
            case 202:
                Toast.makeText(getApplicationContext(),
                        "分享...", Toast.LENGTH_SHORT).show();
                Break;
            case 203:
                Toast.makeText(getApplicationContext(),
                        "删除...", Toast.LENGTH_SHORT).show();
                Break;}
        return true;
    }
}
```

上述代码中,通过 addSubmenu()方法为 menu 菜单添加了 SubMenu 子菜单;使用 add()方法为子菜单连续添加了三个菜单项;在子菜单中添加菜单项的方式和在菜单中添加菜单项的方式完全相同。此外,通过 MenuItem 的 setIcon()方法为子菜单的每个菜单项设置相应的图标;运行代码并单击右侧图标后,界面效果如图 4-12 所示。然后单击"基础操作"菜单项,弹出与之对应的子菜单项,效果界面如图 4-13 所示。

图 4-12 与图 4-13 的菜单项相比,图 4-12 中显示的"基础操作"菜单项视觉效果较差。

接下来使用 SubMenu 的 setIcon()方法为"基础操作"子菜单添加图标,代码如下所示。

```
//添加子菜单
SubMenu subMenu = menu.addSubMenu(0, 2, Menu.NONE, "基础操作");
subMenu.setIcon(android.R.drawable.ic_menu_manage);
```

但是运行代码时,效果仍然与图 4-12 相同,即使用 setIcon()方法也没有成功为子菜单添加相应的图标,原因是 Android 4.0 及以上的版本所提供的 setIcon()方法并不完善,从而造成了图标不显示。

可以通过在 SubMenuDemoActivity 类中添加 setIconEnable()方法来解决图标不显示的问题,代码如下所示。

```
//处理 setIcon()显示不出图标的问题
private void setIconEnable(Menu menu, boolean enable) {
    try {
        Class<?> clazz = Class
                .forName("com.android.internal.view.menu.MenuBuilder");
        Method m = clazz.getDeclaredMethod("setOptionalIconsVisible",
                boolean.class);
        m.setAccessible(true);
        //下面传入参数
        m.invoke(menu, enable);
    } catch (Exception e) {
        e.printStackTrace();}
}
```

然后在 onCreateOptionsMenu()方法中调用 setIconEnable()方法,代码如下所示。

```
//添加子菜单
SubMenu subMenu = menu.addSubMenu(0, 2, Menu.NONE, "基础操作");
subMenu.setIcon(android.R.drawable.ic_menu_manage);
setIconEnable(menu, true);
```

重新运行 SubMenuDemoActivity,单击 后,界面效果如图 4-14 所示。

图 4-12　子菜单弹出

图 4-13　子菜单项

图 4-14　为子菜单增加图标

注意

当调用 setIcon()方法设置图标时,图标显示不出来的原因是:MenuBuilder 的 optionalIconsVisible 属性默认为 false,所以 icon 图标未能显示,需要调用 setOptionalIconsVisible(true)方法改变其状态并将 icon 图标显示出来。由于 MenuBuilder 处于 com.android.internal 包中,无法直接调用,因此在创建 menu 时只能通过反射机制来调用 MenuBuilder 对象的 setOptionalIconsVisible()方法。

在 Menu 中可以包含多个 SubMenu，SubMenu 可以包含多个 MenuItem，但 SubMenu 不能包含 SubMenu，即子菜单不能嵌套。例如，下面语句在运行时会报错：

```
subMenu.addSubMenu("子菜单嵌套");        //编译时通过,运行时报错
```

上面语句虽然能够通过编译，但在运行时报错。

4．ContextMenu 上下文菜单

在 Windows 操作系统中，用户能够在文件上单击右键来执行"打开""复制""剪切"等操作，右键所弹出的菜单就是上下文菜单。在手机中经常通过长按某个视图元素来弹出上下文菜单。

上下文菜单是通过调用 ContextMenu 接口中的方法来实现的。ContextMenu 接口继承了 Menu 接口，如图 4-15 所示，因此可以像操作选项菜单一样为上下文菜单增加菜单项。上下文菜单与选项菜单最大的不同是：选项菜单的拥有者是 Activity，而上下文菜单的拥有者是 Activity 中的 View 对象。每个 Activity 有且只有一个选项菜单，并为整个 Activity 服务。而一个 Activity 通常拥有多个 View，根据需要为某些特定的 View 提供上下文菜单，通过调用 Activity 的 registerForContextMenu()方法将某个上下文菜单注册到指定的 View 上。

图 4-15　Menu、ContextMenu 关系示意图

虽然 ContextMenu 对象的拥有者是 View 对象，但是需要使用 Activity 的 onCreateContextMenu()方法来生成 ContextMenu 对象，该方法的签名如下所示。

【语法】

```
onCreateContextMenu(ContextMenu menu, View v,
            ContextMenu.ContextMenuInfo menuInfo)
```

上述方法与 onCreateOptionsMenu(Menu menu)方法相似，两者不同之处在于：onCreateOptionsMenu()只在用户第一次按菜单键时被调用；而 onCreateContextMenu()

会在用户每一次长按 View 组件时被调用,并且需要为该 View 注册上下文菜单对象。

注意

> 在图 4-15 中,ContextMenuInfo 接口的实例作为 onCreateContextMenu()方法的参数。该接口实例用于视图元素需要向上下文菜单传递一些信息,例如该 View 对应 DB 记录的 ID 等,此时需要使用 ContextMenuInfo。当需要传递额外信息时,需要重写 getContextMenuInfo()方法,并返回一个带有数据的 ContextMenuInfo 实现类对象。限于篇幅,此处不再赘述。

创建上下文菜单的步骤如下。

(1) 通过 registerForContextMenu()方法为 ContextMenu 分配一个 View 对象。
(2) 通过 onCreateContextMenu()创建一个上下文对象。
(3) 重写 onContextItemSelected()方法实现子菜单单击事件的响应处理。

【代码 4-12】 ContextMenuDemoActivity.java

```java
public class ContextMenuDemoActivity extends AppCompatActivity {
    Button contextMenuBtn;
    @Override
    protected void onCreate(Bundle savedInstanceState) {
        super.onCreate(savedInstanceState);
        setContentView(R.layout.activity_contextmenu);
        //显示列表
        contextMenuBtn = (Button) findViewById(R.id.contextMenuBtn);
        //①为按钮注册上下文菜单,长按按钮则弹出上下文菜单
        this.registerForContextMenu(contextMenuBtn);}
    //②生成上下文菜单
    @Override
    public void onCreateContextMenu(ContextMenu menu, View v,
            ContextMenuInfo menuInfo) {
        //观察日志确定每次是否重新调用
        Log.d("ContextMenuDemoActivity", "被创建...");
        menu.setHeaderTitle("文件操作");
        //为上下文添加菜单项
        menu.add(0, 1, Menu.NONE, "发送");
        menu.add(0, 2, Menu.NONE, "重命名");
        menu.add(0, 3, Menu.NONE, "删除");}
    //③响应上下文菜单项
    @Override
    public boolean onContextItemSelected(MenuItem item) {
        switch (item.getItemId()) {
        case 1:
            Toast.makeText(this, "发送...", Toast.LENGTH_SHORT).show();
            break;
        case 2:
            Toast.makeText(this, "重命名...", Toast.LENGTH_SHORT).show();
            break;
        case 3:
            Toast.makeText(this, "删除...", Toast.LENGTH_SHORT).show();
            break;
```

```
            default:
                return super.onContextItemSelected(item);}
        return true;}
}
```

上述代码中,首先在 onCreate()方法中加载了 activity_contextmenu.xml 视图文件,该文件位于 res/layout 文件夹下,其中只包含一个按钮组件,读者可自行查看;然后为按钮注册上下文菜单;接下来通过 onCreateContextMenu() 回调方法为系统创建的 ContextMenu 对象添加菜单项;最后通过 onContextItemSelected()方法实现菜单项事件处理。

运行上述代码并长按"上下文菜单"按钮时,系统会弹出上下文菜单,效果如图 4-16 所示。

图 4-16 ContextMenu 效果图

注意

在运行程序时,通过 LogCat 的输出信息发现,每次唤出上下文菜单时都会调用 onCreateContextMenu()方法。

5. 使用 XML 资源生成菜单

前面介绍的常用菜单都是通过硬编码方式添加菜单项,Android 为开发人员提供了一种更加方便的菜单生成方式,即通过 XML 文件来加载和响应菜单,此种方式易于维护,可读性更强。

使用 XML 资源生成菜单项的步骤如下。

(1) 在 res 目录中创建 menu 子目录。

(2) 在 menu 子目录中创建一个 Menu Resource file(XML 文件),文件名可以随意,Android 会自动为其生成资源 ID,例如,R.menu.context_menu 对应 menu 目录的 context_menu.xml 资源文件,在该 XML 文件中可以提供 menu 所需的菜单项。

(3) 使用 XML 文件的资源 ID(如 R.menu.context_menu),在 Activity 中将 XML 文件中所定义的菜单元素添加到 menu 对象中。

(4) 通过判断菜单项对应的资源 ID(如 R.id.item_send)来实现相应的事件处理。

下面将工程 chapter04 中的 ContextMenuDemoActivity 类文件进行复制,并改名为 XMLContextMenuDemoActivity。接下来在 XMLContextMenuDemoActivity 中使用 XML 资源来生成菜单。

1) 定义菜单资源文件

在 res 目录下创建 menu 子目录,在 menu 目录下创建一个 XML 资源文件,并命名为 context_menu.xml,代码如下所示。

【代码 4-13】 context_menu.xml

```
<?xml version = "1.0" encoding = "utf - 8"?>
```

```xml
<menu xmlns:android="http://schemas.android.com/apk/res/android">
    <group android:id="@+id/group1">
        <item
            android:id="@+id/item_send"
            android:title="发送"/>
        <item
            android:id="@+id/item_rename"
            android:title="重命名"/>
        <item
            android:id="@+id/item_del"
            android:title="删除"/>
    </group>
</menu>
```

在 context_menu.xml 文件中针对 XMLContextMenuDemoActivity 所定义的菜单项进行重写，并为每个菜单项分配了一个可读性较强的 ID。

2）使用 MenuInflater 添加菜单项

Inflater 为 Android 建立了从资源文件到对象的桥梁，MenuInflater 把 XML 菜单资源转换为对象并将其添加到 menu 对象中。在 XMLContextMenuDemoActivity 中重写 onCreateContextMenu()方法，并使用 Activity 的 getMenuInflater()方法获取 MenuInflater 对象，然后将 XML 文件中所定义的菜单元素添加到 menu 对象中，代码如下所示。

```java
//②生成上下文菜单
@Override
public void onCreateContextMenu(ContextMenu menu, View v,
        ContextMenuInfo menuInfo) {
    Log.d("ContextMenuDemoActivity", "被创建...");
    menu.setHeaderTitle("文件操作");
    getMenuInflater().inflate(R.menu.context_menu, menu);
}
```

3）响应菜单项

接下来重写 XMLContextMenuDemoActivity 类的 onContextItemSelected()方法，实现菜单项的事件处理功能，代码如下所示。

```java
//③响应上下文菜单项
@Override
public boolean onContextItemSelected(MenuItem item) {
    switch (item.getItemId()) {
    case R.id.item_send:
        Toast.makeText(this, "发送...", Toast.LENGTH_SHORT).show();
        break;
    case R.id.item_rename:
        Toast.makeText(this, "重命名...", Toast.LENGTH_SHORT).show();
        break;
    case R.id.item_del:
        Toast.makeText(this, "删除...", Toast.LENGTH_SHORT).show();
        break;
```

```
        default:
            return super.onContextItemSelected(item);}
        return true;
}
```

上述代码演示了使用 XML 资源文件生成菜单的优势。Android 不仅为 context_menu.xml 文件生成了资源 ID，还为文件中 group、menu 和 item 等元素来自动生成相应的 ID（与布局文件中所定义的 ID 相同）。菜单项 ID 的创建与管理全部由 Android 系统来完成，无须开发人员花费心思进行定义。运行 XMLContextMenuDemoActivity，效果与图 4-16 完全相同。

使用 XML 生成菜单是在 Android 中创建菜单的推荐方式。实际上，开发人员在代码中对菜单项或分组等操作都能在 XML 资源文件中完成，下面简单介绍一些比较常见的操作。

（1）资源文件实现子菜单。通过在 item 元素中嵌套 menu 子元素来实现子菜单，代码如下所示。

```xml
<item android:title="系统设置">
    <menu>
        <item android:id="@+id/mi_display_setting" android:title="显示设置"/>
        <item android:id="@+id/mi_network_setting" android:title="网络设置"/>
        <!--其他菜单项-->
    </menu>
</item>
```

（2）为菜单项添加图标。

```xml
<item android:id="@+id/mi_exit" android:title="退出"
    android:icon="@drawable/exit"/>
```

（3）设置菜单项的可选策略。使用 android:checkableBehavior 设置一组菜单项的可选策略，可选值为：none、all 或 single。

```xml
<group android:id="..." android:checkableBehavior="all">
    <!-- 菜单项 -->
</group>
```

（4）使用 android:checked 设置特定菜单项。

```xml
<item android:id="..." android:title="sometitle" android:checked="true"/>
```

（5）设置菜单项"可用/不可用"。

```xml
<item android:id="..." android:title="sometitle" android:enabled="false"/>
```

（6）设置菜单项"可见/不可见"。

```xml
<item android:id="..." android:title="sometitle" android:visible="false"/>
```

4.2.2 Toolbar 操作栏

Toolbar 是在 Android 5.0 开始推出的一个 Material Design 风格的导航组件,Google 非常推荐大家使用 Toolbar 来作为 Android 客户端的导航栏,以此来取代之前的 Actionbar。Actionbar 需要固定在 Activity 的顶部,与 Actionbar 相比 Toolbar 明显更灵活,Toolbar 可以放到界面的任意位置。除此之外,在设计 Toolbar 时,Google 也为开发者预留了许多可定制修改的空间,例如:设置导航栏图标、设置 App 的 Logo 图标、支持设置标题和子标题、支持添加一个或多个自定义组件、支持 Action Menu 等。

Toolbar 继承自 ViewGroup 类,Toolbar 的主要方法如表 4-2 所示。

表 4-2 Toolbar 常用方法

方法	功能描述	方法	功能描述
setTitle(int resId)	设置标题	setSubtitleTextColor(int color)	设置子标题字体颜色
setSubtitle(int resId)	设置子标题	setNavigationIcon(Drawable icon)	设置导航栏的图标
setTitleTextColor(int color)	设置标题字体颜色	setLogo(Drawable drawable)	设置 Toolbar 的 Logo 图标

1. Toolbar 的简单应用

首先需要在应用的 build.gradle 文件中添加对 v7 appcompat 库的支持,其语法如下所示。

【语法】

```
compile 'com.android.support:appcompat-v7:24.1.1'
```

完成上述配置后,在 res/values/styles.xml 文件中对< style >元素进行设置,使用 appcompat 中的 NoActionBar 主题来去除 ActionBar 提供的操作栏,代码如下所示。

【语法】

```
< style name = "AppTheme" parent = "Theme.AppCompat.Light.NoActionBar">
```

然后在 Activity 对应的布局文件中添加 Toolbar 组件,语法如下所示。

【语法】

```
< android.support.v7.widget.Toolbar
    android:id = "@ + id/my_toolbar"
    android:layout_width = "match_parent"
    android:layout_height = "?attr/actionBarSize"
    android:background = "?attr/colorPrimary" >
```

接下来,在 Activity 的 onCreate()方法中使用 setSupportActionBar()方法将 Toolbar 设置为 Activity 的操作栏,其语法如下所示。

【语法】 显示 Toolbar 组件

```
@Override
protected void onCreate(Bundle savedInstanceState) {
```

```
        super.onCreate(savedInstanceState);
        setContentView(R.layout.activity_my);
        Toolbar toolbar = (Toolbar) findViewById(R.id.my_toolbar);
        setSupportActionBar(toolbar);
    }
```

以上各步操作对应的完整代码如下所示。

【代码 4-14】 toolbar.xml

```xml
<RelativeLayout xmlns:android = "http://schemas.android.com/apk/res/android"
    xmlns:tools = "http://schemas.android.com/tools"
    xmlns:app = "http://schemas.android.com/apk/res-auto"
    android:id = "@+id/relativeLayoutContainer"
    android:layout_width = "match_parent"
    android:layout_height = "match_parent">
    <android.support.v7.widget.Toolbar
        android:id = "@+id/my_toolbar"
        android:layout_width = "match_parent"
        android:layout_height = "?attr/actionBarSize"
        android:background = "?attr/colorPrimary" />
</RelativeLayout>
```

【代码 4-15】 ToolbarActivity.java

```java
public class ToolbarActivity extends AppCompatActivity {
    @Override
    protected void onCreate(Bundle savedInstanceState) {
        super.onCreate(savedInstanceState);
        setContentView(R.layout.toolbar);
        Toolbar toolbar = (Toolbar) findViewById(R.id.my_toolbar);
        setSupportActionBar(toolbar);
    }
}
```

运行 ToolbarActivity,结果如图 4-17 所示。

2. Toolbar 的综合应用

在图 4-17 中只显示了应用程序的名称。除此之外,Toolbar 还可以包含导航按钮、应用的 Logo、标题和子标题、若干个自定义 View 以及动作菜单等元素,代码如下所示。

图 4-17 Toolbar 的简单使用

【代码 4-16】 toolbar.xml

```xml
<RelativeLayout xmlns:android = "http://schemas.android.com/apk/res/android" …>
    <android.support.v7.widget.Toolbar
        android:id = "@+id/my_toolbar"
        android:layout_width = "match_parent"
        android:layout_height = "?attr/actionBarSize"
```

```xml
        android:background = "?attr/colorPrimary" >
    <TextView
            android:id = "@+id/toolbar_title"
            android:layout_width = "wrap_content"
            android:layout_height = "wrap_content"
            android:layout_gravity = "center"
            android:text = "自定义"
            android:textColor = "#fff"
            android:textSize = "21sp"/>
    </android.support.v7.widget.Toolbar>
</RelativeLayout>
```

上述代码中,在 Toolbar 组件中添加一个 TextView 组件,然后在 Activity 中通过 ID 来获取该 TextView 组件,并为其添加相应的事件处理。

下面在 res/menu 目录中创建一个 XML 布局文件 menu_tool_demo.xml,代码如下所示。

【代码 4-17】 menu_tool_demo.xml

```xml
<menu xmlns:android = "http://schemas.android.com/apk/res/android" …>
    <item android:id = "@+id/toolbar_action"
    android:icon = "@mipmap/ic_search"
    android:title = "Action1"
    app:showAsAction = "ifRoom"/>
    <item android:id = "@+id/action_item1"
        android:title = "item1"
        app:showAsAction = "never" />
    <item android:id = "@+id/action_item2"
        android:title = "item2"
        app:showAsAction = "never" />
</menu>
```

上述代码用于设置 Toolbar 右侧导航栏的内容。接下来,在 Activity 中设置 Toolbar 的标题及字体颜色、应用的图标、导航按钮图标等特征,代码如下所示。

【代码 4-18】 ToolbarActivity.java

```java
public class ToolbarActivity extends AppCompatActivity {
    @Override
    protected void onCreate(Bundle savedInstanceState) {
        super.onCreate(savedInstanceState);
        setContentView(R.layout.toolbar);
        Toolbar toolbar = (Toolbar) findViewById(R.id.my_toolbar);
        toolbar.setTitle("ToolbarDemo");
        setSupportActionBar(toolbar);
        //显示应用的 Logo 并设置图标
        getSupportActionBar().setLogo(R.mipmap.ic_launcher);
        //显示标题和子标题并设置颜色
```

第4章 UI进阶

```
        toolbar.setTitleTextColor(Color.WHITE);
        toolbar.setSubtitle("Android 基础");
        toolbar.setSubtitleTextColor(Color.WHITE);
        //显示导航按钮图标
        toolbar.setNavigationIcon(R.mipmap.ic_drawer_home);}
    //显示 Menu 菜单按钮
    public boolean onCreateOptionsMenu(Menu menu) {
        getMenuInflater().inflate(R.menu.menu_toolbar_demo, menu);
        return true;}
}
```

上述代码中，getSupportActionBar()是用于获取已设定的 Toolba 组件，并通过 setTitle()、setSubtitle()等方法对 Toolbar 进一步进行设置。

运行 ToolbarActivity，结果如图 4-18 所示。

图 4-18　Toolbar 的综合应用

> **注意**
>
> 上述代码中，Toolbar 的 setTitle()方法需要在 setSupportActionBar()方法之前调用，否则无效。

4.3　高级组件

4.3.1　AdapterView 与 Adapter

Java EE 中提供了一种架构模式：MVC（Model View Controller）架构，即模型—视图—控制器三层架构。MVC 架构的实现原理为：数据模型 M 用于存放数据，利用控制器 C 将数据显示在视图 V 中。在 Android 中提供了一种高级组件——AdapterView，其实现过程类似于 MVC 架构。AdapterView 之所以称为高级组件，是因为该组件的使用方式与其他组件不同，不仅需要在界面中使用 AdapterView，还需要通过适配器为其添加所需的数据或组件。

控制层：在 AdapterView 实现的过程中，Adapter 适配器承担了控制层的角色，通过 Adapter 可以将数据源中数据以某种样式（例如 xml 文件）呈现到视图中。

视图层：AdapterView 充当了 MVC 中的视图层，用于将前端显示和后端数据分离，其内容一般是包含多项相同格式资源的列表。

模型层：将数据源当作模型层，其中包括数组、XML 文件等形式的数据。

1．AdapterView 组件

AdapterView 是一组重要的组件，AdapterView 本身是一个抽象类，其所派生的子类的使用方式十分相似，但显示特征有所不同。AdapterView 具有以下特征：

（1）AdapterView 继承了 ViewGroup，其本质上是容器。

（2）AdapterView 可以包括多个"列表项"，并将"列表项"以合适的形式显示出来。

· 191 ·

（3）AdapterView 所显示的"列表项"是由 Adapter 提供的，通过 AdapterView 的 setAdapter()方法来设置 Adapter 适配器。

AdapterView 及其子类的继承关系如图 4-19 所示。

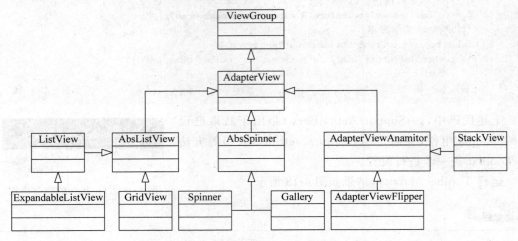

图 4-19　AdapterView 及其子类的继承关系

由图 4-19 可以看出，从 AdapterView 派生出三个子类：AbsListView、AbsSpinner 和 AdapterViewAnimator。这些子类依然是抽象类，在实际运用时需要使用这些类的子类，如 GridView、ListView、Spinner 等，具体如下：ListView 是列表类型；Spinner 是下拉列表，用于为用户提供选择；Gallery 是缩略图，已经被 ScrollView 和 ViewPicker 所取代，但有时也会用到，多用于将子项以中心锁定、水平滚动的列表；GridView 是网格图，以表格形式显示资源，并允许左右滑动。

> 通常将 ListView、GridView、Spinner 和 Gallery 等 AdapterView 子类作为容器，然后使用 Adapter 为容器提供"列表项"，AdapterView 负责采用合适的方式显示这些列表项。

2．Adapter 组件

Adapter 是一个接口，ListAdapter 和 SpinnerAdapter 是 Adapter 的子接口。其中，ListAdapter 为 AbsListView 提供列表项，而 SpinnerAdapter 为 AbsSpinner 提供列表项。Adapter 的继承关系如图 4-20 所示。

大多数 Adapter 实现类都继承自 BaseAdapter 类，而 BaseAdapter 类实现了 ListAdapter 和 SpinnerAdapter 接口，因此 BaseAdapter 及其子类可以为 AbsListView 和 AbsSpinner 提供列表项。

Adapter 的常用子接口及实现类介绍如下：

（1）ListAdapter 接口继承自 Adapter 接口，是 ListView 和 List 数据集合之间的桥梁。ListView 组件能够显示由 ListAdapter 所包装的任何数据。

第4章 UI进阶

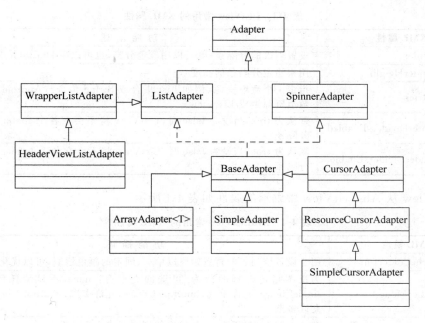

图 4-20 Adapter 的继承关系

（2）BaseAdapter 抽象类是一个能够在 ListView 和 Spinner 中使用的 Adapter 类的父类。提供扩展 BaseAdapter 可以对个列表项进行最大限度的定制。

（3）SimpleCursorAdapter 类适用于简单的纯文字型 ListView，需要将 Cursor 字段和 View 中列表项的 ID 对应起来，如需要实现更复杂的 UI，可以重写其方法来实现。

（4）ArrayAdapter 类是简单易用的 Adapter，通常用于将数组或 List 集合包装成多个列表项。

（5）SimpleAdapter 类是一种简单 Adapter，可以将静态数据在 View 组件中显示。开发人员可以把 List 集合中的数据封装为一个 Map 泛型的 ArrayList。ArrayList 的列表项与 List 集合中的数据相对应。SimpleAdapter 功能强大，使用较为广泛。

> **注意**
>
> Adapter 对象扮演着桥梁的角色，通过桥梁连接着 AdapterView 和所要显示的数据。Adapter 提供了一个连通数据项的途径，将数据集呈现到 View 中。

4.3.2 ListView 列表视图

ListView 列表视图以垂直列表的形式显示所有列表项，在手机应用中使用比较广泛。ListView 通常具有以下两个职责：①将数据填充到布局，以列表的方式来显示数据；②处理用户的选择、单击等操作。

通常创建 ListView 有以下两种方式：①直接使用 ListView 进行创建；②使用 Activity 继承 ListActivity，实现 ListView 对象的获取。ListView 常用的属性如表 4-3 所示。

193

表 4-3 ListView 常用的 XML 属性

XML 属性	功能描述
android:divider	设置列表的分隔条（既可以用颜色分割，也可以用 Drawable 分割）
android:dividerHeight	用来指定分隔条的高度
android:entries	指定一个数组资源（例如 List 集合或 String 集合），Android 将根据该数组资源生成 ListView
android:footerDividersEnabled	默认为 true，当设为 false 时，ListView 将不会在各个 footer 之间绘制分隔条
android:headerDividersEnabled	默认为 true，当设为 false 时，ListView 将不会在各个 header 之间绘制分隔条

ListView 从 AbsListView 中继承的属性如表 4-4 所示。

表 4-4 AbsListView 常用的 XML 属性

XML 属性	功能描述
android:cacheColorHint	用于设置该列表的背景始终以单一、固定的颜色绘制，可以优化绘制过程
android:choiceMode	为视图指定选择的行为，可选的类型有：none，不显示任何选中项；singleChoice，允许单选；multipleChoice，允许多选；multipleChoiceModal，允许多选
android:drawSelectorOnTop	默认为 false，如果为 true，选中的列表项将会显示在上面
android:fastScrollEnabled	用于设置是否允许使用快速滚动滑块。如果设为 true，则将会显示滚动图标，并允许用户拖动该滚动图标进行快速滚动
android:listSelector	设置选中项显示的可绘制对象，可以是图片或者颜色属性
android:scrollingCache	设置在滚动时是否使用绘制缓存，默认为 true。如果为 true，则将使滚动显示更快速，但会占用更多内存
android:smoothScrollbar	默认该属性为 true，列表会使用更精确的基于条目在屏幕上的可见像素高度的计算方法。如果适配器需要绘制可变高的选项，则应该设为 false
android:stackFromBottom	设置 GridView 或 ListView 是否将列表项从底部开始显示
android:textFilterEnabled	设置是否对列表项进行过滤。当设为 true 时，列表会将结果进行过滤
android:transcriptMode	设置该组件的滚动模式。该属性支持如下值：disabled，关闭滚动，默认值；normal，当新条目添加进列表中并且已经准备好显示的时候，列表会自动滑动到底部以显示最新条目；alwaysScroll，列表会自动滑动到底部，无论新条目是否已经准备好显示

如果想对 ListView 的外观、行为进行定制，则需要将 ListView 作为 AdapterView 来使用，通过 Adapter 来控制每个列表的外观和行为。

在 ListView 中，每个 Item 子项既可以是一个字符串，也可以是一个组合控件。通常而言，使用 ListView 需要以下步骤：准备 ListView 所要显示的数据；使用数组或 List 集合存储数据；创建适配器作为列表项数据源；将适配器对象添加到 ListView，并进行展示。

对于简单的 List 列表，直接使用 ArrayAdapter 将数据显示到 ListView 中。如果列表中的内容比较复杂，就需要使用自定义布局来实现 List 列表。接下来分别演示并介绍 ListView 的使用场景。

1. 通过继承 ListActivity 实现 ListView

通过继承 ListActivity 类可以实现 ListView。ListActivity 是 Android 中常用的布局组件之一，通常用于显示可以滚动的列表项。ListActivity 默认布局是由一个位于屏幕中心的全屏列表构成的（默认 ListView 占满全屏），该 ListView 组件本身默认的 ID 为 @id/android:list，所以在 onCreate() 方法中不需要调用 setContentView() 方法进行设置布局，直接调用 getListView() 即可以获取系统默认的 ListView 组件并进行使用。

下述代码通过继承 ListActivity 实现一个简单的 ListView 组件。

【代码 4-19】 ListViewSimpleDemoActivity.java

```java
public class ListViewSimpleDemoActivity extends ListActivity {
    //数据源列表
    private String[] mListStr = { "姓名：张三","性别：男","年龄：25",
            "居住地：青岛","邮箱：itshixun@gmail.com" };
    ListView mListView = null;
    @Override
    protected void onCreate(Bundle savedInstanceState) {
        //获取系统默认的 ListView 组件
        mListView = getListView();
        setListAdapter(new ArrayAdapter<String>(this,
            android.R.layout.simple_list_item_1, mListStr));
        mListView.setOnItemClickListener(new OnItemClickListener() {
            @Override
            public void onItemClick(AdapterView<?> parent, View view,
                    int position, long id) {
                Toast.makeText(ListVewSimpleDemoActivity.this, "您选择了"
                        + mListStr[position], Toast.LENGTH_LONG).show();}
        });
        //设置 ListView 作为显示
        super.onCreate(savedInstanceState);}
}
```

上述代码中，ListViewSimpleDemoActivity 继承了 ListActivity 类，使用 ListActivity 中默认的布局及 ListView 组件，因此无须定义该 Activity 的布局文件，也无须调用 setContentView() 方法设置布局，直接调用 getListView() 获取系统默认的 ID 为 @id/android:list 的 ListView 组件即可。除此之外，代码中还定义了一个 ArrayAdapter 对象，并通过 ListActivity 的 setListAdapter() 方法将其设为 ListView 的适配器对象。

图 4-21 简单的 ListView 视图

运行上述代码，界面效果如图 4-21 所示。

如果需要在 ListActivity 中显示其他组件，如文本框和按钮等组件，可以采用如下步骤：

（1）先定义 Activity 的布局文件，在布局 UI 界面时先增加其他组件，再添加一个 ListView 组件用于展示数据。

（2）在 Activity 中通过 setContentView() 方法来添加布局对象。

创建 ListActivity 的布局文件，代码如下所示。

【代码 4-20】 listview_demo.xml

```xml
<LinearLayout xmlns:android="http://schemas.android.com/apk/res/android" …>
```

```xml
<!-- 添加按钮 -->
<LinearLayout
    android:layout_width = "match_parent"
    android:layout_height = "wrap_content" >
    <EditText
        android:id = "@ + id/addTxt"
        android:layout_width = "212dp"
        android:layout_height = "wrap_content" >
    </EditText>
    <Button
        android:id = "@ + id/addBtn"
        android:layout_width = "83dp"
        android:layout_height = "wrap_content"
        android:text = "添加" >
    </Button>
</LinearLayout>
<!-- 自定义的 ListView -->
<ListView
    android:id = "@id/android:list"
    android:layout_width = "match_parent"
    android:layout_height = "0dip"
    android:layout_weight = "1"
    android:drawSelectorOnTop = "false" />
</LinearLayout>
```

注意

> 在通过继承 ListActivity 来实现 ListView 时，如果用户也定义了一个 ID 为 @id/android:list 的 ListView，与 ListActivity 中的默认 ListView 组件 ID 一致，则使用 setContentView() 方法会指定用户定义的 ListView 作为 ListActivity 的布局，否则会使用系统提供的 ListView 作为 ListActivity 的布局。

下述代码创建一个 Activity 来加载自定义的 XML 布局文件。

【代码 4-21】 ListViewDemoActivity.java

```java
public class ListViewDemoActivity extends ListActivity {
    //数据源列表
    private String[] mListStr = { "姓名：张三","性别：男","年龄：25",
            "居住地：青岛","邮箱：itshixun@gmail.com" };
    ListView mListView = null;
    @Override
    protected void onCreate(Bundle savedInstanceState) {
        super.onCreate(savedInstanceState);
        //设置 Activity 的布局
        setContentView(R.layout.listview_demo);
        //获取 id 为 android:list 的 ListView 组件
        mListView = getListView();
        setListAdapter(new ArrayAdapter<String>(this,
                android.R.layout.simple_list_item_1, mListStr));
```

```
        mListView.setOnItemClickListener(new OnItemClickListener() {
            @Override
            public void onItemClick(AdapterView<?> parent, View view,
                int position, long id) {
                Toast.makeText(ListVewDemoActivity.this,
                    "您选择了" + mListStr[position],
                    Toast.LENGTH_LONG).show();}});}
}
```

运行上述代码,结果如图 4-22 所示。

图 4-22 继承 ListActivity 实现自定义 ListView

2. 在 AppCompatActivity 中使用自定义的 ListView

首先,修改 listview_demo.xml 文件,将 ListView 组件中的 ID 修改为用户自定义的字段,代码如下所示。

【代码 4-22】 listview_demo.xml

```xml
<LinearLayout xmlns:android = "http://schemas.android.com/apk/res/android" …>
    <ListView
        android:id = "@ + id/listview"
        android:layout_width = "match_parent"
        android:layout_height = "0dip"
        android:layout_weight = "1"
        android:drawSelectorOnTop = "false" />
</LinearLayout>
```

下面创建一个 Activity 来加载上述自定义的 XML 布局文件,代码如下所示。

【代码 4-23】 ListViewDemoActivity.java

```java
public class ListViewDemoActivity extends AppCompatActivity{
    //数据源列表
    private String[] mListStr = { "姓名:张三","性别:男","年龄:25",
        "居住地:青岛","邮箱: itshixun@gmail.com" };
    ListView mListView = null;
    @Override
    protected void onCreate(Bundle savedInstanceState) {
```

```
super.onCreate(savedInstanceState);
//设置 Activity 的布局
setContentView(R.layout.listview_demo);
//获取 id 为 listview 的 ListView 组件
mListView = (ListView) findViewById(R.id.listview);
mListView.setAdapter(new ArrayAdapter<String>(this,
        android.R.layout.simple_list_item_1, mListStr));
mListView.setOnItemClickListener(new OnItemClickListener() {
    @Override
    public void onItemClick(AdapterView<?> parent, View view,
                    int position, long id) {
        Toast.makeText(ListVewDemoActivity.this,
                "您选择了" + mListStr[position],
                Toast.LENGTH_LONG).show();}});
}}
```

上述代码中，ListViewDemoActivity 与 ListViewSimpleDemoActivity 基本相同，不同之处在于：ListViewSimpleDemoActivity 没有调用 setContentView()方法，而 ListViewDemoActivity 使用 listview_demo.xml 布局文件来渲染整个布局。

运行 ListVewDemoActivity 时，界面效果如图 4-23 所示。

图 4-23 ListView 视图

3. 复杂 ListView 的使用

前面介绍的两个例子都只展示了文本行，在实际应用中图文混排也是较常见的，即在行中既包括文字又包括图片。图文混排功能需要用户根据需求来自定义 Adapter 适配器。通常实现图文混排的步骤如下：

（1）定义行选项的布局格式；
（2）自定义一个 Adapter，并重写其中的关键方法，如 getCount()、getView()等方法；
（3）注册列表选项的单击事件；
（4）创建 Activity 并加载对应的布局文件。

下述代码通过上述步骤来完成一个图文混排的列表案例。新建行选项的布局文件 item.xml，代码如下所示。

【代码 4-24】 item.xml

```
<RelativeLayout xmlns:android = "http://schemas.android.com/apk/res/android" …>
    <TextView
        android:id = "@+id/itemTxt"
        android:layout_width = "wrap_content"
        android:layout_height = "wrap_content"
        android:layout_alignParentLeft = "true"
        android:layout_marginLeft = "10dp"
```

```xml
            android:layout_marginTop = "10dp"
            android:textColor = "#000"
            android:textSize = "20sp"/>
    <ImageView
            android:id = "@+id/itemImg"
            android:layout_width = "wrap_content"
            android:layout_height = "wrap_content"
            android:layout_alignParentRight = "true"
            android:layout_marginRight = "10dp" />
</RelativeLayout>
```

上述代码主要定义一个 TextView 和一个 ImageView,用于显示列表每行中的文本和图片。然后,创建一个自定义的 TextImageAdapter,代码如下所示。

【代码 4-25】 TextImageAdapter.java

```java
public class TextImageAdapter extends BaseAdapter {
    private Context mContext;
    private List<String> texts;      //展示的文字
    private List<Integer> images;    //展示的图片
    public TextImageAdapter(Context context, List<String> texts,
            List<Integer> images) {
        this.mContext = context;
        this.texts = texts;
        this.images = images;}
    /**元素的个数*/
    public int getCount() {
        return texts.size();}
    public Object getItem(int position) {
        return null;}
    public long getItemId(int position) {
        return 0;}
    //用以生成在 ListView 中展示的一个 View 元素
    public View getView(int position, View convertView, ViewGroup parent) {
        //优化 ListView
        if (convertView == null) {
            convertView = LayoutInflater.from(mContext)
                    .inflate(R.layout.item, null);
            ItemViewCache viewCache = new ItemViewCache();
            viewCache.mTextView = (TextView) convertView
                    .findViewById(R.id.itemTxt);
            viewCache.mImageView = (ImageView) convertView
                    .findViewById(R.id.itemImg);
            convertView.setTag(viewCache);}
        ItemViewCache cache = (ItemViewCache) convertView.getTag();
        //设置文本和图片,然后返回这个 View,用于 ListView 的 Item 的展示
        cache.mTextView.setText(texts.get(position));
        cache.mImageView.setImageResource(images.get(position));
        return convertView;}
    //元素的缓冲类,用于优化 ListView
    private class ItemViewCache {
        public TextView mTextView;
```

```
    public ImageView mImageView;}
}
```

上述代码中创建了 TextImageAdapter 类,用于进行数据的适配与展示,其中该类继承了 BaseAdapter,由于 BaseAdapter 已经实现了 Adapter 的大部分方法,因此在 TextImageAdapter 中只需要实现所需要的部分即可,例如 getCount()和 getView()方法。getCount()方法用于返回 ListView 中文本元素的数量,getView()方法用于生成所要展示的 View 对象。在 ListView 中,每添加一个 View 就会调用一次 Adapter 的 getView()方法,所以有必要对该方法进行优化,代码 4-25 中通过自定义 ItemViewCache 类实现了部分优化。

创建 Activity 的布局文件,代码如下所示。

【代码 4-26】 listview_image.xml

```xml
<LinearLayout xmlns:android = "http://schemas.android.com/apk/res/android" …>
    <ListView
        android:id = "@ + id/list_image"
        android:layout_width = "match_parent"
        android:layout_height = "match_parent"/>
</LinearLayout>
```

接下来创建一个用于展现图文混排的 Activity,并注册单击选择项的事件,代码如下所示。

【代码 4-27】 ListVewImageDemoActivity.java

```java
public class ListVewImageDemoActivity extends AppCompatActivity {
    private  String[] texts = new String[]{"樱花","小鸡","坚果"};//展示的文字
    //展示的图片
    private int[] images = new int[]{R.drawable.cherry_blossom,
            R.drawable.chicken,R.drawable.chestnut};
    ListView mListView = null;
    @Override
    protected void onCreate(Bundle savedInstanceState) {
        super.onCreate(savedInstanceState);                    //设置 ListView 作为显示
        setContentView(R.layout.listview_image);               //设置 Activity 布局
        mListView = (ListView)findViewById(R.id.list_image);
                                                //获取 ID 为 list_image 的 ListView 组件
        //加载适配器
        TextImageAdapter adapter = new TextImageAdapter(this, texts, images);
        mListView.setAdapter(adapter);
        mListView.setOnItemClickListener(new OnItemClickListener() {
            @Override
            public void onItemClick(AdapterView<?> parent, View view,
                int position, long id) {
                Toast.makeText(ListVewImageDemoActivity.this,
                "您选择了" + texts[position], Toast.LENGTH_LONG).show();}});
    }
}
```

上述代码中,首先通过 ListActivity 提供的 getListView()方法来获取 ListView 对象;然后创建一个 TextImageAdapter 适配器对象,并作为 setListAdapter()方法的传入参数,从而把 ListView 和 Adapter 对象进行绑定;最后通过定义内部监听器来实现 ListView 中选择项的单击事件。

运行上述代码时,界面运行效果如图 4-24 所示。

图 4-24　图文混排效果

> 上述几个案例主要介绍了 ListView 的常用功能,限于篇幅还有很多功能没有提到,例如 ListView 的分割部分、headView、footView 以及 ListView 的分页等,请参考贯穿案例中的相应实现。

4.3.3　GridView 网格视图

GridView 用于按行和列的分布方式来显示多个组件。GridView 与 ListView 拥有相同的父类 AbsListView,因此两者有许多相同之处,唯一的区别在于:ListView 只显示一列,GridView 可以显示多列。从这个角度看,ListView 可以视为一个特殊的 GridView,当 GridView 只显示一列时,GridView 就变成了 ListView。GridView 也需要通过 Adapter 来提供显示数据。

GridView 常用的 XML 属性如表 4-5 所示。

表 4-5　GridView 常用的 XML 属性

XML 属性	功 能 描 述
android:numColumns	设置列数,可以设置自动,如 auto_fit
android:columnWidth	设置每一列的宽度
android:stretchMode	设置拉伸模式。none,拉伸被禁用,不允许被拉伸;spacingWidth,列与列之间的间距会被拉伸,因此使用该拉伸模式时,必须指定 columnWidth,而指定 horizontalSpacing 就会无效。columnWidth,每列的宽度相等,只需要指定 numColumns 和 horizontalSpacing 属性。spacingWidthUniform,每列的间距均被拉伸。当拉伸被禁用时不可以被拉伸
android:verticalSpacing	设置各个元素之间的垂直边距
android:horizontalSpacing	设置各个元素之间的水平边距

在使用 GridView 时一般都需要为其指定 numColumns 属性,否则 numColumns 默认为 1。当 numColumns 属性设置为 1 时,意味着该 GridView 只有 1 列,此时功能与 ListView 相同。在实际开发中,创建 GridView 的过程与 ListView 相似,步骤如下:

(1) 在布局文件中使用<GridView>元素来定义 GridView 组件;
(2) 自定义一个 Adapter,并重写其中的关键方法,如 getCount()、getView()等方法;
(3) 注册列表选项的单击事件;
(4) 创建 Activity 并加载对应的布局文件。

下面通过一个简单示例演示 GridView 的用法。新建布局文件 gridview_demo.xml,代码如下所示。

· 201 ·

【代码 4-28】 gridview_demo.xml

```xml
<GridView xmlns:android = "http://schemas.android.com/apk/res/android"
    android:id = "@+id/gridview"
    android:layout_width = "fill_parent"
    android:layout_height = "fill_parent"
    android:columnWidth = "90dp"
    android:gravity = "center"
    android:horizontalSpacing = "10dp"
    android:numColumns = "auto_fit"
    android:stretchMode = "columnWidth"
    android:verticalSpacing = "10dp" >
</GridView>
```

上述代码比较很简单,整个布局文件中只有一个 GridView。通过 columnWidth 属性设置列宽为 90dp;将属性 numColumns 设为 auto_fit,Android 会自动计算手机屏幕的大小,以决定每行展示几个元素;将属性 stretchMode 设为 columnWidth,则根据列宽自动缩放;horizontalSpacing 属性用于定义列之间的间隔;verticalSpacing 用于定义行之间的间隔。

然后,自定义一个 Adapter 适配器用于适配 GridView,代码如下所示。

【代码 4-29】 ImageAdapter.java

```java
public class ImageAdapter extends BaseAdapter {
    private Context mContext;
    //一组 Image 的 Id
    private int[] mThumbIds;
    public ImageAdapter(Context context) {
        this.mContext = context;}
    @Override
    public int getCount() {
        return mThumbIds.length;}
    @Override
    public Object getItem(int position) {
        return mThumbIds[position];}
    @Override
        public long getItemId(int position) {
        return 0;}
    @Override
    public View getView(int position, View convertView, ViewGroup parent) {
        //定义一个 ImageView,显示在 GridView 里
        ImageView imageView;
        if (convertView == null) {
            imageView = new ImageView(mContext);
            imageView.setLayoutParams(new GridView.LayoutParams(200, 200));
            imageView.setScaleType(ImageView.ScaleType.CENTER_CROP);
            imageView.setPadding(8, 8, 8, 8);
        } else {
            imageView = (ImageView) convertView;}
        imageView.setImageResource(mThumbIds[position]);
        return imageView;}
}
```

第4章 UI进阶

上述代码中采用了自定义 Adapter 的方式,与前面自定义 Adapter 的方式相似,以"九宫格"的方式展示图片,每幅图片大小为 200×200。

最后,创建 GridViewDemoActivity,并加载相应的布局文件,用于显示使用 GridView 布局的界面,代码如下所示。

【代码 4-30】 GridViewDemoActivity.java

```
public class GridViewDemoActivity extends AppCompatActivity {
    @Override
    protected void onCreate(Bundle savedInstanceState) {
        super.onCreate(savedInstanceState);
        setContentView(R.layout.gridview_demo);
        GridView gridView = (GridView)findViewById(R.id.gridview);
        ImageAdapter imageAdapter = new ImageAdapter(this,mThumbIds);
        gridView.setAdapter(imageAdapter);
        //单击 GridView 元素的响应
        gridView.setOnItemClickListener(new OnItemClickListener() {
            @Override
            public void onItemClick(AdapterView<?> parent, View view,
                    int position, long id) {
                //弹出单击的 GridView 元素的位置
                Toast.makeText(GridViewDemoActivity.this,mThumbIds[position],
                    Toast.LENGTH_SHORT).show(); }});
    }
    //展示图片
    private int[] mThumbIds = {
        R.drawable.flg_1, R.drawable.flg_2,
        R.drawable.flg_3, R.drawable.flg_4,
        R.drawable.flg_5, R.drawable.flg_6,
        R.drawable.flg_7, R.drawable.flg_8,
        R.drawable.flg_9 };
}
```

上述代码中,定义了一组国旗图片,并通过 setOnItemClickListener()方法实现了 gridView 中的图片单击事件,当单击一个图片时,会显示该图片所存储的位置。

运行上述代码时,界面运行效果如图 4-25 所示。

4.3.4 TabHost

TabHost 可以很方便地在窗口中放置多个标签页,每个标签页所显示的区域与其外部容器大小相同,通过叠放标签页可以在容器中放置更多组件。TabHost 是一种比较实用的组件,在应用中比较常见,例如手机通讯记录中包括"未接电话""已接电话"等 Tab 页。

图 4-25 九宫格

在使用 TabHost 时,通常需要与 TabWidget、TabSpec 组件结合使用,具体功能如下:

(1) TabWidget 组件用于显示 TabHost 标签页中上部和下部的按钮,单击按钮时切换选项卡；

(2) TabSpec 代表选项卡界面，通过将 TabSpec 添加到 TabHost 中实现选项卡的添加。

TabHost 仅仅是一个简单的容器，通过以下方法来创建、添加选项卡：

(1) newTabSpec(String tag)方法用于创建选项卡；

(2) addTab(tabSpec)方法用于添加选项卡。

使用 TabHost 有两种形式：继承 TabActivity 和不继承 TabActivity。

1. 继承 TabActivity 使用 TabHost

当继承 TabActivity 时，使用 TabHost 的步骤如下。

(1) **定义布局**：在 XML 文件中使用 TabHost 组件，并在其中定义一个 FrameLayout 选项卡内容；

(2) **创建 TabActivity**：用于显示选项卡组件的 Activity，需要继承自 TabActivity；

(3) **获取组件**：通过 getTabHost()方法获取 TabHost 对象；

(4) **创建选项卡**：通过 TabHost 来创建一个选项卡。

下述代码通过一个简单示例演示 TabHost 的用法。新建布局文件 tabhost_demo1.xml 代码如下所示。

【代码 4-31】 tabhost_demo1.xml

```xml
<TabHost xmlns:android="http://schemas.android.com/apk/res/android"
    android:id="@android:id/tabhost" …>
    <LinearLayout
        android:layout_width="match_parent"
        android:layout_height="match_parent"
        android:orientation="vertical" >
        <TabWidget
            android:id="@android:id/tabs"
            android:layout_width="match_parent"
            android:layout_height="wrap_content"
            android:orientation="horizontal"/>
        <FrameLayout
            android:id="@android:id/tabcontent"
            android:layout_width="match_parent"
            android:layout_height="match_parent"
            android:layout_weight="1" >
            <LinearLayout
                android:id="@+id/content1"
                android:layout_width="match_parent"
                android:layout_height="match_parent"
                android:orientation="vertical" >
                <TextView android:text="内容1"
                    android:layout_width="wrap_content"
                    android:layout_height="wrap_content"/>
            </LinearLayout>
```

```xml
                <LinearLayout
                    android:id = "@ + id/content2"
                    android:layout_width = "match_parent"
                    android:layout_height = "match_parent"
                    android:orientation = "vertical" >
                     <TextView android:text = "内容 2"
                            android:layout_width = "wrap_content"
                            android:layout_height = "wrap_content"/>
                </LinearLayout>
                <LinearLayout
                    android:id = "@ + id/content3"
                    android:layout_width = "match_parent"
                    android:layout_height = "match_parent"
                    android:orientation = "vertical" >
                     <TextView android:text = "内容 3"
                            android:layout_width = "wrap_content"
                            android:layout_height = "wrap_content"/>
                </LinearLayout>
            </FrameLayout>
        </LinearLayout>
</TabHost>
```

上述布局文件解释如下：布局文件中的根元素为 TabHost，其中其 ID 必须引用 Android 系统自带的 ID，即 android:id＝@android:id/tabhost；使用 TabHost 一定要有 TabWidget 和 FramLayout 两个控件；TabWidget 必须使用系统 ID，即@android:id/tabs；FrameLayout 作为标签内容的基本框架，也必须使用系统 ID，即@android:id/tabcontent。

接下来创建 TabHostDemo1Activity，代码如下所示。

【代码 4-32】 TabHostDemo1Activity.java

```java
public class TabHostDemo1Activity extends TabActivity{
    @Override
    protected void onCreate(Bundle savedInstanceState) {
        super.onCreate(savedInstanceState);
        setContentView(R.layout.tabhost_demo1);
        TabHost tabHost = getTabHost();
        //添加第 1 个标签
        TabSpec page1 = tabHost.newTabSpec("tab1")    //创建新标签
                .setIndicator("标签 1")                //设置标签内容
                .setContent(R.id.content1);
        tabHost.addTab(page1);
        //添加第 2 个标签
        TabSpec page2 = tabHost.newTabSpec("tab2")
                .setIndicator("标签 2")
                .setContent(R.id.content2);
        tabHost.addTab(page2);
        //添加第 3 个标签
        TabSpec page3 = tabHost.newTabSpec("tab3")
                .setIndicator("标签 3")
                .setContent(R.id.content3);
        tabHost.addTab(page3);}
}
```

上述代码解释如下：通过调用从 TabActivity 继承而来的 getTabHost()方法来获取布局文件中的 TabHost 组件；调用 TabHost 组件的 newTabSpec(tag) 方法创建一个选项卡，其中参数 tag 是一个字符串，即选项卡的唯一标识；使用 TabHost.TabSpec 的 setIndicator()方法来设置新选项卡的名称；使用

图 4-26　继承 TabActivity 使用 TabHost

TabHost.TabSpec 的 setContent()方法来设置选项卡的内容，可以是视图组件、Activity 或 Fragment；使用 tabHost.add(tag)方法将选项卡添加到 TabHost 组件中，其中传入的 tag 参数是选项卡的唯一标识。

运行上述代码，界面效果如图 4-26 所示。

在实际应用中，有时会改变选项卡标签的高度，在代码中通过 getTabWidget()方法来获取 TabWidget 对象，然后使用该对象的 getChildAt()方法来获得指定的标签，最后对该标签中内容的位置进行设置，代码如下所示。

【示例】　改变选项卡标签的高度

```
TabWidget mTabWidget = tabHost.getTabWidget();
for (int i = 0; i < mTabWidget.getChildCount(); i++) {
    mTabWidget.getChildAt(i).getLayoutParams().height = 50;    //设置选项卡的宽度
    mTabWidget.getChildAt(i).getLayoutParams().width = 60;     //设置选项卡的高度
}
```

2．不继承 TabActivity 使用 TabHost

当不继承 TabActivity 时，使用 TabHost 的步骤如下。
（1）**定义布局**：在 XML 文件中使用 TabHost 组件；
（2）**创建 TabActivity**：用于显示选项卡组件的 Activity，需要继承 TabActivity；
（3）**获取组件**：通过 findViewById()方法获取 TabHost 对象；
（4）**创建选项卡**：通过 TabHost 来创建一个选项卡。

新建布局文件 tabhost_demo2.xml，代码如下所示。

【代码 4-33】　tabhost_demo2.xml

```xml
<LinearLayout xmlns:android = "http://schemas.android.com/apk/res/android" …>
    <TabHost
        android:layout_width = "wrap_content"
        android:layout_height = "wrap_content"
        android:id = "@+id/tabHost"
        android:layout_weight = "1">
        <LinearLayout
            android:layout_width = "match_parent"
            android:layout_height = "match_parent"
            android:orientation = "vertical">
            <TabWidget
```

```xml
            android:id = "@android:id/tabs"
            android:layout_width = "match_parent"
            android:layout_height = "wrap_content"></TabWidget>
        <FrameLayout
            android:id = "@android:id/tabcontent"
            android:layout_width = "match_parent"
            android:layout_height = "match_parent"
            android:layout_weight = "1">
            <LinearLayout
                android:id = "@ + id/content_1"
                android:layout_width = "match_parent"
                android:layout_height = "match_parent"
                android:orientation = "vertical">
                <TextView
                    android:layout_width = "wrap_content"
                    android:layout_height = "wrap_content"
                    android:textSize = "25sp"
                    android:text = "内容 1"
                    android:id = "@ + id/textView" />
            </LinearLayout>
            <LinearLayout
                android:id = "@ + id/content_2"
                android:layout_width = "match_parent"
                android:layout_height = "match_parent"
                android:orientation = "vertical">
                <TextView
                    android:layout_width = "wrap_content"
                    android:layout_height = "wrap_content"
                    android:textSize = "25sp"
                    android:text = "内容 2"
                    android:id = "@ + id/textView2" />
            </LinearLayout>
            <LinearLayout
                android:id = "@ + id/content_3"
                android:layout_width = "match_parent"
                android:layout_height = "match_parent"
                android:orientation = "vertical">
                <TextView
                    android:layout_width = "wrap_content"
                    android:layout_height = "wrap_content"
                    android:textSize = "25sp"
                    android:text = "内容 3"
                    android:id = "@ + id/textView3" />
            </LinearLayout>
        </FrameLayout>
    </LinearLayout>
</TabHost>
</LinearLayout>
```

布局文件 tabhost_demo2.xml 与 tabhost_demo1.xml 相比，TabHost 组件的 ID 是用户自定义的，不再使用 Android 系统自带的 ID。

创建 TabHostDemo2Activity，代码如下所示。

【代码 4-34】 TabHostDemo2Activity.java

```java
public class TabHostDemo2Activity extends AppCompatActivity{
    protected void onCreate(Bundle savedInstanceState) {
        super.onCreate(savedInstanceState);
        setContentView(R.layout.tabhost_demo2);
        TabHost tabHost = (TabHost) findViewById(R.id.tabHost);
        tabHost.setup();
        tabHost.addTab(tabHost.newTabSpec("tab1").setIndicator("标签1")
                .setContent(R.id.content_1));
        tabHost.addTab(tabHost.newTabSpec("tab2").setIndicator("标签2")
                .setContent(R.id.content_2));
        tabHost.addTab(tabHost.newTabSpec("tab3").setIndicator("标签3")
                .setContent(R.id.content_3));
        TabWidget mTabWidget = tabHost.getTabWidget();          //设置选项卡的高度和宽度
        for (int i = 0; i < mTabWidget.getChildCount(); i++) {
            mTabWidget.getChildAt(i).getLayoutParams().height = 80;   //设置选项卡的高度
            mTabWidget.getChildAt(i).getLayoutParams().width = 60;    //设置选项卡的宽度
        }
    }
}
```

与代码 TabHostDemo1Activity 相比，获取 TabHost 组件不再使用 getTabHost()方法，而是使用 findViewById 来获取。使用 addTab()方法来添加选项卡使用 newTabSpec(tag)方法创建一个选项卡。

运行上述代码，界面如图 4-27 所示。

图 4-27　不继承 TabActivity 使用 TabHost

TabActivity 在 Android 3.0 以后已过时，推荐使用"不继承 TabActivity 的方式"使用 TabHost。

4.3.5　WebView

WebView 是 Android 系统中一种特殊的 View，通常用于在 App 中显示网页或开发用户自己的浏览器。WebView 提供了网页的前进和后退，以及页面的放大、缩小和搜索等功能。前端开发者还可以使用 Web 检查器（Web Inspector）来调试 HTML、CSS、Javascript 等代码。

WebView 控件功能强大，除了具有一般 View 的属性和方法外，还可以对 URL 请求、页面加载、渲染以及页面的交互进行处理。WebView 具有以下几点功能：

（1）WebChromeClient 可以辅助 WebView 实现与浏览器的交互动作，例如 WebView 关闭和隐藏、WebView 获得焦点、页面加载进展、JavaScript 确认框和警告框、JavaScript 操作超时等。通过 WebView 的 setWebChromeClient()方法可以对 WebChromeClient 进行

第4章 UI进阶

设置。

（2）WebViewClient 主要帮助 WebView 处理各种通知、请求事件等，例如表单因错误需要重新提交、页面开始加载及加载完成、资源加载中、接收到 HTTP 认证请求处理、页面键盘响应、页面中的 URL 打开处理等。通过 WebView 的 setWebViewClient()方法可以对 WebViewClient 进行设置。

（3）使用 WebSettings 可以对 WebView 进行配置和管理，例如是否允许对文件进行操作、缓存的设置、页面是否支持放大和缩小、是否允许使用数据库 API、字体及文字编码设置、是否允许 javascript 脚本运行、是否允许图片自动加载、是否允许数据及密码保存等。通过 getSetting()方法返回一个 WebSettings 对象。

（4）通过 addJavascriptInterface()方法将 Java 对象绑定到 WebView 中，以便 JavaScript 从页面中控制 Java 对象，从而实现 WebView 与 HTML 页面的交互。

注意

> 在 Android 4.3 及以前版本中，WebView 内部采用 Webkit 渲染引擎；在 Android 4.4 以上版本中，采用 chromium 渲染引擎来渲染 View 的内容。

下面通过一个简单的案例演示 WebView 的基本用法，步骤如下：①在 AndroidManifest.xml 中配置访问权限；②在布局文件中创建 WebView 元素；③在代码中加载网页。

以加载百度首页为例，演示 WebView 的步骤的实现。

（1）由于访问网络需要网络访问权限，所以需要在 AndroidManifest.xml 配置文件中配置相应的权限，代码如下所示。

```xml
<uses-permission android:name="android.permission.INTERNET" />
```

（2）在 res/layout 文件夹下创建一个名为 webview_demo.xml 的布局文件，代码如下所示。

【代码 4-35】 webview_demo.xml

```xml
<LinearLayout xmlns:android="http://schemas.android.com/apk/res/android" …>
    <WebView
        android:id="@+id/webView"
        android:layout_width="match_parent"
        android:layout_height="match_parent" />
</LinearLayout>
```

上述代码比较简单，在 LinearLayout 布局中添加一个 ID 为 webView 的组件，通过该组件对网页进行加载。

（3）创建一个名为 WebViewDemoActivity 的 Activity 类，代码如下所示。

【代码 4-36】 WebViewDemoActivity.java

```java
public class WebViewDemoActivity extends AppCompatActivity{
    //定义 WebView 类型的变量
    WebView webView;
```

```java
    @Override
    protected void onCreate(Bundle savedInstanceState) {
        super.onCreate(savedInstanceState);
        setContentView(R.layout.webview_demo);
        //获取 webview 对象,并加载百度首页
        webView = (WebView)findViewById(R.id.webView);
        webView.loadUrl("http://www.baidu.com");
    }
}
```

上述代码中,通过 webView 对象的 loadUrl()方法加载百度的首页。运行上述代码时,效果如图 4-28 所示。在加载网页内容时,除了使用 WebView 的 loadUrl()方法进行加载外,还可以使用 loadData()或 loadDataWithBaseURL()方法将 HTML 代码片段或本地存储的 HTML 页面内容显示出来。接下来将上面的实例进行调整,调整后代码如下所示。

【代码 4-37】 WebViewDemoActivity.java

```java
public class WebViewDemoActivity extends AppCompatActivity {
    //定义 WebView 类型的变量
    WebView webView;
    @Override
    protected void onCreate(Bundle savedInstanceState) {
        super.onCreate(savedInstanceState);
        setContentView(R.layout.webview_demo);
        //获取 WebView 对象,并加载百度首页
        webView = (WebView)findViewById(R.id.webView);
        StringBuffer htmlBuffer = new StringBuffer();
        htmlBuffer.append("<html>");
        htmlBuffer.append("<body>请单击<a href=\"http://www.baidu.com\">
            百度</a></body>");
        htmlBuffer.append("</html>");
        webView.loadDataWithBaseURL("",htmlBuffer.toString(),"text/html",
            "UTF-8","");
    }
}
```

上述代码中使用了 loadDataWithBaseURL()方法来加载 HTML 代码片段并显示。界面效果如图 4-29 所示。

loadDataWithBaseURL()方法用于将指定的数据加载到 WebView 中,由于本身机制的原因,该方法不能加载来自网络的内容。loadDataWithBaseURL()方法签名如下所示。

【语法】

```java
public void loadDataWithBaseURL(String baseUrl, String data, String mimeType,
    String encoding, String historyUrl)
```

其中:baseUrl 为基础目录,data 中的文件路径可以是相对于 baseUrl 的相对目录,如果为空,则作用和 about:blank 相同;data 是被加载的内容,通常为 HTML 代码片段;mimeType 用于指定资源的媒体类型(即 MIME Type),可以取值 text/html、image/jpeg等;encoding 用于设置网页的编码格式,可以取值 utf-8、gbk 等;historyUrl 用作历史记录的字段,可以设置为 null。

图 4-28　WebView 视图　　　　　图 4-29　自定义视图

4.4　贯穿任务实现

4.4.1　实现【任务 4-1】

礼品分为多种类型，每种礼品又有多种款式，按照"GIFT-EMS 礼记"项目的实际需求，在移动端每类礼品只会显示一个礼品。当用户浏览销售的礼品时，首先需要在礼品类型列表中选择一个类型，而在浏览礼品攻略时，首先需要查看攻略的列表。下面编写礼品类型和礼品攻略共用的列表界面。礼品类型和攻略列表界面的布局文件对应的代码如下所示。

【任务 4-1】 list_level2.xml

```xml
<RelativeLayout xmlns:android = "http://schemas.android.com/apk/res/android" …>
    <include
        android:id = "@+id/titleBarRelativeLayout"
        layout = "@layout/title_bar" />
    <ListView
        android:id = "@+id/listView"
        android:layout_width = "match_parent"
        android:layout_height = "wrap_content"
        android:layout_below = "@id/titleBarRelativeLayout" >
    </ListView>
    <include
        android:id = "@+id/cornerMenuRelativeLayout"
        layout = "@layout/corner_menu" />
</RelativeLayout>
```

在列表布局中，包含一个 ListView 用于显示礼品类型和攻略。接下来编写 ListView 对应的子项布局，代码如下所示。

【任务 4-1】 list_level2_item.xml

```xml
<RelativeLayout xmlns:android="http://schemas.android.com/apk/res/android" …>
    <ImageView
        android:id="@+id/imageView"
        android:layout_width="match_parent"
        android:layout_height="200dp"
        android:scaleType="centerCrop" />
    <TextView
        android:id="@+id/textView"
        android:layout_width="match_parent"
        android:layout_height="wrap_content"
        android:layout_alignBottom="@+id/imageView"
        android:background="#80000000"
        android:padding="5dp"
        android:textSize="20sp" />
</RelativeLayout>
```

列表界面的 ListView 的每个子项中,包含一个 ImageView 用于显示礼品类型或攻略的图片,还有一个 TextView 显示名称。编写列表界面的 Activity,代码如下。

【任务 4-1】 ListLevel2Activity.java

```java
package com.qst.giftems.activity;
…//省略 import
/**二级页面,礼品类型和攻略共用一个.*/
public class ListLevel2Activity extends BaseActivity {
    public static final int TYPE_GIFT = 50;
    public static final int TYPE_STRATEGY = 100;
    private int type;
    private ArrayList list = new ArrayList();
    private GiftTypeListViewAdapter giftTypeListViewAdapter
        = new GiftTypeListViewAdapter();
    private StrategyListViewAdapter strategyListViewAdapter
        = new StrategyListViewAdapter();
    private ListView listView;
    public ListLevel2Activity() {
        super(R.layout.list_level2, "");}
    @Override
    protected void onCreate(Bundle savedInstanceState) {
        super.onCreate(savedInstanceState);
        listView = (ListView) findViewById(R.id.listView);
        type = getIntent()..getIntExtra("type", 0);
        if (type == TYPE_GIFT) {
            titleTextView.setText("礼品中心");
            //TODO 后续章节改为从服务器获取礼品数据
            int[] pics = { R.drawable.gift_type_1, R.drawable.gift_type_2,
                R.drawable.gift_type_3, R.drawable.gift_type_4,
                R.drawable.gift_type_5, R.drawable.gift_type_6,
                R.drawable.gift_type_7 };
            for (int i = 1; i <= 7; i++) {
                GiftType gt = new GiftType();
                gt.id = i;
```

```java
                    gt.name = "礼品类型 " + i;
                    gt.pic = pics[i - 1] + "";
                    gt.orderNum = i;
                    list.add(gt);}
                listView.setAdapter(giftTypeListViewAdapter);
            } else if (type == TYPE_STRATEGY) {
                titleTextView.setText("礼品攻略");
                //TODO 后续章节改为从服务器获取礼品攻略数据
                int[] pics = { R.drawable.strategy_1, R.drawable.strategy_2,
                        R.drawable.strategy_3, R.drawable.strategy_4 };
                for (int i = 1; i <= 4; i++) {
                    Strategy s = new Strategy();
                    s.id = i;
                    s.title = "送礼攻略 " + i;
                    s.remark = "备注备注备注备注备注备注备注备注备注备注备注备注备注备";
                    s.pic1 = pics[i - 1] + "";
                    list.add(s);}
                listView.setAdapter(strategyListViewAdapter);}
        }
        private class GiftTypeListViewAdapter extends BaseAdapter {
            public int getCount() {
                return list.size();}
            @Override
            public boolean areAllItemsEnabled() {
                return false; }
            public Object getItem(int position) {
                return position; }
            public long getItemId(int position) {
                return position; }
            public View getView(int position, View convertView, ViewGroup parent) {
                final Context context = getApplicationContext();
                final GiftType giftType = (GiftType) list.get(position);
                if (giftType == null)
                    return null;
                if (convertView == null)
                    convertView = LayoutInflater.from(context).inflate(
                            R.layout.list_level2_item, null);
                ImageView imageView = (ImageView) convertView
                        .findViewById(R.id.imageView);
                TextView textView = (TextView) convertView
                        .findViewById(R.id.textView);
                textView.setText(giftType.name);
                imageView.setImageBitmap(readBitMapFromResource(
                        ListLevel2Activity.this, Integer.parseInt(giftType.pic)));
                convertView.setOnClickListener(new OnClickListener() {
                    @Override
                    public void onClick(View v) {
                        Intent intent = new Intent(context, GiftActivity.class);
                        intent.putExtra("giftTypeId", giftType.id);
                        startActivity(intent); }});
                return convertView;}
        }
        private class StrategyListViewAdapter extends BaseAdapter {
```

```java
            public int getCount() {
                return list.size();}
            @Override
            public boolean areAllItemsEnabled() {
                return false;}
            public Object getItem(int position) {
                return position;}
            public long getItemId(int position) {
                return position;}
            public View getView(int position, View convertView, ViewGroup parent) {
                final Context context = getApplicationContext();
                final Strategy strategy = (Strategy) list.get(position);
                if (strategy == null)
                    return null;
                if (convertView == null)
                    convertView = LayoutInflater.from(context).inflate(
                        R.layout.list_level2_item, null);
                ImageView imageView = (ImageView) convertView
                    .findViewById(R.id.imageView);
                TextView textView = (TextView) convertView
                    .findViewById(R.id.textView);
                textView.setText(strategy.title);
                imageView.setImageBitmap(readBitMapFromResource(
                    ListLevel2Activity.this, Integer.parseInt(strategy.pic1)));
                convertView.setOnClickListener(new OnClickListener() {
                    @Override
                    public void onClick(View v) {
                        Intent intent = new Intent(context, StrategyActivity.class);
                        intent.putExtra("strategyId", strategy.id);
                        startActivity(intent);}});
                return convertView;}}
}
```

上述代码中，在 onCreate() 方法中构造了礼品类型和礼品攻略的数据，后续章节中将改为从服务器端进行获取。根据需要显示礼品类型或礼品攻略，指定 listView 所使用的 Adapter，在 Adapter 中使用 list_level2_item 布局来创建 listView 的子项，并显示对应的图片和名称。

需要注意，由于 Android 系统对应用程序的进程所使用的内存大小有限制，当界面上需要显示大量图片或很大的图片时，如果不做任何处理，很容易出现 OOM（Out of Memory）错误。上述代码中通过调用 ImageView 的 setImageBitmap() 方法来显示图片，该方法比 setImageResource() 节省内存。setImageBitmap() 方法会将指定的 Bitmap 对象显示在 ImageView 中，因此需要将图片资源转为 Bitmap。在 BaseActivity 中添加 readBitMapFromResource() 方法，实现将图片资源转为 Bitmap 的操作，代码如下所示。

【任务 4-1】 BaseActivity.java

```java
package com.qst.giftems.activity;
...//省略 import
public abstract class BaseActivity extends Activity {
    ...//省略其他属性和方法
```

```
/**将图片资源转为 Bitmap
 * @param context
 *         Context
 * @param resId
 *         图片资源 ID
 * @return Bitmap */
public static Bitmap readBitMapFromResource(Context context, int resId) {
    BitmapFactory.Options opt = new BitmapFactory.Options();
    opt.inPreferredConfig = Bitmap.Config.RGB_565;
    InputStream is = context.getResources().openRawResource(resId);
    return BitmapFactory.decodeStream(is, null, opt);}
}
```

运行项目,从应用首页进入"礼品中心",如图 4-30 所示。
返回首页,进入"礼品攻略"界面,如图 4-31 所示。

图 4-30 "礼品类型"列表界面　　　　　图 4-31 "礼品攻略"列表界面

4.4.2 实现【任务 4-2】

本任务完成"礼品"详情界面,对应的布局文件代码如下所示。

【任务 4-2】 gift.xml

```
<RelativeLayout xmlns:android = "http://schemas.android.com/apk/res/android" …
    android:focusable = "true"
    android:focusableInTouchMode = "true" >
    <include
        android:id = "@ + id/titleBarRelativeLayout"
        layout = "@layout/title_bar" />
    <RelativeLayout
        android:id = "@ + id/bottomRelativeLayout"
        android:layout_width = "match_parent"
        android:layout_height = "80dp"
        android:layout_alignParentBottom = "true"
        android:background = "@color/title_bottom"
```

```xml
        android:gravity = "center_vertical" >
        <LinearLayout
            android:layout_width = "match_parent"
            android:layout_height = "40dp"
            android:orientation = "horizontal"
            android:paddingLeft = "10dp"
            android:paddingRight = "50dp" >
            <Button
                android:id = "@+id/buyButton"
                style = "@style/button"
                android:layout_width = "80dp"
                android:layout_height = "wrap_content"
                android:text = "购买" />
            …省略"找人送我"、"加入购物袋"两个<Button>
        </LinearLayout>
</RelativeLayout>
<ScrollView
    android:id = "@+id/giftScrollView"
    android:layout_width = "match_parent"
    android:layout_height = "match_parent"
    android:layout_above = "@id/bottomRelativeLayout"
    android:layout_below = "@id/titleBarRelativeLayout" >
    <LinearLayout
        android:layout_width = "match_parent"
        android:layout_height = "match_parent"
        android:orientation = "vertical" >
        <!-- 轮播显示礼品的多个图片 -->
        <ViewFlipper
            android:id = "@+id/picViewFlipper"
            android:layout_width = "match_parent"
            android:layout_height = "200dp"
            android:autoStart = "true"
            android:background = "#000000"
            android:flipInterval = "5000" />
        <View
            android:layout_width = "match_parent"
            android:layout_height = "1dp"
            android:layout_margin = "10dp"
            android:background = "@color/title_bottom" />
        <!-- 礼品名称 -->
        <TextView
            android:id = "@+id/nameTextView"
            style = "@style/text3"
            android:layout_width = "match_parent"
            android:layout_height = "wrap_content"
            android:layout_marginTop = "5dp"
            android:ellipsize = "end" />
        <!-- 礼品多个款式的图片 -->
        <RelativeLayout
            android:id = "@+id/styleRelativeLayout"
            android:layout_width = "match_parent"
            android:layout_height = "wrap_content"
            android:layout_marginTop = "10dp"
```

```xml
            android:orientation = "horizontal" >
         <!-- 款式选择框 -->
         <ImageView
             android:id = "@ + id/styleSelectedImageView"
             android:layout_width = "wrap_content"
             android:layout_height = "wrap_content"
             android:src = "@drawable/style_select"
             android:visibility = "gone" />
</RelativeLayout>
<LinearLayout
         android:id = "@ + id/buyLinearLayout"
         android:layout_width = "match_parent"
         android:layout_height = "wrap_content"
         android:layout_marginTop = "10dp"
         android:gravity = "center_vertical"
         android:orientation = "horizontal" >
         <TextView
             android:id = "@ + id/quantityTextView"
             style = "@style/text1"
             android:layout_width = "wrap_content"
             android:layout_height = "wrap_content"
             android:layout_marginLeft = "20dp"
             android:text = "购买数量" />
         <LinearLayout
             android:layout_width = "wrap_content"
             android:layout_height = "wrap_content"
             android:layout_marginLeft = "5dp"
             android:layout_marginRight = "5dp"
             android:orientation = "vertical" >
             <ImageView
                 android:id = "@ + id/quantityIncreaseImageView"
                 android:layout_width = "30dp"
                 android:layout_height = "30dp"
                 android:src = "@drawable/up" />
             <EditText
                 android:id = "@ + id/quantityEditText"
                 style = "@style/text1"
                 android:layout_width = "30dp"
                 android:layout_height = "30dp"
                 android:background = "@drawable/background_edit"
                 android:inputType = "number"
                 android:text = "1" />
             <ImageView
                 android:id = "@ + id/quantityReduceImageView"
                 android:layout_width = "30dp"
                 android:layout_height = "30dp"
                 android:src = "@drawable/down" />
         </LinearLayout>
         <TextView
             android:id = "@ + id/priceTextView0"
             style = "@style/text1"
             android:layout_width = "wrap_content"
             android:layout_height = "wrap_content"
```

```xml
                android:text = "套,价格" />
            <TextView
                android:id = "@ + id/priceTextView"
                style = "@style/text3"
                android:layout_width = "wrap_content"
                android:layout_height = "wrap_content"
                android:layout_marginLeft = "5dp"
                android:text = "0" />
            <TextView
                style = "@style/text1"
                android:layout_width = "wrap_content"
                android:layout_height = "wrap_content"
                android:layout_marginLeft = "5dp"
                android:text = "元" />
        </LinearLayout>
        <TextView
            android:id = "@ + id/introTextView"
            style = "@style/text1"
            android:layout_width = "match_parent"
            android:layout_height = "wrap_content" />
        <LinearLayout
            android:id = "@ + id/loadingLinearLayout"
            android:layout_width = "match_parent"
            android:layout_height = "wrap_content"
            android:gravity = "center"
            android:orientation = "horizontal" >
            <ProgressBar
                android:layout_width = "wrap_content"
                android:layout_height = "wrap_content" />
            <TextView
                style = "@style/text1"
                android:textSize = "12sp"
                android:layout_width = "wrap_content"
                android:layout_height = "wrap_content"
                android:layout_marginLeft = "10dp"
                android:textColor = "#999999"
                android:text = "正在加载礼品详情" />
        </LinearLayout>
        <!-- 礼品介绍,HTML 格式 -->
        <com.qst.giftems.activity.SimpleWebView
            android:id = "@ + id/remarkWebView"
            android:layout_width = "match_parent"
            android:layout_height = "wrap_content" >
        </com.qst.giftems.activity.SimpleWebView>
        </LinearLayout>
    </ScrollView>
    <include
        android:id = "@ + id/cornerMenuRelativeLayout"
        layout = "@layout/corner_menu" />
</RelativeLayout>
```

上述布局文件中,在上部使用 ViewFlipper 轮播方式显示礼品的款式,中部列出每款礼品的第一幅图片,单击并选择该款式,下部是购买的数量等,在窗体的最下方使用 WebView

第4章 UI进阶

显示礼品的详情,礼品详情为 HTML 格式的内容。

编写"礼品"界面对应的 Activity,代码如下所示。

【任务 4-2】 GiftActivity.java

```java
/**礼品*/
public class GiftActivity extends BaseActivity {
    int giftTypeId;
    int giftId;
    Gift gift;
    GiftStyle selectedGiftStyle;
    RelativeLayout container;
    ViewFlipper picViewFlipper;
    TextView nameTextView;
    RelativeLayout styleRelativeLayout;
    LinearLayout buyLinearLayout;
    EditText quantityEditText;
    ImageView quantityIncreaseImageView;
    ImageView quantityReduceImageView;
    TextView priceTextView;
    TextView introTextView;
    ImageView styleSelectedImageView;
    LinearLayout loadingLinearLayout;
    WebView remarkWebView;
    Button buyButton;
    Button anotherPayButton;
    Button shoppingBagButton;
    OnClickListener styleOnClickListener = new OnClickListener() {
        @Override
        public void onClick(View v) {
            GiftStyle style = (GiftStyle) v.getTag();
            selectStyle(style);}};
    public GiftActivity() {
        super(R.layout.gift, "礼品");}
    @Override
    protected void onCreate(Bundle savedInstanceState) {
        super.onCreate(savedInstanceState);
        final Context context = getApplicationContext();
        giftTypeId = getIntent().getIntExtra("giftTypeId", 0);
        giftId = getIntent().getIntExtra("giftId", 0);
        container = (RelativeLayout) findViewById(R.id.container);
        …//省略获取各组件的 findViewById()语句
        picViewFlipper.setInAnimation(context, R.anim.in_from_right);
        picViewFlipper.setOutAnimation(context, R.anim.out_to_left);
        WebSettings webSettings = remarkWebView.getSettings();
        webSettings.setLayoutAlgorithm(LayoutAlgorithm.SINGLE_COLUMN);
        webSettings.setUseWideViewPort(true);
        webSettings.setLoadWithOverviewMode(true);
        collectView.setOnClickListener(new OnClickListener() {
            @Override
            public void onClick(View v) {
                if (!checkLogin())
                    return;
                //TODO 后续章节完成收藏功能
            }});
        quantityEditText.addTextChangedListener(new TextWatcher() {
```

```java
            @Override
            public void onTextChanged(CharSequence s, int start, int before,
                    int count) {
                try {
                    int quantity = Integer.parseInt(quantityEditText.getText()
                            .toString());
                    priceTextView
                            .setText((selectedGiftStyle.discount == 0
                                    ? selectedGiftStyle.price
                                    : selectedGiftStyle.discount) * quantity + "");
                } catch (Exception e) {
                    priceTextView.setText("0");}}
            @Override
            public void beforeTextChanged(CharSequence s, int start, int count,
                    int after) {
            }
            @Override
            public void afterTextChanged(Editable s) {
            }});
    quantityIncreaseImageView.setOnClickListener(new OnClickListener() {
        @Override
        public void onClick(View v) {
            try {
                int quantity = Integer.parseInt(quantityEditText.getText()
                        .toString());
                quantityEditText.setText(quantity + 1 + "");
            } catch (Exception e) {
                quantityEditText.setText("1");}}});
    quantityReduceImageView.setOnClickListener(new OnClickListener() {
        @Override
        public void onClick(View v) {
            try {
                int quantity = Integer.parseInt(quantityEditText.getText()
                        .toString());
                if (quantity > 1)
                    quantityEditText.setText(quantity - 1 + "");
                else
                    quantityEditText.setText("1");
            } catch (Exception e) {
                quantityEditText.setText("1");}}});
    buyButton.setOnClickListener(new OnClickListener() {
        @Override
        public void onClick(View v) {
            if (!checkLogin())
                return;
            if (selectedGiftStyle == null) {
                showMessage("请先选择一个礼品款式");
                return;}
            int orderCount;
            try {
                orderCount = Integer.parseInt(quantityEditText.getText()
                        .toString());
                if (orderCount == 0) {
                    showMessage("请输入购买数量");
```

```java
                        return;}
                } catch (NumberFormatException e) {
                    e.printStackTrace();
                    showMessage("请输入购买数量");
                    return;}
                //TODO 后续章节完成购买功能
            }
        });
        anotherPayButton.setOnClickListener(new OnClickListener() {
            @Override
            public void onClick(View v) {
                if (!checkLogin())
                    return;
                if (selectedGiftStyle == null) {
                    showMessage("请先选择一个礼品款式");
                    return; }
                int orderCount;
                try {
                    orderCount = Integer.parseInt(quantityEditText.getText()
                            .toString());
                    if (orderCount == 0) {
                        showMessage("请输入购买数量");
                        return; }
                } catch (NumberFormatException e) {
                    e.printStackTrace();
                    showMessage("请输入购买数量");
                    return; }
                //TODO 后续章节完成购买功能
            }
        });
        shoppingBagButton.setOnClickListener(new OnClickListener() {
            @Override
            public void onClick(View v) {
                if (!checkLogin())
                    return;
                if (selectedGiftStyle == null) {
                    showMessage("请先选择一个礼品款式");
                    return; }
                int orderCount;
                try {
                    orderCount = Integer.parseInt(quantityEditText.getText()
                            .toString());
                    if (orderCount == 0) {
                        showMessage("请输入购买数量");
                        return; }
                } catch (NumberFormatException e) {
                    e.printStackTrace();
                    showMessage("请输入购买数量");
                    return; }
                //TODO 后续章节完成购买功能
        }});
        //TODO 后续章节改为从服务器获取礼品数据
        gift = new Gift();
```

```java
gift.id = 1;
gift.name = "测试礼品";
gift.remark = "礼品介绍";
gift.likes = 10;
gift.sales = 100;
gift.collected = false;
int[] pics = { R.drawable.gift_1, R.drawable.gift_2, R.drawable.gift_3 };
for (int i = 1; i <= 3; i++) {
    GiftStyle gs = new GiftStyle();
    gs.id = i;
    gs.name = "款式 " + i;
    gs.price = 199;
    gs.discount = 169;
    gs.remark = "款式介绍";
    gs.pic1 = pics[i - 1] + "";
    gs.pic2 = R.drawable.gift_0 + "";
    gs.pic3 = R.drawable.gift_0 + "";
    gs.orderNumber = i;
    gs.gift = gift;
    gift.styles.add(gs);}
collectView.setVisibility(View.VISIBLE);
collectView.setBackgroundResource(gift.collected ? R.drawable.collect2
        : R.drawable.collect1);
selectStyle(gift.styles.get(0));
nameTextView.setText(gift.name);
styleRelativeLayout.removeAllViews();
for (int s = 0; s < gift.styles.size(); s++) {
    GiftStyle st = gift.styles.get(s);
    ImageView iv = new ImageView(context);
    iv.setId(st.id);
    iv.setTag(st);
    iv.setImageBitmap(readBitMapFromResource(GiftActivity.this,
            Integer.parseInt(st.pic1)));
    RelativeLayout.LayoutParams params = new RelativeLayout.LayoutParams(
            120, 120);
    params.leftMargin = 5;
    if (s > 0)
        params.addRule(RelativeLayout.RIGHT_OF,
                gift.styles.get(s - 1).id);
    styleRelativeLayout.addView(iv, params);
    iv.setOnClickListener(styleOnClickListener);
    TextView tv = new TextView(context);
    tv.setTextAppearance(context, R.style.text1);
    tv.setText(st.name);
    tv.setGravity(Gravity.CENTER);
    tv.setEllipsize(TruncateAt.END);
    RelativeLayout.LayoutParams params2 = new RelativeLayout.LayoutParams(
            RelativeLayout.LayoutParams.WRAP_CONTENT,
            RelativeLayout.LayoutParams.WRAP_CONTENT);
    params2.addRule(RelativeLayout.BELOW, st.id);
    params2.addRule(RelativeLayout.ALIGN_LEFT, st.id);
    params2.addRule(RelativeLayout.ALIGN_RIGHT, st.id);
    styleRelativeLayout.addView(tv, params2);}
```

```java
        styleRelativeLayout.addView(styleSelectedImageView);}
@Override
protected void onDestroy() {
    super.onDestroy();
    remarkWebView.removeAllViews();
    remarkWebView.destroy();}
void selectStyle(GiftStyle style) {
    if (selectedGiftStyle == style)
        return;
    selectedGiftStyle = style;
    Context context = getApplicationContext();
    try {
        int quantity = Integer.parseInt(quantityEditText.getText()
            .toString());
        priceTextView
            .setText((selectedGiftStyle.discount == 0
                ? selectedGiftStyle.price
                : selectedGiftStyle.discount) * quantity + "");
    } catch (Exception e) {
        priceTextView.setText("0");}
    buyLinearLayout.setVisibility(View.VISIBLE);
    RelativeLayout.LayoutParams params = (RelativeLayout.LayoutParams)
        styleSelectedImageView.getLayoutParams();
    params.addRule(RelativeLayout.ALIGN_TOP, selectedGiftStyle.id);
    params.addRule(RelativeLayout.ALIGN_BOTTOM, selectedGiftStyle.id);
    params.addRule(RelativeLayout.ALIGN_LEFT, selectedGiftStyle.id);
    params.addRule(RelativeLayout.ALIGN_RIGHT, selectedGiftStyle.id);
    styleRelativeLayout.requestLayout();
    styleSelectedImageView.setVisibility(View.VISIBLE);
    picViewFlipper.removeAllViews();
    if (!StringUtils.isEmpty(style.pic1)) {
        ImageView iv = new ImageView(context);
        iv.setScaleType(ScaleType.CENTER_CROP);
        iv.setImageBitmap(readBitMapFromResource(GiftActivity.this,
            Integer.parseInt(style.pic1)));
        picViewFlipper.addView(iv, ViewFlipper.LayoutParams.MATCH_PARENT,
            ViewFlipper.LayoutParams.MATCH_PARENT);}
    if (!StringUtils.isEmpty(style.pic2)) {
        ImageView iv = new ImageView(context);
        iv.setScaleType(ScaleType.CENTER_CROP);
        iv.setImageBitmap(readBitMapFromResource(GiftActivity.this,
            Integer.parseInt(style.pic2)));
        picViewFlipper.addView(iv, ViewFlipper.LayoutParams.MATCH_PARENT,
            ViewFlipper.LayoutParams.MATCH_PARENT); }
    if (!StringUtils.isEmpty(style.pic3)) {
        ImageView iv = new ImageView(context);
        iv.setScaleType(ScaleType.CENTER_CROP);
        iv.setImageBitmap(readBitMapFromResource(GiftActivity.this,
            Integer.parseInt(style.pic3)));
        picViewFlipper.addView(iv, ViewFlipper.LayoutParams.MATCH_PARENT,
            ViewFlipper.LayoutParams.MATCH_PARENT); }
    if (!StringUtils.isEmpty(style.pic4)) {
        ImageView iv = new ImageView(context);
```

```
            iv.setScaleType(ScaleType.CENTER_CROP);
            iv.setImageBitmap(readBitMapFromResource(GiftActivity.this,
                    Integer.parseInt(style.pic4)));
            picViewFlipper.addView(iv, ViewFlipper.LayoutParams.MATCH_PARENT,
                    ViewFlipper.LayoutParams.MATCH_PARENT); }
        if (!StringUtils.isEmpty(style.pic5)) {
            ImageView iv = new ImageView(context);
            iv.setScaleType(ScaleType.CENTER_CROP);
            iv.setImageBitmap(readBitMapFromResource(GiftActivity.this,
                    Integer.parseInt(style.pic5)));
            picViewFlipper.addView(iv, ViewFlipper.LayoutParams.MATCH_PARENT,
                    ViewFlipper.LayoutParams.MATCH_PARENT); }
loadingLinearLayout.setVisibility(View.VISIBLE);
remarkWebView.setVisibility(View.GONE);
remarkWebView.setWebViewClient(new WebViewClient() {
    @Override
    public void onPageFinished(WebView view, String url) {
        loadingLinearLayout.setVisibility(View.GONE);
        remarkWebView.setVisibility(View.VISIBLE); }});
remarkWebView.loadDataWithBaseURL("", style.remark, "text/html",
        "UTF-8", ""); }
}
```

上述代码中,在 onCreate()方法中构造了礼品的数据(后续章节后改为从服务器获取),并将礼品的图片、名称等内容展示在界面上。

运行项目,从首页进入"礼品中心",单击某个礼品类型后进入"礼品"详情界面,如图 4-32 所示。

图 4-32 "礼品"详情界面

4.4.3 实现【任务 4-3】

本任务完成"礼品攻略"详情界面。对应的布局文件代码如下所示。

【任务 4-3】 strategy.xml

```xml
<RelativeLayout xmlns:android="http://schemas.android.com/apk/res/android" …>
    <include
        android:id="@+id/titleBarRelativeLayout"
        layout="@layout/title_bar" />
    <ScrollView
        android:layout_width="match_parent"
        android:layout_height="match_parent"
        android:layout_below="@id/titleBarRelativeLayout" >
        <LinearLayout
            android:layout_width="match_parent"
            android:layout_height="match_parent"
            android:orientation="vertical" >
            <ViewFlipper
                android:id="@+id/viewFlipper"
                android:layout_width="match_parent"
                android:layout_height="200dp"
                android:autoStart="true"
                android:flipInterval="5000"
                android:inAnimation="@anim/in_from_right"
                android:outAnimation="@anim/out_to_left" >
            </ViewFlipper>
            <TextView
                android:id="@+id/nameTextView"
                style="@style/text3"
                android:layout_width="match_parent"
                android:layout_height="wrap_content"
                android:layout_marginTop="10dp"
                android:padding="10dp"
                android:scrollbars="vertical"
                android:singleLine="true"
                android:textSize="16sp" />
            <TextView
                android:id="@+id/remarkTextView"
                style="@style/text1"
                android:layout_width="match_parent"
                android:layout_height="wrap_content"
                android:gravity="left"
                android:padding="10dp"
                android:scrollbars="vertical"
                android:singleLine="false" />
            <LinearLayout
                android:id="@+id/loadingLinearLayout"
                android:layout_width="match_parent"
```

```xml
                android:layout_height = "wrap_content"
                android:gravity = "center"
                android:orientation = "horizontal" >
                <ProgressBar
                    android:layout_width = "wrap_content"
                    android:layout_height = "wrap_content" />
                <TextView
                    style = "@style/text1"
                    android:textSize = "12sp"
                    android:layout_width = "wrap_content"
                    android:layout_height = "wrap_content"
                    android:layout_marginLeft = "10dp"
                    android:textColor = "#999999"
                    android:text = "正在加载攻略详情" />
            </LinearLayout>
            <com.qst.giftems.activity.SimpleWebView
                android:id = "@+id/contentWebView"
                android:layout_width = "match_parent"
                android:layout_height = "wrap_content" >
            </com.qst.giftems.activity.SimpleWebView>
        </LinearLayout>
    </ScrollView>
    <include
        android:id = "@+id/cornerMenuRelativeLayout"
        layout = "@layout/corner_menu" />
</RelativeLayout>
```

上述布局中,在上部使用 ViewFlipper 轮播方式显示多个礼品攻略,中部列出该攻略的标题和内容,最下方使用 WebView 显示攻略的详情,攻略详情为 HTML 格式的内容。

编写攻略界面对应的 Activity,代码如下所示。

【任务 4-3】 StrategyActivity.java

```java
/** 礼品攻略*/
public class StrategyActivity extends BaseActivity {
    int strategyId;
    Strategy strategy;
    ViewFlipper viewFlipper;
    TextView nameTextView;
    TextView remarkTextView;
    LinearLayout loadingLinearLayout;
    WebView contentWebView;
    public StrategyActivity() {
        super(R.layout.strategy, "礼品攻略");}
    @Override
    protected void onCreate(Bundle savedInstanceState) {
        super.onCreate(savedInstanceState);
        Context context = getApplicationContext();
        strategyId = getIntent().getIntExtra("strategyId", 0);
        viewFlipper = (ViewFlipper) findViewById(R.id.viewFlipper);
        nameTextView = (TextView) findViewById(R.id.nameTextView);
        remarkTextView = (TextView) findViewById(R.id.remarkTextView);
```

```java
loadingLinearLayout = (LinearLayout) findViewById(
        R.id.loadingLinearLayout);
contentWebView = (WebView) findViewById(R.id.contentWebView);
viewFlipper.setInAnimation(context, R.anim.in_from_right);
viewFlipper.setOutAnimation(context, R.anim.out_to_left);
WebSettings webSettings = contentWebView.getSettings();
webSettings.setLayoutAlgorithm(LayoutAlgorithm.SINGLE_COLUMN);
webSettings.setUseWideViewPort(true);
webSettings.setLoadWithOverviewMode(true);
//TODO 后续章节改为从服务器获取礼品数据
strategy = new Strategy();
strategy.id = 1;
strategy.title = "测试礼品攻略";
strategy.remark = "礼品攻略礼品攻略礼品攻略礼品攻略礼品攻略礼品攻略礼品
        攻略礼品攻略礼品攻略礼品攻略礼品攻略";
strategy.pic1 = R.drawable.strategy_1 + "";
strategy.pic2 = R.drawable.strategy_2 + "";
strategy.pic3 = R.drawable.strategy_3 + "";
strategy.pic4 = R.drawable.strategy_4 + "";
strategy.content = "<div style=\"font-size:80pt;background-color:green;
        color:red\">这是 HTML 内容</div>";
if (!StringUtils.isEmpty(strategy.pic1)) {
    ImageView v = new ImageView(context);
    viewFlipper.addView(v);
    v.setImageBitmap(readBitMapFromResource(StrategyActivity.this,
            Integer.parseInt(strategy.pic1)));}
if (!StringUtils.isEmpty(strategy.pic2)) {
    ImageView v = new ImageView(context);
    viewFlipper.addView(v);
    v.setImageBitmap(readBitMapFromResource(StrategyActivity.this,
            Integer.parseInt(strategy.pic2)));}
if (!StringUtils.isEmpty(strategy.pic3)) {
    ImageView v = new ImageView(context);
    viewFlipper.addView(v);
    v.setImageBitmap(readBitMapFromResource(StrategyActivity.this,
            Integer.parseInt(strategy.pic3)));}
if (!StringUtils.isEmpty(strategy.pic4)) {
    ImageView v = new ImageView(context);
    viewFlipper.addView(v);
    v.setImageBitmap(readBitMapFromResource(StrategyActivity.this,
            Integer.parseInt(strategy.pic4)));}
if (!StringUtils.isEmpty(strategy.pic5)) {
    ImageView v = new ImageView(context);
    viewFlipper.addView(v);
    v.setImageBitmap(readBitMapFromResource(StrategyActivity.this,
            Integer.parseInt(strategy.pic5)));}
viewFlipper.setAutoStart(viewFlipper.getChildCount() > 1);
nameTextView.setText(strategy.title);
remarkTextView.setText(strategy.remark);
loadingLinearLayout.setVisibility(View.VISIBLE);
contentWebView.setVisibility(View.GONE);
contentWebView.setWebViewClient(new WebViewClient() {
    @Override
```

```
            public void onPageFinished(WebView view, String url) {
                loadingLinearLayout.setVisibility(View.GONE);
                contentWebView.setVisibility(View.VISIBLE);}});
        contentWebView.loadDataWithBaseURL("", strategy.content, "text/html",
            "UTF-8", ""); }
    @Override
    protected void onDestroy() {
        super.onDestroy();
        contentWebView.removeAllViews();
        contentWebView.destroy();}
}
```

上述代码中,在onCreate()方法中构造了礼品攻略的数据(后续章节后改为从服务器获取),并将攻略的图片、名称等内容展示在界面上。

运行项目,从首页进入礼品攻略模块,单击某个礼品攻略后进入"礼品攻略"详情界面,如图4-33所示。

图4-33 "礼品攻略"详情界面

4.4.4 实现【任务4-4】

本任务完成收礼人列表界面。首先编写布局文件,代码如下所示。

【任务4-4】 list_user_address.xml

```
<RelativeLayout xmlns:android="http://schemas.android.com/apk/res/android" …>
```

第4章 UI进阶

```xml
<include
    android:id = "@ + id/titleBarRelativeLayout"
    layout = "@layout/title_bar" />
<RelativeLayout
    android:id = "@ + id/bottomRelativeLayout"
    android:layout_width = "match_parent"
    android:layout_height = "80dp"
    android:layout_alignParentBottom = "true"
    android:background = "@color/title_bottom"
    android:gravity = "center" >
    <Button
        android:id = "@ + id/okButton"
        style = "@style/button"
        android:layout_width = "150dp"
        android:layout_height = "40dp" />
</RelativeLayout>
<ListView
    android:id = "@ + id/listView"
    android:layout_width = "match_parent"
    android:layout_height = "wrap_content"
    android:layout_above = "@id/bottomRelativeLayout"
    android:layout_below = "@id/titleBarRelativeLayout" >
</ListView>
<TextView
    android:id = "@ + id/nodataTextView"
    style = "@style/text3"
    android:layout_width = "match_parent"
    android:layout_height = "100dp"
    android:layout_below = "@id/titleBarRelativeLayout"
    android:text = "您还没有登记收礼人" />
<include
    android:id = "@ + id/cornerMenuRelativeLayout"
    layout = "@layout/corner_menu" />
</RelativeLayout>
```

上述布局文件中，使用一个 ListView 显示用户已登记的收礼人列表。接下来，编写 ListView 的子项所需的布局文件，代码如下所示。

【任务 4-4】 list_user_address_item.xml

```xml
<RelativeLayout xmlns:android = "http://schemas.android.com/apk/res/android" …
    style = "@style/item"
    android:padding = "2dp" >
    <LinearLayout
        android:id = "@ + id/llll"
        android:layout_width = "match_parent"
        android:layout_height = "wrap_content"
        android:orientation = "vertical"
        android:padding = "10dp" >
        <LinearLayout
```

· 229 ·

```xml
            android:layout_width = "match_parent"
            android:layout_height = "wrap_content"
            android:gravity = "center_vertical"
            android:orientation = "horizontal" >
            <TextView
                android:id = "@+id/nameTextView"
                style = "@style/text3"
                android:layout_width = "wrap_content"
                android:layout_height = "wrap_content"
                android:gravity = "left"
                android:textSize = "16sp" />
            <TextView
                style = "@style/text1"
                android:layout_width = "wrap_content"
                android:layout_height = "wrap_content"
                android:layout_marginLeft = "20dp"
                android:text = "手机：" />
            <TextView
                android:id = "@+id/mobileTextView"
                style = "@style/text1"
                android:layout_width = "wrap_content"
                android:layout_height = "wrap_content" />
        </LinearLayout>
        <TextView
            android:id = "@+id/pcaTextView"
            style = "@style/text1"
            android:layout_width = "match_parent"
            android:layout_height = "wrap_content"
            android:layout_marginTop = "5dp"
            android:gravity = "left" />
        <TextView
            android:id = "@+id/addressTextView"
            style = "@style/text1"
            android:layout_width = "match_parent"
            android:layout_height = "wrap_content"
            android:layout_marginTop = "5dp"
            android:gravity = "left"
            android:textSize = "10sp" />
    </LinearLayout>
    <ImageView
        android:id = "@+id/deleteImageView"
        android:layout_width = "30dp"
        android:layout_height = "30dp"
        android:layout_alignParentRight = "true"
        android:layout_alignParentTop = "true"
        android:layout_marginRight = "5dp"
        android:layout_marginTop = "5dp"
        android:src = "@drawable/close"
        android:visibility = "gone"/>
    <ImageView
```

```xml
        android:id = "@+id/selectedImageView"
        android:layout_width = "wrap_content"
        android:layout_height = "wrap_content"
        android:layout_alignBottom = "@id/llll"
        android:layout_alignLeft = "@id/llll"
        android:layout_alignRight = "@id/llll"
        android:layout_alignTop = "@id/llll"
        android:scaleType = "fitXY"
        android:src = "@drawable/style_select"
        android:visibility = "gone"/>
</RelativeLayout>
```

上述布局中,显示了收礼人的姓名、电话、地址等信息,其中 deleteImageView 用于单击删除收礼人,selectedImageView 用于当选择收礼人时凸显边框。

编写收礼人列表 Activity,代码如下所示。

【任务 4-4】 UserAddressListActivity.java

```java
public class UserAddressListActivity extends BaseActivity {
    public static final int SELECT_USER_ADDRESS_RESULT_CODE = 10;
    ArrayList<UserAddress> userAddresses = new ArrayList<UserAddress>();
    int currentUserAddressId;
    UserAddress selectedUserAddress;
    View selectedUserAddressItemView;
    ListView listView;
    TextView nodataTextView;
    RelativeLayout bottomRelativeLayout;
    Button okButton;
    public UserAddressListActivity() {
        super(R.layout.list_user_address, "已登记收礼人");}
    @Override
    protected void onCreate(Bundle savedInstanceState) {
        super.onCreate(savedInstanceState);
        listView = (ListView) findViewById(R.id.listView);
        nodataTextView = (TextView) findViewById(R.id.nodataTextView);
        okButton = (Button) findViewById(R.id.okButton);
        nodataTextView.setVisibility(View.GONE);
        if (getCallingActivity() != null) {
            okButton.setText("确定选择收礼人");
            okButton.setOnClickListener(new OnClickListener() {
                @Override
                public void onClick(View v) {
                    if (selectedUserAddress == null) {
                        showMessage("请选择收礼人");
                        return;}
                    Intent intent = new Intent();
                    intent.putExtra("userAddressId", selectedUserAddress.id);
                    intent.putExtra("name", selectedUserAddress.name);
                    intent.putExtra("mobile", selectedUserAddress.mobile);
                    intent.putExtra("provinceId",
                            selectedUserAddress.provinceId);
                    intent.putExtra("cityId", selectedUserAddress.cityId);
                    intent.putExtra("areaId", selectedUserAddress.areaId);
```

```java
                    intent.putExtra("address", selectedUserAddress.address);
                    setResult(SELECT_USER_ADDRESS_RESULT_CODE, intent);
                    finish();}});
        } else {
            okButton.setText("登记新收礼人");
            okButton.setOnClickListener(new OnClickListener() {
                @Override
                public void onClick(View v) {
                    Intent intent = new Intent(getApplicationContext(),
                            UserAddressActivity.class);
                    startActivity(intent); }}); }
        listView.setAdapter(new UserAddressListViewAdapter());
        //TODO 后续章节改为从服务器获取数据
        for (int i = 1; i <= 10; i++) {
            UserAddress ua = new UserAddress();
            ua.id = i;
            ua.userId = 1;
            ua.name = "姓名 " + i;
            ua.mobile = "1234567890" + i;
            ua.provinceId = 1;
            ua.cityId = 1;
            ua.areaId = 1;
            ua.address = "地址地址地址地址地址地址地址地址地址" + i;
            ua.postcode = "123456";
            ua.info = "";
            ua.provinceName = "北京市";
            ua.cityName = "北京市";
            ua.areaName = "东城区";
            userAddresses.add(ua);}
        ((BaseAdapter) listView.getAdapter()).notifyDataSetChanged();}
    private class UserAddressListViewAdapter extends BaseAdapter {
        public int getCount() {
            return userAddresses.size();}
        @Override
        public boolean areAllItemsEnabled() {
            return false; }
        public Object getItem(int position) {
            return userAddresses.get(position); }
        public long getItemId(int position) {
            return position; }
        public View getView(int position, View convertView, ViewGroup parent) {
            UserAddress userAddress = userAddresses.get(position);
            if (userAddress == null)
                return null;
            if (convertView == null) {
                convertView = LayoutInflater
                        .from(getApplicationContext())
                        .inflate(R.layout.list_user_address_item, parent, false);}
            convertView.setTag(userAddress);
            TextView nameTextView = (TextView) convertView
                    .findViewById(R.id.nameTextView);
            TextView mobileTextView = (TextView) convertView
                    .findViewById(R.id.mobileTextView);
```

```java
            TextView pcaTextView = (TextView) convertView
                    .findViewById(R.id.pcaTextView);
            TextView addressTextView = (TextView) convertView
                    .findViewById(R.id.addressTextView);
            ImageView deleteImageView = (ImageView) convertView
                    .findViewById(R.id.deleteImageView);
            nameTextView.setText(userAddress.name);
            mobileTextView.setText(userAddress.mobile);
            pcaTextView.setText(userAddress.provinceName + " "
                    + userAddress.cityName + " " + userAddress.areaName);
            addressTextView.setText(userAddress.address);
            deleteImageView.setTag(userAddress);
            if (getCallingActivity() != null) {
                deleteImageView.setVisibility(View.GONE);
                convertView.setOnClickListener(onClickListener1);
            } else {
                deleteImageView.setVisibility(View.VISIBLE);
                deleteImageView.setOnClickListener(deleteOCL);
                convertView.setOnClickListener(onClickListener2); }
            convertView.setTag(userAddress);
            return convertView; }
    OnClickListener onClickListener1 = new OnClickListener() {
        @Override
        public void onClick(View v) {
            ImageView selectedImageView = (ImageView) v
                    .findViewById(R.id.selectedImageView);
            UserAddress item = (UserAddress) v.getTag();
            if (selectedImageView.getVisibility() == View.GONE) {
                if (selectedUserAddress != null) {
                    selectedUserAddressItemView.findViewById(
                            R.id.selectedImageView)
                            .setVisibility(View.GONE); }
                selectedImageView.setVisibility(View.VISIBLE);
                selectedUserAddress = item;
                selectedUserAddressItemView = v;
            } else {
                selectedImageView.setVisibility(View.GONE);
                selectedUserAddress = null;
                selectedUserAddressItemView = null; }}};
    OnClickListener onClickListener2 = new OnClickListener() {
        @Override
        public void onClick(View v) {
            UserAddress item = (UserAddress) v.getTag();
            Intent intent = new Intent(getApplicationContext(),
                    UserAddressActivity.class);
            intent.putExtra("userAddressId", item.id);
            intent.putExtra("name", item.name);
            intent.putExtra("mobile", item.mobile);
            intent.putExtra("provinceId", item.provinceId);
            intent.putExtra("cityId", item.cityId);
            intent.putExtra("areaId", item.areaId);
            intent.putExtra("address", item.address);
            startActivity(intent);}};
```

```
                OnClickListener deleteOCL = new OnClickListener() {
                    @Override
                    public void onClick(View v) {
                        final int userAddressId = ((UserAddress) v.getTag()).id;
                        Dialogs.showSimpleDialog(UserAddressListActivity.this,
                                "确定删除这条收礼人信息?", true, new OnOkListener() {
                                    @Override
                                    public void onOk() {
                                        currentUserAddressId = userAddressId;
                                        //TODO 后续章节改为从服务器删除数据
                                        Iterator<UserAddress> it = userAddresses
                                                .iterator();
                                        while (it.hasNext()) {
                                            UserAddress item = it.next();
                                            if (item.id == currentUserAddressId)
                                                it.remove();}
                                        ((BaseAdapter) listView.getAdapter())
                                                .notifyDataSetChanged();}});}};
}
```

上述代码中,onCreate()方法中定义了一个收礼人集合并显示在 ListView 中。本界面除了显示收礼人列表外,还可以为其他需要选择收礼人的界面提供选择功能,因此 onCreate()方法中根据 getCallingActivity()的返回值判断是否是通过 startActivityForResult()方法启动的当前 Activity。当通过 startActivityForResult()方法启动时,不显示删除功能,但是单击某个收礼人条目时会显示选择框,否则仅显示删除功能而不显示选择框,单击收礼人时跳转到收礼人信息编辑界面,并通过 Intent 传送当前单击收礼人的信息。

 注意

有关 Intent 的具体使用参见第 5 章内容。

运行项目,进入"个人中心",单击"登记新收礼人"按钮,打开"已登记收礼人"列表界面,单击某个收礼人的删除图标,则可以删除此收礼人,效果如图 4-34 所示。

图 4-34 "已登记收礼人"列表界面

4.4.5 实现【任务 4-5】

本任务完成收礼人编辑界面。首先编写布局文件，代码如下所示。

【任务 4-5】 user_address.xml

```xml
<RelativeLayout xmlns:android="http://schemas.android.com/apk/res/android" …>
    <include
        android:id="@+id/titleBarRelativeLayout"
        layout="@layout/title_bar" />
    <RelativeLayout
        android:id="@+id/bottomRelativeLayout"
        android:layout_width="match_parent"
        android:layout_height="80dp"
        android:layout_alignParentBottom="true"
        android:background="@color/title_bottom"
        android:gravity="center" >
        <Button
            android:id="@+id/okButton"
            style="@style/button"
            android:layout_width="100dp"
            android:layout_height="40dp"
            android:text="保存" />
    </RelativeLayout>
    <LinearLayout
        android:layout_width="match_parent"
        android:layout_height="wrap_content"
        android:layout_above="@id/bottomRelativeLayout"
        android:layout_below="@id/titleBarRelativeLayout"
        android:orientation="vertical"
        android:padding="10dp" >
        <TextView
            style="@style/text1"
            android:layout_width="match_parent"
            android:layout_height="wrap_content"
            android:gravity="left"
            android:text="请输入收礼人个人信息和详细地址：" />
        <LinearLayout
            android:layout_width="match_parent"
            android:layout_height="wrap_content"
            android:layout_marginTop="10dp"
            android:orientation="horizontal" >
            <TextView
                style="@style/text1"
                android:layout_width="wrap_content"
                android:layout_height="wrap_content"
                android:text="收礼人：" />
            <EditText
                android:id="@+id/nameEditText"
                style="@style/text1"
```

```xml
            android:layout_width = "80dp"
            android:layout_height = "35dp"
            android:background = "@drawable/background_edit" >
        </EditText>
        <TextView
            style = "@style/text1"
            android:layout_width = "wrap_content"
            android:layout_height = "wrap_content"
            android:layout_marginLeft = "10dp"
            android:text = "手机号: " />
        <EditText
            android:id = "@+id/mobileEditText"
            style = "@style/text1"
            android:layout_width = "130dp"
            android:layout_height = "35dp"
            android:background = "@drawable/background_edit"
            android:inputType = "phone" >
        </EditText>
    </LinearLayout>
    <LinearLayout
        android:layout_width = "match_parent"
        android:layout_height = "wrap_content"
        android:layout_marginTop = "10dp"
        android:orientation = "horizontal" >
        <TextView
            style = "@style/text1"
            android:layout_width = "wrap_content"
            android:layout_height = "wrap_content"
            android:layout_gravity = "center_vertical"
            android:text = "地址: " />
        <Spinner
            android:id = "@+id/provinceSpinner"
            android:layout_width = "wrap_content"
            android:layout_height = "wrap_content" >
        </Spinner>
        <Spinner
            android:id = "@+id/citySpinner"
            android:layout_width = "wrap_content"
            android:layout_height = "wrap_content" >
        </Spinner>
        <Spinner
            android:id = "@+id/areaSpinner"
            android:layout_width = "wrap_content"
            android:layout_height = "wrap_content" >
        </Spinner>
    </LinearLayout>
    <EditText
        android:id = "@+id/addressEditText"
        style = "@style/text1"
        android:layout_width = "match_parent"
```

```xml
            android:layout_height = "80dp"
            android:background = "@drawable/background_edit"
            android:gravity = "left|top"
            android:hint = "请输入详细的街道地址"
            android:padding = "10dp"
            android:singleLine = "false" >
        </EditText>
    </LinearLayout>
    <include
        android:id = "@ + id/cornerMenuRelativeLayout"
        layout = "@layout/corner_menu" />
</RelativeLayout>
```

上述布局文件中，添加了收礼人各个属性的输入框，包括姓名、手机号、省市区、详细地址等。然后编写相应的"登记新收礼人"Activity，代码如下所示。

【任务 4-5】 UserAddressActivity.java

```java
public class UserAddressActivity extends BaseActivity {
    UserAddress userAddress = new UserAddress();
    EditText nameEditText;
    EditText mobileEditText;
    Spinner provinceSpinner;
    Spinner citySpinner;
    Spinner areaSpinner;
    EditText addressEditText;
    Button okButton;
    public UserAddressActivity() {
        super(R.layout.user_address, "我的收礼人");}
    @Override
    protected void onCreate(Bundle savedInstanceState) {
        super.onCreate(savedInstanceState);
        nameEditText = (EditText) findViewById(R.id.nameEditText);
        …//省略获取各组件的 findViewById()语句
        okButton = (Button) findViewById(R.id.okButton);
        ArrayAdapter < Areas.Province > paa = new ArrayAdapter < Areas.Province >(
                getApplicationContext(), R.layout.spinner_item, Areas.provinces);
        provinceSpinner.setAdapter(paa);
        int userAddressId = getIntent().getIntExtra("userAddressId", 0);
        userAddress.id = userAddressId;
        if (userAddressId == 0) {
            titleTextView.setText("登记新收礼人");
        } else {
            titleTextView.setText("修改收礼人信息");
            userAddress.name = getIntent().getStringExtra("name");
            userAddress.mobile = getIntent().getStringExtra("mobile");
            userAddress.provinceId = getIntent().getIntExtra("provinceId", 0);
            userAddress.cityId = getIntent().getIntExtra("cityId", 0);
            userAddress.areaId = getIntent().getIntExtra("areaId", 0);
```

```java
            userAddress.address = getIntent().getStringExtra("address");
            nameEditText.setText(userAddress.name);
            mobileEditText.setText(userAddress.mobile);
            addressEditText.setText(userAddress.address);
            //设置省市区下拉菜单选中收礼人的省市区
            int pp = 0, cc = 0, aa = 0;
            if (userAddress.provinceId != 0) {
                for (int i = 0; i < provinceSpinner.getCount(); i++) {
                    Areas.Province p = Areas.provinces.get(i);
                    if (p.id == userAddress.provinceId) {
                        //provinceSpinner.setSelection(i);
                        pp = i;
                        ArrayAdapter<Areas.City> caa
                                = new ArrayAdapter<Areas.City>(
                                getApplicationContext(), R.layout.spinner_item,
                                p.cities);
                        citySpinner.setAdapter(caa);
                        if (userAddress.cityId != 0) {
                            for (int j = 0; j < citySpinner.getCount(); j++) {
                                Areas.City c = p.cities.get(j);
                                if (c.id == userAddress.cityId) {
                                    //citySpinner.setSelection(j);
                                    cc = j;
                                    ArrayAdapter<Areas.Area> aaa
                                            = new ArrayAdapter<Areas.Area>(
                                            getApplicationContext(),
                                            R.layout.spinner_item, c.areas);
                                    areaSpinner.setAdapter(aaa);
                                    if (userAddress.areaId != 0) {
                                        for (int k = 0; k < areaSpinner
                                                .getCount(); k++) {
                                            Areas.Area a = c.areas.get(k);
                                            if (a.id == userAddress.areaId) {
                                                //areaSpinner.setSelection(k);
                                                aa = k;
                                                break;}}}
                                    break;}}}
                        break; }}}
                provinceSpinner.setSelection(pp);
                citySpinner.setSelection(cc);
                areaSpinner.setSelection(aa);}
        provinceSpinner.setOnItemSelectedListener(new OnItemSelectedListener() {
            @Override
            public void onItemSelected(AdapterView<?> parent, View view,
                    int position, long id) {
                Areas.Province p = (Areas.Province) provinceSpinner
                        .getSelectedItem();
                if (p.id == userAddress.provinceId)
                    return;
                ArrayAdapter<Areas.City> caa = new ArrayAdapter<Areas.City>(
                        getApplicationContext(),
                        R.layout.spinner_item,
                        ((Areas.Province) provinceSpinner.getSelectedItem())
```

```java
                    .cities);
            citySpinner.setAdapter(caa);}
        @Override
        public void onNothingSelected(AdapterView<?> parent) {
        }});
citySpinner.setOnItemSelectedListener(new OnItemSelectedListener() {
        @Override
        public void onItemSelected(AdapterView<?> parent, View view,
                int position, long id) {
            Areas.City c = (Areas.City) citySpinner.getSelectedItem();
            if (c.id == userAddress.cityId)
                return;
            ArrayAdapter<Areas.Area> aaa = new ArrayAdapter<Areas.Area>(
                    getApplicationContext(), R.layout.spinner_item,
                    ((Areas.City) citySpinner.getSelectedItem()).areas);
            areaSpinner.setAdapter(aaa);}
        @Override
        public void onNothingSelected(AdapterView<?> parent) {
        }});
okButton.setOnClickListener(new OnClickListener() {
        @Override
        public void onClick(View v) {
            if (!checkLogin())
                return;
            String name = nameEditText.getText().toString();
            if (StringUtils.isEmpty(name)) {
                showMessage("请输入收礼人姓名");
                return; }
            String mobile = mobileEditText.getText().toString();
            if (StringUtils.isEmpty(mobile)) {
                showMessage("请输入收礼人手机号码");
                return; }
            int provinceId = 0;
            int cityId = 0;
            int areaId = 0;
            Areas.Province province = (Areas.Province) provinceSpinner
                    .getSelectedItem();
            if (province != null) {
                provinceId = province.id;
                Areas.City city = (Areas.City) citySpinner
                        .getSelectedItem();
                if (city != null) {
                    cityId = city.id;
                    Areas.Area area = (Areas.Area) areaSpinner
                            .getSelectedItem();
                    if (area != null)
                        areaId = area.id; }}
            String address = addressEditText.getText().toString();
            userAddress.name = name;
            userAddress.mobile = mobile;
            userAddress.provinceId = provinceId;
            userAddress.cityId = cityId;
            userAddress.areaId = areaId;
```

```
            userAddress.address = address;
            //TODO 后续章节完成保存数据到服务器的功能
        }});}
}
```

上述代码在 onCreate()方法中通过 getIntent()获取前一个 Activity 传过来的收礼人信息,并显示在界面中,其中省市区是三个 Spinner 控件,需要根据当前收礼人的省市区设置默认值。

运行项目,进入收礼人列表界面,单击某个收礼人后会进入该收礼人信息编辑界面,如图 4-35 所示。

图 4-35 收礼人信息编辑界面

4.4.6 实现【任务 4-6】

本任务完成"我的收藏"界面。首先编写布局文件,代码如下所示。

【任务 4-6】 list_favorite.xml

```xml
<RelativeLayout xmlns:android = "http://schemas.android.com/apk/res/android" …>
    xmlns:tools = "http://schemas.android.com/tools"
    <include
        android:id = "@+id/titleBarRelativeLayout"
        layout = "@layout/title_bar" />
    <ListView
        android:id = "@+id/listView"
        android:layout_width = "match_parent"
        android:layout_height = "wrap_content"
        android:layout_above = "@id/bottomRelativeLayout"
        android:layout_below = "@id/titleBarRelativeLayout" >
    </ListView>
    <TextView
        android:id = "@+id/nodataTextView"
```

第4章 UI进阶

```xml
            style = "@style/text3"
            android:layout_width = "match_parent"
            android:layout_height = "100dp"
            android:layout_below = "@id/titleBarRelativeLayout"
            android:text = "您还没有收藏礼品" />
    <include
            android:id = "@ + id/cornerMenuRelativeLayout"
            layout = "@layout/corner_menu" />
</RelativeLayout>
```

上述布局文件中,使用一个 ListView 显示用户已收藏的礼品列表。编写 ListView 的子项对应的布局文件,代码如下所示。

【任务 4-6】 list_favorite_item.xml

```xml
<LinearLayout xmlns:android = "http://schemas.android.com/apk/res/android" …>
    <ImageView
        android:id = "@ + id/stylePicImageView"
        android:layout_width = "100dp"
        android:layout_height = "100dp"
        android:padding = "3dp" />
    <LinearLayout
        android:layout_width = "match_parent"
        android:layout_height = "match_parent"
        android:orientation = "vertical" >
        <RelativeLayout
            android:layout_width = "match_parent"
            android:layout_height = "0dp"
            android:layout_weight = "1"
            android:gravity = "center_vertical"
            android:paddingLeft = "5dp" >
            <TextView
                android:id = "@ + id/giftNameTextView"
                style = "@style/text3"
                android:layout_width = "wrap_content"
                android:layout_height = "match_parent"
                android:textSize = "14sp" />
            <ImageView
                android:id = "@ + id/deleteImageView"
                android:layout_width = "30dp"
                android:layout_height = "30dp"
                android:layout_alignParentRight = "true"
                android:layout_marginRight = "5dp"
                android:layout_marginTop = "5dp"
                android:src = "@drawable/close" />
        </RelativeLayout>
        <LinearLayout
            android:layout_width = "match_parent"
            android:layout_height = "0dp"
            android:layout_weight = "1"
            android:gravity = "center_vertical"
            android:orientation = "horizontal"
            android:paddingLeft = "5dp" >
```

```xml
<TextView
    style = "@style/text1"
    android:layout_width = "wrap_content"
    android:layout_height = "match_parent"
    android:text = "价格: " />
<TextView
    android:id = "@+id/priceTextView"
    style = "@style/text1"
    android:layout_width = "wrap_content"
    android:layout_height = "match_parent"
    android:textColor = "@color/selected" />
<TextView
    android:id = "@+id/ccc"
    style = "@style/text1"
    android:layout_width = "wrap_content"
    android:layout_height = "match_parent"
    android:text = "元" />
        </LinearLayout>
    </LinearLayout>
</LinearLayout>
```

上述布局代码中,定义了礼品的图片、名称、价格等信息,其中 deleteImageView 用于单击后取消收藏这个礼品。然后编写对应的"我的收藏"Activity,代码如下所示。

【任务 4-6】 FavorityActivity.java

```java
public class FavoriteActivity extends BaseActivity {
    ArrayList<Gift> gifts = new ArrayList<Gift>();
    Gift selectedGift;
    View selectedGiftItemView;
    int currentGiftId;
    TextView nodataTextView;
    ListView listView;
    public FavoriteActivity() {
        super(R.layout.list_favorite, "我的收藏");}
    @Override
    protected void onCreate(Bundle savedInstanceState) {
        super.onCreate(savedInstanceState);
        nodataTextView = (TextView) findViewById(R.id.nodataTextView);
        listView = (ListView) findViewById(R.id.listView);
        listView.setAdapter(new GiftListViewAdapter());
        nodataTextView.setVisibility(View.GONE);
        //TODO 后续章节改为从服务器获取数据
        for (int i = 1; i <= 10; i++) {
            Gift gift = new Gift();
            gift.id = i;
            gift.name = "测试礼品 " + i;
            gift.remark = "礼品介绍礼品介绍";
            gift.likes = 10;
            gift.sales = 100;
            gift.collected = false;
            int[] pics = { R.drawable.gift_1, R.drawable.gift_2,
                R.drawable.gift_3 };
```

```java
            for (int j = 1; j <= 3; j++) {
                GiftStyle gs = new GiftStyle();
                gs.id = j;
                gs.name = "款式" + j;
                gs.price = 199;
                gs.discount = 169;
                gs.remark = "款式介绍";
                gs.pic1 = pics[j - 1] + "";
                gs.pic2 = R.drawable.gift_0 + "";
                gs.pic3 = R.drawable.gift_0 + "";
                gs.orderNumber = i;
                gs.gift = gift;
                gift.styles.add(gs);}
        gifts.add(gift);}
    ((BaseAdapter) listView.getAdapter()).notifyDataSetChanged();
}
private class GiftListViewAdapter extends BaseAdapter {
    public int getCount() {
        return gifts.size();}
    @Override
    public boolean areAllItemsEnabled() {
        return false;}
    public Object getItem(int position) {
        return gifts.get(position);}
    public long getItemId(int position) {
        return position;}
    public View getView(int position, View convertView, ViewGroup parent) {
        Gift gift = gifts.get(position);
        if (gift == null)
            return null;
        if (convertView == null) {
            convertView = LayoutInflater.from(getApplicationContext())
                    .inflate(R.layout.list_favorite_item, parent, false);
            convertView.setOnClickListener(onClickListener);}
        convertView.setTag(gift);
        ImageView stylePicImageView = (ImageView) convertView
                .findViewById(R.id.stylePicImageView);
        TextView giftNameTextView = (TextView) convertView
                .findViewById(R.id.giftNameTextView);
        ImageView deleteImageView = (ImageView) convertView
                .findViewById(R.id.deleteImageView);
        TextView priceTextView = (TextView) convertView
                .findViewById(R.id.priceTextView);
        stylePicImageView.setImageBitmap(readBitMapFromResource(
                FavoriteActivity.this,
                Integer.parseInt(gift.styles.get(0).pic1)));
        giftNameTextView.setText(gift.name);
        priceTextView.setText(gift.styles.get(0).discount + "");
        deleteImageView.setTag(gift);
        deleteImageView.setOnClickListener(deleteOCL);
```

```java
            return convertView;}
OnClickListener onClickListener = new OnClickListener() {
    @Override
    public void onClick(View v) {
        final Gift gift = (Gift) v.getTag();
        Intent intent = new Intent(getApplicationContext(),
            GiftActivity.class);
        intent.putExtra("giftId", gift.id);
        startActivity(intent);}};
OnClickListener deleteOCL = new OnClickListener() {
    @Override
    public void onClick(View v) {
        final Gift gift = (Gift) v.getTag();
        Dialogs.showSimpleDialog(FavoriteActivity.this,
            "您确定不再收藏这个礼品吗?", true, new OnOkListener() {
                @Override
                public void onOk() {
                    currentGiftId = gift.id;
                    Iterator<Gift> it = gifts.iterator();
                    while (it.hasNext()) {
                        Gift item = it.next();
                        if (item.id == currentGiftId)
                            it.remove();}
                    ((BaseAdapter) listView.getAdapter())
                        .notifyDataSetChanged();}});}};
}
```

上述代码在 onCreate()方法中模拟了多个收藏的礼品并显示在 ListView 中。

运行项目,进入"个人中心",单击"我的收藏夹"按钮,打开收藏礼品列表界面,单击某个礼品的删除图标,则可以取消收藏此礼品。运行效果如图 4-36 所示。

图 4-36 收藏礼品列表界面

第4章 UI进阶

本章总结

小结

- Fragment 允许将 Activity 拆分成多个完全独立封装的可重用的组件,每个组件拥有自己的生命周期和 UI 布局。
- 创建 Fragment 需要实现三个方法:onCreate()、onCreateView()和 onPause()。
- Fragment 的生命周期与 Activity 的生命周期相似,具有以下状态:活动状态、暂停状态、停止状态和销毁状态。
- Android 中提供的菜单有如下几种:选项菜单、子菜单、上下文菜单和图标菜单等。
- 在 Android 中提供了一种高级控件,其实现过程就类似于 MVC 架构,该控件就是 AdapterView。
- ListView(列表视图)是手机应用中使用非常广泛的组件,以垂直列表的形式显示所有列表项。
- GridView 用于在界面上按行、列分布的方式显示多个组件。GridView 与 ListView 拥有相同的父类 AbsListView,且都是列表项,两者唯一的区别在于:ListView 只显示一列,GridView 则可以显示多列。

Q&A

问题:简述 Fragment 的生命周期。

回答:Fragment 的生命周期与 Activity 的生命周期类似,也存在如下四个状态。

(1) 活动状态:当前状态 Fragment 位于前台,用户可见,并且可以获取焦点;

(2) 暂停状态:其他 Activity 位于前台,该 Fragment 仍然可见,但不能获取焦点;

(3) 停止状态:该 Fragment 不可见,失去焦点;

(4) 销毁状态:该 Fragment 被完全删除或该 Fragment 所在的 Activity 结束。

章节练习

习题

1. 在 Android 中使用 Menu 时可能需要重写的方法有_____(多选)。

    A. onCreateOptionsMenu()　　　　B. onCreateMenu()

    C. onOptionsItemSelected()　　　　D. onItemSelected()

2. 自定义 Adapter 需要重写的方法有_____(多选)。

    A. getCount()　　　　　　　　　　B. getItem()

    C. getItemId()　　　　　　　　　　D. getView()

3. 下面对自定义 style 的方式正确的是_____。

A.

```
<resources>
    <style name="myStyle">
        <item name="android:layout_width">match_parent</item>
    </style>
</resources>
```

B.

```
<style name="myStyle">
    <item name="android:layout_width">match_parent</item>
</style>
```

C.

```
<resources>
    <item name="android:layout_width">match_parent</item>
</resources>
```

D.

```
<resources>
    <style name="android:layout_width">match_parent</style>
</resources>
```

上机

训练目标：ListView 应用。

培养能力	熟练使用 ListView 实现列表功能		
掌握程度	★★★★★	难度	中
代码行数	400	实施方式	重复编码
结束条件	熟练使用 ListView 实现列表功能		
参考训练内容			
通过使用 ListView 来完成用户的列表功能，并且在每个 item 上显示用户的删除和更新按钮，同时通过假数据的方式实现对应的业务功能			

第5章 Intent与BroadcastReceiver

本章任务是完成"GIFT-EMS 礼记"的用户日程相关功能：
- 【任务 5-1】 完成用户日程界面。
- 【任务 5-2】 完成用户日程编辑界面。
- 【任务 5-3】 完成用户日程提醒功能。

知 识 点	Listen（听）	Know（懂）	Do（做）	Revise（复习）	Master（精通）
Intent 原理及使用	★	★	★	★	★
BroadCastReceiver	★	★	★	★	
Handler 消息传递	★	★	★	★	★
AsyncTask 使用	★	★	★		

5.1 Intent 意图

　　Intent 是 Android 应用内不同组件之间的通信载体，当在 Android 应用中连接不同的组件时，通常需要借助于 Intent 来实现。使用 Intent 可以激活 Android 的三个核心组件：

Activity、Service 和 BroadcastReceiver。通过 Intent 可以启动一个 Activity，也可以启动一个 Service，还可以发送一条广播消息来触发系统中的 BroadcastReceiver。Android 三大组件之间的通信都以 Intent 为载体，使用 Intent 封装当前组件在启动目标组件时所需的信息，实现应用程序间的交互机制，因此 Intent 通常翻译成"意图"。

5.1.1 Intent 原理及分类

Intent 消息传递机制既可以在应用程序中使用，也可以在应用程序之间使用。在 Android 中。通过 Intent 机制来协助应用程序间的交互，Intent 负责对应用中的一次行为操作所涉及的数据和附加信息进行描述，Android 根据 Intent 的描述找到相应的组件，并将 Intent 传递给目标组件来完成组件的调用。因此，Intent 起着媒体中介的作用，专门提供组件之间相互调用的信息，实现调用者与被调用者之间的解耦。

Intent 也可以在系统范围内广播消息，应用程序通过注册一个 Broadcast Receiver 来监听和响应广播的 Intent，从而实现基于内部的、系统的或者第三方应用程序的事件驱动的应用程序。Android 系统通过广播 Intent 来发布系统事件，如网络连接状态或电池电量的改变事件等；Android 系统应用程序（如拨号程序和 SMS 管理器）通过注册监听特定的广播 Intent（例如"来电"或者"收到 SMS 消息"）来做出相应的响应。同样，开发者也可以通过注册监听相同 Intent 的 BroadcastReceiver 来替换本地应用程序。

为了组件之间的解耦和无缝地替换应用程序元素，Android 架构鼓励通过 Intent 来传播意图，包括在同一个应用程序内的传播。除此之外，Intent 还提供一个简易扩展应用程序功能模型的机制。

综上所述，Android 使用 Intent 来封装程序的"调用意图"，无论是启动一个 Activity 组件或 Service 组件，Android 都使用统一的 Intent 对象来封装这种"启动意图"，从而实现 Activity、Service 和 BroadcastReceiver 之间的通信。Intent 与三大组件之间的关系如图 5-1 所示。

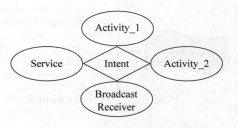

图 5-1 Intent 与三大组件关系图

使用 Intent 启动 Activity、Service 和 BroadcastReceiver 三大组件所使用的机制略有不同。

（1）当启动 Activity 组件时，通常需要调用 startActivity（Intent intent）或 startActivityForResult(Intent intent, int requestCode)方法，其中 Intent 参数用于封装目标 Activity 所需信息。

（2）当启动 Service 组件时，通常需要调用 startService(Intent intent)或 bindService(Intent intent, ServiceConnection conn, int flags)方法，其中 Intent 参数用于封装目标 Service 所需信息。

（3）当触发 BroadcastReceiver 组件时，通过调用 sendBroadcast(Intent intent)等方法来发送广播信息，其中 Intent 参数用于封装目标 BroadcastReceiver 所需信息。

通过上述描述可以看出，Intent 用于封装当前组件在启动目标组件时所需的信息，系统通过该信息找到对应的组件，完成组件之间的调用。

根据 Intent 所描述的信息，可以将 Intent 意图分为以下两类。
- **显式 Intent**：明确指定需要启动或触发组件的类名，对于显式 Intent 而言，Android 系统无须对该 Intent 做任何解析，系统直接找到指定的目标组件，然后启动该组件即可。
- **隐式 Intent**：只指定了需要启动或触发的组件应满足的条件，对于隐式 Intent 而言，Android 系统需要 Intent 进行解析，并得到启动组件所需要的条件，然后在系统中查找与之匹配的目标组件，如果找到符合该条件的组件，就启动相应的目标组件。

在 Android 中，通过 IntentFilter 来判断所调用的组件是否符合隐式 Intent，即通过 IntentFilter 来声明所调用组件的满足条件，从而声明最终需要处理哪些隐式 Intent。

本章重点介绍使用 Intent 启动 Activity 和 BroadcastReceiver 这两种组件的过程，启动 Service 则在本书第 8 章再进行详细介绍。

5.1.2　Intent 属性

Intent 对象其实就是一个信息的捆绑包，通过 Intent 的属性来设定相应的启动目标。通常 Intent 对象中包含 Component、Action 等属性，下面分别进行介绍。

1. Component 组件

Component 组件为目标组件，需要接收一个 ComponentName 对象，而 ComponentName 对象的构造方法有以下几种方式。

- **ComponentName（String pkg, String className）**：用于创建 pkg 包下的 className 所对应的组件。其中参数 pkg 代表应用程序的包名，参数 className 代表组件的类名。
- **ComponentName（Context context, String className）**：用于创建 context 上下文中 className 所对应的组件。
- **ComponentName（Context context, Class <?> className）**：用于创建 context 上下文中 className 所对应的组件。

上述三个构造方法本质上是相同的，用于创建一个 ComponentName 对象，通过包名和类名就可以确定唯一的组件类，应用程序可以根据给定的组件类去启动相应的组件。

除此之外，Intent 还有如下三个方法用于指定待启动组件的包名和类名：setClass（Context ctx, Class <?> cls）；setClassName（Context ctx, String className）；setClassName（String pkg, String className）。

通过 Intent 的 Component 属性来明确指定所要启动的组件，称为显式 Intent；而没有指定 Component 属性的 Intent 被称为隐式 Intent。

当指定 Component 属性时，将直接使用该属性所指定的组件，而 Intent 的其他属性都是可选的。例如，在某个 Activity 中启动 SecondActivity 时，代码编写方式如下所示。

【示例】　创建 ComponentName 对象

```
button1.setOnClickListener(newOnClickListener(){
@Override
public void onClick(View v) {
```

```
            //创建一个意图对象
            Intent intent = new Intent();
            //创建组件,通过组件来响应
            ComponentName component =
                new ComponentName(MainActivity.this, SecondActivity.class);
            intent.setComponent(component);
            startActivity(intent);}});
```

使用 setClass()对上述代码进行改造,代码如下所示。

【示例】 使用 setClass()方法指定待启动组件

```
Intent intent = new Intent();
/**
 * setClass()方法的第一个参数是一个 Context 对象
 * Activity 是 Context 类的子类,即所有的 Activity 对象都是 Context 的子对象
 * setClass()方法的第二个参数是一个 Class 对象,是被启动的 Activity 类的 class 对象
 **/
intent.setClass(MainActivity.this,SecondActivity.class);
startActivity(intent);
```

上述代码还可以继续简化,直接使用 Intent()构造方法指定启动组件,该方式代码最精简,在实际编程中经常用到,代码如下所示。

【示例】 使用 Intent()构造方法指定启动组件

```
Intent intent = new Intent(MainActivity.this,SecondActivity.class);
startActivity(intent);
```

2. Action 动作

在日常生活中描述一个意愿或愿望时,总是有一个动词,例如:我想"吃"一碗香喷喷的大米饭,我要"写"一封情书等。在 Intent 中,Action 用来描述具体的动作,执行者依照 Action 动作指示接收相关输入、表现对应行为、产生符合的输出,附加的 Action 越多,匹配越精确。

在 Android 中,Action 是一个字符串,用于描述一个 Android 应用程序的组件。一个 Intent Filter 中可以包含一个或多个 Action,当在 AndroidManifest.xml 中定义 Activity 时,在<intent-filter>节点中指定一个 Action 列表用于标识 Activity 所能接收的"动作"。

在 Intent 类中提供了大量的标准 Action 常量,启动 Activity 的系统标准 Action 如表 5-1 所示。

表 5-1 启动 Activity 的系统标准 Action

Action 常量	字 符 串	描 述
ACTION_MAIN	android.intent.action.MAIN	应用程序入口
ACTION_VIEW	android.intent.action.VIEW	最常见的动作,视图要求以最合理的方式查看 Intent 的 URI 中所提供的数据。不同的应用程序将会根据 URI 模式来处理视图请求。一般情况下,"http:"地址将会打开浏览器;"tel:"地址将会打开拨号程序以拨打该号码;"geo:"地址会在 Google 地图应用程序中显示出来,而联系人信息将会在联系人管理器中显示出来

续表

Action 常量	字符串	描述
ACTION_EDIT	android.intent.action.EDIT	请求一个 Activity,要求该 Activity 可以编辑 Intent 的数据 URI 中的数据
ACTION_PICK	android.intent.action.PICK	启动一个子 Activity,可以从 Intent 的数据 URI 指定的 ContentProvider 中选择一个项。当关闭的时候,返回所选择的项的 URI。启动的 Activity 与选择的数据有关,例如,传递 content://contacts/people 将会调用本地联系人列表
ACTION_DIAL	android.intent.action.DIAL	打开一个拨号程序,要拨打的号码由 Intent 的数据 URI 预先提供。默认情况下,这是由本地 Android 电话拨号程序进行处理的。拨号程序可以规范化大部分号码样式,例如,tel:555-1234 和 tel:(212)555-1212 都是有效号码
ACTION_CALL	android.intent.action.CALL	打开一个电话拨号程序,并立即使用 Intent 的数据 URI 所提供的号码拨打一个电话,此动作只应用于代替本地电话的 Activity
ACTION_SEND	android.intent.action.SEND	启动一个 Activity,该 Activity 会发送 Intent 中指定的数据。接收人需要由解析的 Activity 来选择。使用 setType 可以设置要传输的数据的 MIME 类型。数据本身应该根据其类型使用 EXTRA_TEXT 或者 EXTRA_STREAM 存储为 extra。对于 E-mail,本地 Android 应用程序也可以使用 EXTRA_EMAIL、EXTRA_CC、EXTRA_BCC 和 EXTRA_SUBJECT 键来接收 extra。应该只使用 ACTION_SEND 动作向远程接收人(而不是设备上的另外一个应用程序)发送数据
ACTION_SENDTO	android.intent.action.SENDTO	启动一个 Activity 来向 Intent 的数据 URI 所指定的联系人发送一条消息
ACTION_ANSWER	android.intent.action.ANSWER	打开一个处理来电的 Activity,通常这个动作是由本地电话拨号程序处理
ACTION_INSERT	android.intent.action.INSERT	打开一个子 Activity 能在 Intent 的数据 URI 指定的游标处插入新项的 Activity。当作为子 Activity 调用时,应该返回一个指向新插入项的 URI
ACTION_DELETE	android.intent.action.DELETE	启动一个 Activity,允许删除 Intent 的数据 URI 中指定的数据
ACTION_ALL_APPS	android.intent.action.ACTION_ALL_APPS	打开一个列出所有已安装应用程序的 Activity,通常此操作由启动器处理
ACTION_SEARCH	android.intent.action.SEARCH	通常用于启动特定的搜索 Activity。如果没有在特定的 Activity 上触发,就会提示用户从所有支持搜索的应用程序中做出选择。可以使用 SearchManager.QUERY 键把搜索词作为一个 Intent 的 extra 中的字符串来提供

Android 中预先定义了一些标准 Action,主要针对一些系统级的事件,这些 Action 都对应于 Intent 类中的常量,其值和意义总结如下。

- **ACTION_BOOT_COMPLETED**:系统启动完成广播,用于指定应用的初始化,例如装载闹钟等。
- **ACTION_TIME_CHANGED**:时间改变广播,用于改变系统时间。

- **ACTION_DATE_CHANGED**：日期改变广播，用于改变日期。
- **ACTION_TIME_TICK**：每分钟改变一次时间，不能在 Manifest 中注册接收，只能通过 context.registerReceiver()显式注册接收。
- **ACTION_TIMEZONE_CHANGED**：时区改变广播，用于改变时区。
- **ACTION_BATTERY_LOW**：电量低广播，显示 Low battery warning 系统对话框。
- **ACTION_PACKAGE_ADDED**：添加包广播，一个新的应用包已被安装，显示内容包含包的名称。
- **ACTION_PACKAGE_REMOVED**：删除包广播，提醒应用包已被卸载。

注意

> 上述列表只列举了部分 Action 常量，读者如有兴趣可以进一步查阅 Android API 关于 Intent 的说明。

3. Category 类别

Category 属性用来描述动作的类别，在< intent-filter >元素中进行声明，Intent 类中提供了标准的 Category 常量及对应的字符串，如表 5-2 所示。

表 5-2　标准 Category

Category 常量	字　符　串	描　　述
CATEGORY_DEFAULT	android.intent.category.DEFAULT	默认的 Category
CATEGORY_BROWSABLE	android.intent.category.BROWSABLE	指定 Activity 能被浏览器安全调用
CATEGORY_TAB	android.intent.category.TAB	指定 Activity 能作为 TabActivity 的 Tab 页
CATEGORY_LAUNCHER	android.intent.category.LAUNCHER	Activity 显示顶级程序列表中
CATEGORY_INFO	android.intent.category.INFO	用于提供包信息
CATEGORY_HOME	android.intent.category.HOME	设置该 Activity 随系统启动而运行
CATEGORY_PREFERENCE	android.intent.category.PREFERENCE	该 Activity 是参数面板
CATEGORY_TEST	android.intent.category.TEST	该 Activity 是一个测试
CATEGORY_CAR_DOCK	android.intent.category.CAR_DOCK	指定移动设备（如手机）被插入汽车底座时运行该 Activity
CATEGORY_DESK_DOCK	android.intent.category.DESK_DOCK	指定移动设备（如手机）被插入桌面底座时运行该 Activity
CATEGORY_CAR_MODE	android.intent.category.CAR_MODE	设置该 Activity 可以在车载环境下使用

Category 属性为 Action 增加额外的附加类别信息。CATEGORY_LAUNCHER 意味着在加载程序的时候 Activity 出现在最上面，而 CATEGORY_HOME 表示页面跳转到 HOME 界面。

下述示例通过 Action 和 Category 属性的联合使用来模拟实现"返回主页"的功能。相应的 XML 布局代码如下所示。

【代码 5-1】 activity_home.xml

```xml
<LinearLayout xmlns:android="http://schemas.android.com/apk/res/android" …>
    <Button android:id="@+id/homeBtn"
        android:layout_marginTop="20dp"
        android:text="返回首页"
        android:layout_width="wrap_content"
        android:layout_height="wrap_content"
        android:layout_gravity="center_horizontal"/>
</LinearLayout>
```

上述界面较为简单,只定义了一个 ID 为 homeBtn 的 Button 组件。接下来创建界面布局对应的 HomeActivity,代码如下所示。

【代码 5-2】 HomeActivity.java

```java
/** 回到主页 */
public class HomeActivity extends AppCompatActivity {
    Button homeButton;
    @Override
    protected void onCreate(Bundle savedInstanceState) {
        super.onCreate(savedInstanceState);
        setContentView(R.layout.activity_home);
        //初始化
        homeButton = (Button) findViewById(R.id.homeBtn);
        //注册事件
        homeButton.setOnClickListener(new View.OnClickListener() {
            @Override
            public void onClick(View view) {
                //创建 Intent 对象
                Intent intent = new Intent();
                //为 Intent 设置 Action 和 Category 属性
                intent.setAction(Intent.ACTION_MAIN);
                intent.addCategory(Intent.CATEGORY_HOME);
                //启动 Activity
                startActivity(intent);}});
    }
}
```

上述代码中,将 Intent 的 Action 属性设为 Intent.ACTION_MAIN,Category 属性设为 Intent.CATEGORY_HOME,以满足该 Intent 对应 Activity 为 Android 系统的 Home 桌面的要求。运行上述代码,如图 5-2 所示,单击"返回首页"按钮可以返回 Home 页面。

图 5-2 Action 和 Category 简单应用

4. Data 数据

Data 属性通常用与 Action 属性结合使用,为 Intent 提供可操作的数据。Data 属性接收一个 URI 对象,其对应的字符串格式如下:

【语法】

scheme://host:port/path

【示例】 URI 字符串

http://www.baidu.com

其中,http 为 scheme 部分;www.baidu.com 为 host 部分;由于是 80 端口,所以 port 部分被省略。下面示例演示 Data 属性和 Action 属性的结合使用,调用浏览器打开一个指定网页,代码如下所示。

【代码 5-3】 activity_uri.xml

```xml
<?xml version = "1.0" encoding = "utf-8"?>
<LinearLayout xmlns:android = "http://schemas.android.com/apk/res/android"
    android:layout_width = "match_parent"
    android:layout_height = "match_parent"
    android:orientation = "vertical" >
    <Button android:id = "@+id/uriBtn"
        android:layout_marginTop = "20dp"
        android:text = "打开百度"
        android:layout_width = "wrap_content"
        android:layout_height = "wrap_content"
        android:layout_gravity = "center_horizontal"/>
</LinearLayout>
```

上述界面较为简单,只定义了一个 ID 为 uriBtn 的 Button 组件。页面布局对应的 UriActivity 代码如下所示。

【代码 5-4】 UriActivity.java

```java
/** 结合 Data 属性和 Action 属性打开指定的网页 */
public class UriActivity extends AppCompatActivity{
    //定义 Button 变量
    Button uriBtn;
    @Override
    protected void onCreate(Bundle savedInstanceState) {
        super.onCreate(savedInstanceState);
        setContentView(R.layout.activity_uri);
        uriBtn = (Button)findViewById(R.id.uriBtn);
        uriBtn.setOnClickListener(new View.OnClickListener() {
            @Override
            public void onClick(View view) {
                //打开网页
                Intent intent = new Intent();
                intent.setAction(Intent.ACTION_VIEW);
                Uri data = Uri.parse("http://www.baidu.com");
                //利用 Data 属性
                intent.setData(data);
                startActivity(intent);}});}
}
```

第5章 Intent与BroadcastReceiver

运行上述Activity，当单击"打开百度"按钮时，将会通过浏览器打开百度首页，界面如图5-3所示。此应用程序中需要展示一个页面，没有必要自己去实现一个浏览器，通过调用系统的浏览器来打开该网页即可。

图5-3 Data简单应用

由上述示例可以看出，使用隐式Intent，开发人员不仅可以启动本程序中的其他Activity，还可以启动其他程序中的Activity，使得Android多个应用程序之间共享功能成为可能。

5. Type 数据类型

Type属性用于指定Data属性URI所对应的MIME类型，该类型可以是自定义的MIME类型，只要符合特定格式的字符串即可，例如text/html。

Data属性与Type属性的关系比较微妙，两个属性之间能够相互覆盖，例如：

- 如果为Intent先设置Data属性，再设置Type属性，那么Type属性将会覆盖Data属性；
- 如果为Intent先设置Type属性，再设置Data属性，那么Data属性将会覆盖Type属性；
- 如果希望Intent既有Data属性也有Type属性，则应该调用Intent的setDataAndType()方法。

注意

限于篇幅，Data属性与Type属性的关系此处不再赘述，读者可以自行验证。

6. Extras 扩展信息

Extras属性是一个Bundle对象，通常用于在多个Activity之间交换数据。其中Bundle与Map非常类似，可以存入多组键值对，在Intent中通过Bundle类型的Extras属性来封装数据，从而实现组件之间的数据传递。Extras属性的使用过程如下所示。

【示例】 使用 Extras 属性

```
Bundle bundle = new Bundle();
bundle.putString("test", "this is a test");
Intent intent = new Intent(MainActivity.this,SecondActivity.class);
intent.putExtras(bundle);
startActivity(intent);
```

上述代码中,在 MainActivity 中通过 Intent 的 Extras 属性来存储数据,然后将其传递到另一个 SecondActivity,接下来在 SecondActivity 中通过 getExtras()方法获得 Bundle 对象并进行取值,代码如下所示。

```
Bundle bundle = this.getIntent().getExtras();
String test = bundle.getString("test");
```

由上述代码可知,要想获取传递的值,利用 Intent 对象的 Extras 属性即可。

7. Flags 标志位

Flag 属性用于为 Intent 添加一些额外的控制标志,通过 Intent 的 addFlags()方法为 Intent 添加控制标志。Intent 类中定义了多个 Flag 常量,常用的 Flag 值的用法如下。

(1) **FLAG_ACTIVITY_CLEAR_TOP**:假设当前 Activity 栈是 A、B、C、D,如果通过 Intent 从 D 跳转到 B,且 Intent 中添加了 FLAG_ACTIVITY_CLEAR_TOP 标记,则栈情况变为 A、B,即在 Activity 栈中已经存在 B 时,将会把 B 之上的 Activity 从栈中弹出并销毁。如果 Intent 中没有添加 FLAG_ACTIVITY_CLEAR_TOP 标记,则会把 B 再次压入栈中,栈情况将会变为 A、B、C、D、B。

(2) **FLAG_ACTIVITY_NEW_TASK**:假设当前 Activity 栈是 A、B、C,通过 Intent 从 C 跳转到 D,并且在 Intent 中添加了 FLAG_ACTIVITY_NEW_TASK 标记,如果在 AndroidManifest.xml 中对 D 这个 Activity 的声明中添加了 Task affinity,并且和栈 1 的 affinity 不同,则系统首先会查找有没有和 D 的 Task affinity 相同的栈存在。如果存在,则将 D 压入该栈;如果不存在,则会新建一个 D 的 affinity 的栈并将其压入。如果 D 的 Task affinity 默认没有设置,或者和栈 1 的 affinity 相同,则会将其压入栈 1,Activity 栈变成 A、B、C、D,此时与没有 FLAG_ACTIVITY_NEW_TASK 标记的效果一样。注意,如果试图从非 Activity 的途径启动一个 Activity(例如从一个 Service 中启动一个 Activity),则 Intent 必须要添加 FLAG_ACTIVITY_NEW_TASK 标记。

(3) **FLAG_ACTIVITY_NO_HISTORY**:假设当前 Activity 栈情况为 A、B、C,通过 Intent 从 C 跳转到 D,并对 Intent 添加了 FLAG_ACTIVITY_NO_HISTORY 标记,则此时界面显示 D 的内容,但是 D 并不会压入栈中,如果按返回键返回到 C,Activity 栈依然是 A、B、C;如果从 D 跳转到 E,栈的情况变为 A、B、C、E,此时按返回键会回到 C,因为 D 根本就没有被压入栈中。

(4) **FLAG_ACTIVITY_SINGLE_TOP**:和上面(3)类似。如果 Intent 添加了 FLAG_ACTIVITY_SINGLE_TOP 标记,并且 Intent 的目标 Activity 就是栈顶的 Activity,那么不会将新建的实例压入栈中。

第5章 Intent与BroadcastReceiver

> **注意**
>
> 为了便于读者更好地理解Flags属性，请结合前面章节的Activity启动方式加以理解和验证。

5.1.3 使用Intent启动Activity

通过调用Context的startActivity()方法可以创建并显示目标Activity，该方法需要传入一个Intent类型的参数，代码如下所示。

```
startActivity(myIntent);
```

startActivity()方法会查找并启动一个与Intent参数相匹配的Activity。因此，通过Intent来显式地指定所要启动的Activity，或者包含一个目标Activity必须执行的动作。在后一种情况中，运行时将会使用一个称为"Intent解析"的过程来动态选择Activity。

如果使用startActivity()方法启动Activity，则在新启动的Activity完成之后，原Activity不会接收到任何信息。如果希望跟踪来自子Activity的反馈，则可以使用startActivityForResult()方法来启动Activity。下面通过4种情况来讲解上述两个方法的用法。

1. 显式Intent启动Activity

当一个应用程序由多个相互关联的Activity组成时，Activity之间需要经常切换，可以通过Intent来显式地指定要打开的Activity，即使用Intent对象来指定要打开的Activity的类名，然后调用startActivity()方法启动Activity。

【示例】 显式Intent启动Activity

```
Intent intent = new Intent(MyActivity.this, MyOtherActivity.class);
startActivity(intent);
```

在调用startActivity()之后，新的Activity(即MyOtherActivity)将会被创建、启动和恢复运行，且移动到Activity栈的顶部。当调用MyOtherActivity的finish()方法结束或按下设备的返回按钮时，系统将关闭该Activity并从栈中移除。每次调用startActivity()方法都会创建一个新的Activity并添加到栈中，而按下后退按钮或调用finish()时则依次删除栈顶的Activity。

下面通过两个Activity演示界面之间的切换，代码如下所示。

【代码5-5】 activity_1.xml

```xml
<LinearLayout xmlns:android = "http://schemas.android.com/apk/res/android" …>
    <TextView
        android:text = "这是第一个Activity"
        android:layout_width = "wrap_content"
        android:layout_height = "wrap_content"
```

```xml
        android:id = "@ + id/textView"
        android:textSize = "24sp" />
    <TextView
        android:text = "您的爱好是："
        android:layout_width = "match_parent"
        android:layout_height = "wrap_content"
        android:id = "@ + id/textView2"
        android:textSize = "24sp" />
    <CheckBox
        android:text = "唱歌"
        android:layout_width = "match_parent"
        android:layout_height = "wrap_content"
        android:id = "@ + id/checkBox"
        android:textSize = "20sp" />
        …//省略"跳舞""运动"和"读书"三个<CheckBox>
    <Button
        android:text = "提交"
        android:layout_width = "wrap_content"
        android:layout_height = "wrap_content"
        android:id = "@ + id/button"
        android:layout_gravity = "center" />
</LinearLayout>
```

【代码 5-6】 activity_2.xml

```xml
<LinearLayout xmlns:android = "http://schemas.android.com/apk/res/android" …>
    <TextView
        android:text = "这是第二个 Activity"
        android:layout_width = "wrap_content"
        android:layout_height = "wrap_content"
        android:id = "@ + id/textView3"
        android:textSize = "24sp" />
    <LinearLayout
        android:orientation = "horizontal"
        android:layout_width = "match_parent"
        android:layout_height = "wrap_content">
        <TextView
            android:text = "您的爱好是："
            android:layout_width = "wrap_content"
            android:layout_height = "wrap_content"
            android:id = "@ + id/textView4"
            android:textSize = "18sp" />
        <TextView
            android:layout_width = "wrap_content"
            android:layout_height = "wrap_content"
            android:id = "@ + id/tx_aihao"
            android:layout_weight = "1"
            android:textSize = "18sp" />
    </LinearLayout>
    <Button
        android:text = "返回"
        android:layout_width = "wrap_content"
        android:layout_height = "wrap_content"
```

```xml
            android:id = "@ + id/back"
            android:layout_gravity = "center" />
</LinearLayout>
```

【代码 5-7】 Activity_1.java

```java
public class Activity_1 extends AppCompatActivity{
    private CheckBox checkBox,checkBox2,checkBox3,checkBox4;
    private List<CheckBox> checkBoxs = new ArrayList<CheckBox>();
    private Button button;
    private String content = "";
    @Override
    protected void onCreate( Bundle savedInstanceState) {
        super.onCreate(savedInstanceState);
        setContentView(R.layout.activity_1);
        checkBox = (CheckBox) findViewById(R.id.checkBox);
        checkBox2 = (CheckBox) findViewById(R.id.checkBox2);
        checkBox3 = (CheckBox) findViewById(R.id.checkBox3);
        checkBox4 = (CheckBox) findViewById(R.id.checkBox4);
        button = (Button) findViewById(R.id.button);
        //添加到集合中
        checkBoxs.add(checkBox);
        checkBoxs.add(checkBox2);
        checkBoxs.add(checkBox3);
        checkBoxs.add(checkBox4);
        button.setOnClickListener(new View.OnClickListener() {
            @Override
            public void onClick(View v) {
                getValues(v);
                Intent intent = new Intent(Activity_1.this,Activity_2.class);
                startActivity(intent);}});}
    public void getValues(View v) {
        for (CheckBox cbx : checkBoxs) {
            if (cbx.isChecked()) {
                content += cbx.getText() + " ";}}
        if ("".equals(content)) {
            content = "请您选择您的爱好";}}
}
```

【代码 5-8】 Activity_2.java

```java
public class Activity_2 extends AppCompatActivity{
    private TextView tx;
    private Button bt;
    @Override
    protected void onCreate(Bundle savedInstanceState) {
        super.onCreate(savedInstanceState);
        setContentView(R.layout.acticity_2);
        tx = (TextView) findViewById(R.id.tx_aihao);
        bt = (Button) findViewById(R.id.back);
        bt.setOnClickListener(new View.OnClickListener() {
            @Override
            void onClick(View v) {
```

```
                finish();}});}
}
```

上述代码中,单击 Activity_1 中的"提交"按钮,调用 startActivity(intent),实现从 Activity_1 到 Activity_2 的跳转,如图 5-4 所示。通过单击 Activity_2 中的"返回"按钮,调用 finish()方法关闭当前 Activity,并返回到上一个 Activity。

图 5-4 显式 Intent 启动 Activity

2. 隐式 Intent 启动 Activity

隐式 Intent 提供了一种机制,可以使匿名的应用程序组件响应动作请求。当系统启动一个可执行给定动作的 Activity 时,不需要指明所要启动的某个应用程序中具体的 Activity。例如,当用户在应用程序中拨打电话时,可以使用一个隐式的 Intent 来请求执行在电话号码(表示为一个 URI)上的动作(拨号),代码如下所示。

【示例】 隐式 Intent 启动 Activity

```
Intent intent = new Intent(Intent.ACTION_DIAL, Uri.parse("tel:555 - 2368"));
startActivity(intent);
```

上述代码中,Android 会解析这个 Intent,并启动一个与之匹配的新 Activity,该 Activity 提供了电话拨号的动作(如果是手机设备,通常都会带有电话应用程序)。

在构建一个隐式的 Intent 时,需要指定一个所要执行的动作;除此之外,还可以提供执行该动作所需的数据 URI。通过向 Intent 添加 Extra 的方式来向目标 Activity 发送额外的数据。

当使用 Intent 启动 Activity 时,Android 将其解析为在最适合的数据类型上执行所需动作的类,而不必提前指定由哪个应用程序提供此功能。

当多个 Activity 都能够执行指定的动作时,会向用户呈现各种选项供用户手动选择,本章后续小节将详细介绍,Intent 解析过程是通过 Intent Filter 来实现。

常见的使用 Intent 来启动内置应用程序的情况有以下 4 种。

1) 启动浏览器

在 Activity 启动内置浏览器时,需要创建一个使用 ACTION_VIEW Action、URI 为 URL 网址的 Intent 对象,代码如下所示。

第5章 Intent与BroadcastReceiver

【示例】 启动浏览器

```
Intent i = new Intent(Intent.ACTION_VIEW,Uri.parse("http://www.sohu.com"));
startActivity(i);
```

2）启动地图

启动内置 Google 地图时，也是使用 ACTION_VIEW Action，URI 为 GPS 坐标值，代码如下所示。

【示例】 启动 Google 地图

```
Intent i = new Intent(Intent.ACTION_VIEW,Uri.parse(
        "geo:25.04692437135412,121.5161783959678"));
startActivity(i);
```

3）打电话

启动拨号器程序时，使用 ACTION_DIAL Action，URI 为电话号码，代码如下所示。

【示例】 启动拨号程序进行打电话

```
Intent i = new Intent(Intent.ACTION_DIAL,Uri.parse("tel:+1234567"));
startActivity(i);
```

4）发送电子邮件

在 Activity 中可以启动内置电子邮件工具来发送邮件，使用 ACTION_SENDTO Action，URI 为收件者的电子邮件地址，代码如下所示。

【示例】 发送电子邮件

```
Intent i = new Intent(Intent.ACTION_SENDTO,Uri.parse("mailto:qst@163.com"));
startActivity(i);
```

在应用程序中可以使用第三方应用的 Activity 和 Service，但是无法确保用户设备上已经安装了特定的应用程序，因此在调用 startActivity() 之前应该通过解析来确定该应用程序是否存在。解析过程具体如下：调用 Intent 的 resolveActivity() 方法，并向该方法传入包管理器，通过对包管理器进行查询来确定是否有 Activity 启动并响应该 Intent，示例代码如下所示。

【示例】 使用 Intent 的 resolveActivity() 方法进行确认

```
//创建隐式 Intent 来启动新的 Activity
Intent intent = new Intent(Intent.ACTION_DIAL, Uri.parse("tel:555-2368"));
//检查这个 Activity 是否存在
PackageManager pm = getPackageManager();
ComponentName cn = intent.resolveActivity(pm);
if (cn == null) {
    //如果这个 Activity 不存在,则指向 the Google Play Store
    Uri marketUri = Uri.parse("market://search?q=pname:com.myapp.packagename");
    Intent marketIntent = new Intent(Intent.ACTION_VIEW).setData(marketUri);
    //如果在 Google Play Store 中有则下载,否则报错
    if (marketIntent.resolveActivity(pm) != null){
        startActivity(marketIntent);
```

```
    }else{
        Log.d(TAG, "Market client not available.");}
}else{
    startActivity(intent);}
```

如果没有找到 Activity,则可以选择禁用相关的功能或者引导用户下载安装合适的应用程序。

3. 传递数据给其他 Activity

通过 Intent 的 putExtra()或 putExtras()方法可以向目标 Activity 传递数据。其中,putExtras()方法用于向 Intent 中批量添加数据,此时通常先将数据批量添加到 Bundle 对象中,然后再调用 Intent 的 putExtras()方法直接传递该 Bundle 对象即可,示例代码如下所示。

【示例】 使用 putExtras()方法批量传递数据

```
Intent intent = new Intent();
Bundle bundle = new Bundle();         //该类用作携带数据
bundle.putString("name","中华文明");
bundle.putString("address","青岛");
intent.putExtras(bundle);
```

使用 putExtra()方法也可以向 Intent 中添加数据,但该方法需要将数据一个一个地添加到 Intent 中,示例代码如下。

【示例】 使用 putExtra()方法单个传递数据

```
Intent intent = new Intent();
intent.putExtra("name","中华文明");
```

下述代码对 Activity_1 和 Activity_2 进行调整,演示如何使用 Intent 在两个 Activity 之间传递数据。

【代码 5-9】 Activity_1.java

```
public class Activity_1 extends AppCompatActivity{
    …//省略
    @Override
    protected void onCreate(Bundle savedInstanceState) {
        super.onCreate(savedInstanceState);
        setContentView(R.layout.activity_1);
        …//省略
        button.setOnClickListener(new View.OnClickListener() {
            @Override
            public void onClick(View v) {
                int i = 0;
                //将选中的喜好放到 bundle 中
                for (CheckBox cbx : checkBoxs) {
                    if (cbx.isChecked()) {
                        bundle.putString("" + i, cbx.getText().toString());
                        i++;}}
```

```
        //喜好的个数也放到 bundle 中
        bundle.putInt("num",i);
        Intent intent = new Intent(Activity_1.this,Activity_2.class);
        intent.putExtras(bundle);
        startActivity(intent);});}
}
```

上述代码中,将被选中的爱好添加到 bundle 对象中,然后再使用 Intent 的 putExtras()方法直接将此 bundle 对象传递到 Activity_2 中。

【代码 5-10】 Activity_2.java

```
public class Activity_2 extends AppCompatActivity{
    private TextView tx;
    private Button bt;
    @Override
    protected void onCreate(Bundle savedInstanceState) {
        super.onCreate(savedInstanceState);
        setContentView(R.layout.acticity_2);tx = (TextView) findViewById(R.id.tx_aihao);
        Intent intent = getIntent();
        //先获取用户的喜好个数
        int num = intent.getIntExtra("num",0);
        String str = "";
        //遍历喜好的内容
        for (int i = 0;i < num;i++){
            str += intent.getStringExtra("" + i) + " ";
        }
        tx.setText(str);}      //显示爱好
}
```

上述代码中,通过 getIntent()方法获取 Activity_1 发送的数据,然后再运用 for 循环调用 getStringExtra()方法遍历获取 String 类型的信息,再显示在 Activity_2 界面中。

运行上述代码,结果如图 5-5 所示。

图 5-5　Activity 之间的跳转与数据传递

4. 从 Activity 返回数据

通过 startActivity()方法新启动的 Activity 与原 Activity 相互独立,在关闭时不会返回任何信息。当需要返回数据时,可以使用 startActivityForResult()方法启动一个 Activity,新启动的 Activity 在关闭时可以原 Activity 返回数据。与其他 Activity 一样,新启动的 Activity 也必须在 AndroidManifest.xml 文件中注册,被注册的任何 Activity 都可以用作目标 Activity,包括系统 Activity 或第三方应用程序 Activity。

当目标 Activity 结束时,会触发 Activity 的 onActivityResult()事件处理方法来返回结果。startActivityForResult()方法特别适用于从一个 Activity 向另一个 Activity 提供数据输入的情况,如登录、注册等功能。

1）启动一个目标 Activity

startActivityForResult()方法需要传入 Intent 参数,用于显式或隐式决定启动哪个 Activity,除此之外还需要传入一个请求码,用于唯一标识返回结果的目标 Activity。

下述代码用于显式启动一个目标 Activity,并设置相应的唯一标识,代码如下所示。

【示例】 显示启动 Activity

```
Intent intent = new Intent(this, MyOtherActivity.class);
startActivityForResult(intent, 1);
```

【示例】 隐式启动 Activity

下述代码是通过隐式 Intent 启动一个目标 Activity 来选取联系人,并设置相应的唯一标识。

```
Uri uri = Uri.parse("content://contacts/people");
Intent intent = new Intent(Intent.ACTION_PICK, uri);
startActivityForResult(intent, 2);
```

2）从目标 Activity 中返回数据

在目标 Activity 中调用 finish()方法之前,通过 setResult()方法向原 Activity 返回一个结果,设置传递到上一个界面的数据。setResult()方法是一个重载方法,其定义如下。

【语法】

```
public final void setResult(int resultCode)
public final void setResult(int resultCode, Intent data)
```

其中,参数 resultCode 用于设置目标 Activity 以何种方式返回,通常为 Activity.RESULT_OK 或者 Activity.RESULT_CANCELED。在某些环境下,当 OK 和 CANCELED 不足以精确描述返回结果时,用户可以使用自己的响应码(response code)来处理应用程序的特定选择,setResult()方法的 resultCode 参数支持使用其他任意的整数值。参数 data 是目标 Activity 所要返回的 Intent 数据载体。Intent 作为结果返回时,通常包含某段内容(如选择的联系人、电话号码或媒体文件等)的 URI 和用于返回的一组附加信息(Extra)。

下面演示在目标 Activity 的 onCreate()方法中,单击 OK 按钮或 Cancel 按钮返回不同的结果,代码如下所示。

【示例】 从目标 Activity 中返回结果

```
Button okButton = (Button) findViewById(R.id.ok_button);
okButton.setOnClickListener(new View.OnClickListener() {
    public void onClick(View view) {
        long selected_horse_id = listView.getSelectedItemId();
        Uri selectedHorse = Uri.parse("content://horses/" + selected_horse_id);
        Intent result = new Intent(Intent.ACTION_PICK, selectedHorse);
```

第5章 Intent与BroadcastReceiver

```
        setResult(Activity.RESULT_OK, result);
        finish();}});
Button cancelButton = (Button) findViewById(R.id.cancel_button);
cancelButton.setOnClickListener(new View.OnClickListener() {
    public void onClick(View view) {
        setResult(Activity.RESULT_CANCELED);
        finish();}});
```

当用户通过硬件返回键关闭 Activity，或者在调用 finish() 方法之前没有调用 setResult() 方法时，resultCode 将被设为 RESULT_CANCELED，且返回结果为 null。

3）处理从目标 Activity 返回的数据

当目标 Activity 关闭时，触发并调用 Activity 的 onActivityResult() 事件处理方法。通过重写 onActivityResult() 方法来处理从目标 Activity 返回的结果，该方法的语法格式如下。

【语法】

```
onActivityResult(int requestCode, int resultCode, Intent data)
```

requestCode 是在启动目标 Activity 时所使用的请求码。

resultCode 表示从目标 Activity 返回的状态码，该值可以是任何整数值，但通常使用 Activity.RESULT_OK 或者 Activity.RESULT_CANCELED。

data 是状态码对应的返回数据，根据目标 Activity 的不同，可能会包含代表选定内容的 URI。另外，目标 Activity 也可以通过 Intent 的 Extra 形式返回数据。

下面实例演示 Activity 的 onActivityResult 事件处理过程。

【代码 5-11】 OnActivityResultActivity.java

```java
public class OnActivityResultActivity extends AppCompatActivity {
    private Button button = null;
    private Button button1 = null;
    private TextView text = null;
    private static final int Mars = 0;
    private static final int Moon = 1;
    @Override
    protected void onCreate(Bundle savedInstanceState) {
        super.onCreate(savedInstanceState);
        setContentView(R.layout.onactivityresult_layout);
        text = (TextView) findViewById(R.id.txv1);
        button = (Button) findViewById(R.id.btn1);
        button.setOnClickListener(new View.OnClickListener() {
            @Override
            public void onClick(View v) {
                Intent intent = new Intent(OnActivityResultActivity.this,
                    MarsActivity.class);
                String content = "地球来的消息:我是来自地球上的Tom,火星的朋友你好.";
                intent.putExtra("FromEarth", content);
                startActivityForResult(intent, Mars);}});
        button1 = (Button) findViewById(R.id.btn2);
        button1.setOnClickListener(new View.OnClickListener() {
```

```java
        @Override
        public void onClick(View v) {
            Intent intent = new Intent(OnActivityResultActivity.this,
                    MoonActivity.class);
            String content = "地球来的消息:我是来自地球上的 Tom,月球的朋友你好.";
            intent.putExtra("FromEarth", content);
            startActivityForResult(intent, Moon);}});}
    @Override
    protected void onActivityResult(int requestCode, int resultCode,
            Intent data){
        switch (requestCode) {
        case Mars:
            Bundle MarsBuddle = data.getExtras();
            String MarsMessage = MarsBuddle.getString("FromMars");
            text.setText(MarsMessage);
            break;
        case Moon:
            Bundle MoonBuddle = data.getExtras();
            String MoonMessage = MoonBuddle.getString("FromMoon");
            text.setText(MoonMessage);
            break;}}
}
```

【代码 5-12】 MoonActivity.java

```java
public class MoonActivity extends AppCompatActivity{
    private Button button   = null;
    @Override
    public void onCreate(Bundle savedInstanceState) {
        super.onCreate(savedInstanceState);
        setContentView(R.layout.moon_layout);
        Intent EarthIntent = getIntent();
        String EarthMessage = EarthIntent.getStringExtra("FromEarth");
        button = (Button) findViewById(R.id.btn3);
        button.setOnClickListener(new View.OnClickListener() {
            @Override
            public void onClick(View v) {
                Intent intent = new Intent(MoonActivity.this,
                        OnActivityResultActivity.class);
                String passString = "月球来的消息:我是月球的 Lucy,非常欢迎你来月球";
                intent.putExtra("FromMoon", passString);
                setResult(RESULT_OK, intent);
                finish();}});
        TextView textView = (TextView) findViewById(R.id.txv2);
        textView.setText(EarthMessage);}
}
```

【代码 5-13】 MarsActivity.java

```java
public class MarsActivity extends AppCompatActivity{
    private Button button = null;
    @Override
    public void onCreate(Bundle savedInstanceState) {
```

```
        super.onCreate(savedInstanceState);
        setContentView(R.layout.mars_layout);
        Intent EarthIntent = getIntent();
        String EarthMessage = EarthIntent.getStringExtra("FromEarth");
        button = (Button) findViewById(R.id.btn4);
        button.setOnClickListener(new View.OnClickListener() {
            @Override
            public void onClick(View v) {
                Intent intent = new Intent(MarsActivity.this,
                    OnActivityResultActivity.class);
                String passString = "火星来的消息:我是火星Jack,非常高兴你能来火星";
                intent.putExtra("FromMars", passString);
                setResult(RESULT_OK, intent);
                finish();}});
        TextView textView = (TextView) findViewById(R.id.txv3);
        textView.setText(EarthMessage);}
}
```

运行上述代码后,结果如图 5-6 所示。

图 5-6　onActivityResult 基本运用

5.1.4　Intent Filter 过滤器

Intent Filter 表示意图的过滤器,用于描述指定的组件可以处理哪些意图。对于 Activity、Service 和 BroadcastReceiver,只有设置了 Intent Filter,才能被隐式 Intent 调用。当安装应用程序时,Android 系统会解析每个组件的 Intent Filter,从而确定这些组件可以处理哪些 Intent。当有 Intent 发生时,Android 根据 Intent Filter 的配置信息,从中找到可以处理该 Intent 的组件。

在 Intent Filter 中,可以包含 Intent 对象的 ACTION、DATA 和 CATEGORY 三个属性。隐式 Intent 必须通过以上三项测试才能传递到所匹配的组件中。当需要组件支持隐式 Intent 时,必须在 AndroidManifest.xml 中配置<intent-filter>元素。下面通过简单示例介绍<intent-filter>的用法。

【示例】　Action 测试

```xml
<intent-filter>
    <action android:name = "com.example.project.SHOW_CURRENT" />
    <action android:name = "com.example.project.SHOW_RECENT" />
    <action android:name = "com.example.project.SHOW_PENDING" />
    ...
</intent-filter>
```

上述代码中,一个 Intent 对象只能命名一个< action >,而一个 Intent 过滤器则可以包含多个< action >;一个 Intent 至少要匹配对应 Intent 过滤器中的一个< action >;当 Intent 对象或者过滤器没有指定< action >时,测试情况如下:

(1) 如果一个 Intent 过滤器没有指定任何< action >,则不会匹配任何 Intent,即所有的 Intent 都不会通过此测试。

(2) 当一个 Intent 对象没有指定任何< action >,而相应的过滤器中至少一个< action >时,将自动通过此测试。

当需要通过 **Category** 测试时,Intent 对象中包含的每个< category >必须匹配 Filter 中的一个。Intent Filter 可以列出额外的< category >,但是不能漏掉 Intent 对象包含的任意一个< category >。

【示例】 Category 测试

```
< intent - filter >
    < category android:name = "android.intent.category.DEFAULT" />
    < category android:name = "android.intent.category.BROWSABLE" />
    ...
</ intent - filter >
```

原则上,一个没有任何< category >的 Intent 总是通过此测试,但是,Android 对所有传入 startActivity()中的隐式 Intent,都认为至少包含了一个< category >,即 android.intent.category.DEFAULT。因此,当 Activity 接收隐式 Intent 时,必须包含 android.intent.category.DEFAULT。

与< action >和< category >相似,< data >也是 Intent Filter 中的子节点。在 Intent Filter 中,可以包含多个< data >节点,也可以没有< data >节点,代码如下所示。

【示例】 Data 测试

```
< intent - filter >
    < data android:mimeType = "video/mpeg" android:scheme = "http://" ... />
    < data android:mimeType = "audio/mpeg" android:scheme = "http://" ... />
    ...
</ intent - filter >
```

每个< data >元素可以指定 URI 和 data type(MIME media type)属性。URI 属性由 schema、host、port 和 path 组成,语法格式如下。

【语法】

```
schema://host:port/path
```

【示例】 URI

```
content://com.example.project:200/folder/subfolder/etc
```

在上述示例中,schema 为"content://";host 为 com.example.project;port 为 200;path 为 folder/subfolder/etc。

主机 host 和 port 一起组成了 URI 验证,如果没有指定 host,port 将被忽略。

< data >节点的属性均为可选,当使用 authority 时必须指定 scheme;当使用 path 时必

须指定 scheme、authority、host 和 port；当 Intent 对象中的 URI 和 Intent Filter 比较时，可以进行局部比较。例如，当 filter 只指定了 scheme 属性时，所包含该 scheme 的 URI 都会匹配；当 filter 指定了 scheme 和 authority 时，包含 scheme 和 authority 的元素将会进行匹配；当 filter 指定了 scheme、authority 和 path 时，只有同时包含 scheme、authority 和 path 的元素才会匹配。path 允许使用通配符进行匹配。

<data>节点的 type 属性用于指定 data 的 MIME 类型，允许使用"*"通配符作为子类型，例如"text/*"或"audio/*"等形式。

5.2 BroadcastReceiver

BroadcastReceiver 是广播接收器，用于接收系统和应用中的广播。在应用程序之间，广播是一种广泛运用的传输信息的机制。BroadcastReceiver 是一种对广播进行过滤接收并响应的组件，该组件本质上就是一个全局监听器，用于监听系统全局的广播消息。

BroadcastReceiver 自身并不提供用户图形界面，但是当收到某个通知时，BroadcastReceiver 可以通过启动 Activity 进行响应，或者通过 NotificationMananger 来提醒用户，也可以启动 Service 等。使用 BroadcastReceiver 可以非常方便地实现系统中不同组件之间的通信。

1．广播接收机制

在 Android 中有各种各样的广播，如电池的使用状态、电话的接收和短信的接收等都会产生一个广播，开发者也可以对广播进行监听并做出相应的逻辑处理。

在应用程序中，如果有一个 Intent 需要多个 Activity 进行处理，可以采用 BroadcastReceiver 将 Intent 广播到多个 Activity 中。由于 BroadcastReceiver 组件本质上就是一个全局监听器，其广播接收机制变相采用了事件处理机制，如图 5-7 所示。

图 5-7　BroadcastReceiver 机制

与事件处理机制类似，实现广播和接收 Intent 的步骤如下。

（1）**定义 BroadcastReceiver 广播接收器**：创建一个 BroadcastReceiver 的子类，并重写 onReceive()方法，该方法是广播接收处理方法，在接收到广播后进行相应的逻辑处理。

（2）**注册 BroadcastReceiver 广播接收器**：用于接收消息并对该消息进行响应。

（3）**发送广播**：该过程将消息内容和用于过滤的信息封装起来并进行广播。

(4) 执行：满足过滤条件的广播接收器接收广播信息并执行 onReceive()方法。

(5) 销毁：广播接收器不使用时将被销毁。

BroadcastReceiver 处理流程如图 5-8 所示。

BroadcastReceiver 广播接收器的生命周期相对比较短暂，只有 10s 左右，如果在 onReceive()方法中处理超过 10s 的事情，就会报错。每当接收到广播时，会重新创建一个 BroadcastReceiver 对象并调用 onReceive()方法，方法执行完后所创建的 BroadcastReceiver 对象就会被销毁。如果 onReceive()方法在 10s 内没有执行完毕，Android 则认为该程序无响应。因此，在 BroadcastReceiver 中不能处理耗时较长的操作，否则会弹出 ANR(Application No Response)的对话框。

当需要完成比较耗时的任务时，不能在 BroadcastReceiver 中使用子线程来完成处理，因为 BroadcastReceiver 的生命周期很短，子线程可能还没有结束，BroadcastReceiver 就先结束了。当 BroadcastReceiver 结束时，其所在进程就属于空进程，没有任何活动组件的进程，在系统需要内存时容易被优先杀死。如果 BroadcastReceiver 的宿主进程被杀死，那么正在工作的子线程也会一起被杀死，因此采用子线程来处理是不可靠的。所以对于耗时较长的任务，需要将 Intent 发送给 Service，由 Service 来完成相应的处理。

图 5-8　BroadcastReceiver 处理流程

2. 使用 BroadcastReceiver

下面以接收短信应用为例，演示 BroadcastReceiver 的使用，实现步骤如下。

(1) 定义一个 BroadcastReceiver 的子类，并重写 onReceive()方法，在接收到广播后进行相应的逻辑处理。

(2) 在 AndroidManifest.xml 文件中注册广播接收器对象，并指明触发 BroadcastReceiver 事件的条件。

(3) 在 AndroidManifest.xml 中添加接收和发送短信权限。

创建一个名为 SMSBroadcastReceiver 广播接收器，代码如下所示。

【代码 5-14】 SMSBroadcastReceiver.java

```java
public class SMSBroadcastReceiver extends BroadcastReceiver {
    private static MessageListener messageListener;
    public SMSBroadcastReceiver() {
        super();}
    @Override
    public void onReceive(Context context, Intent intent) {
        //用来获取短信内容
        Object [] pdus = (Object[]) intent.getExtras().get("pdus");
        for(Object pdu:pdus){
```

```
            SmsMessage smsMessage = SmsMessage.createFromPdu((byte[])pdu);
            String sender = smsMessage.getDisplayOriginatingAddress();
            String content = smsMessage.getMessageBody();
            long date = smsMessage.getTimestampMillis();
            Date timeDate = new Date(date);
            SimpleDateFormat simpleDateFormat
                    = new SimpleDateFormat("yyyy-MM-dd HH:mm:ss");
            String time = simpleDateFormat.format(timeDate);
            MessageListener.OnReceived(content);}}
    //回调接口
    public interface MessageListener {
        public void OnReceived(String message);}
    public void setOnReceivedMessageListener(MessageListener messageListener) {
        this.MessageListener = messageListener;}
}
```

然后，在 AndroidManifest.xml 文件中注册 SMSBroadcastReceiver 类，代码如下所示。

【代码 5-15】 在 AndroidManifest.xml 中注册 SMSBroadcastReceiver 类

```
<receiver android:name = ".activity.SMSBroadcastReceiver">
    <intent-filter android:priority = "1000" >
        <action android:name = "android.provider.Telephony.SMS_RECEIVED" />
    </intent-filter>
</receiver>
```

在 AndroidManifest.xml 中添加短信的收发权限，其代码如下所示。

【代码 5-16】 在 AndroidManifest.xml 中添加短信的收发权限

```
<uses-permission android:name = "android.permission.RECEIVE_SMS" />
<uses-permission android:name = "android.permission.SEND_SMS" /
```

运行上述代码，运行结果如图 5-9 所示。

图 5-9　接收短信内容并显示在程序界面

5.3 Handler 消息传递机制

出于性能优化的考虑,Android UI 操作并不是线程安全的,如果有多个线程并发操作 UI 组件,可能导致线程安全性问题。如果在一个 Activity 中有多个线程去更新 UI,并且没有使用锁机制,可能会导致界面混乱;如果使用锁机制,虽然可以避免该问题却会导致性能下降。因此,Android 中规定只允许 UI 线程修改 Activity 的 UI 组件。当程序第一次启动时,Android 会同时启动一个主线程(main thread),用于负责处理与 UI 相关的事件(如用户按钮事件等),并把事件分发到对应的组件进行处理后再绘制界面,此时主线程又称为 UI 线程。当在新启动的线程中更新 UI 组件时,需要借助 Handler 的消息传递机制来实现。

5.3.1 Handler 简介

在 Android 系统中,Handler 是一种用于在线程之间传递消息的机制。使用 Handler 可以在一个线程中发出消息,在另一个线程中接收消息并进行处理。Handler 类中包含发送、接收和处理消息的方法,其中用于接收消息的方法如表 5-3 所示。

表 5-3 Handler 常用方法

方　　法	功　能　描　述
handleMessage(Message msg)	通过重写该方法来处理消息
hasMessage(int what)	检查消息队列中是否包含 what 所指定的消息
hasMessage(int what,Object object)	检查队列中是否有指定的 what 和指定对象的消息
obtainMessage()	用于获取消息,具有多个重载方法
sendEmptyMessage(int what)	用于发送空消息
sendEmptyMessageDelayed(int what,long delayMillis)	用于在指定的时间之后发送空消息
sendMessage(Message msg)	立即发送消息
sendMessageDelayed(Message msg,long delayMillis)	用于在指定的时间之后发送空消息

下述代码通过一个简单的实例来演示 Handler 的用法,首页创建相应的 XML 布局文件,代码如下所示。

【代码 5-17】 main.xml

```
< LinearLayout xmlns:android = "http://schemas.android.com/apk/res/android" …>
    < ImageView
        android:layout_width = "match_parent"
        android:layout_height = "match_parent"
        android:id = "@ + id/show"
        android:layout_gravity = "center_horizontal"
        android:layout_marginTop = "30dp"
        android:layout_marginLeft = "10dp"
        android:layout_marginRight = "10dp"/>
</LinearLayout >
```

上述布局代码代码中,仅包含一个 ImageView 组件,用于显示 Handler 的处理结果。接下来创建 HandlerActivity,代码如下所示。

第5章 Intent与BroadcastReceiver

【代码5-18】 HandlerActivity.java

```java
public class HandlerActivity extends AppCompatActivity {
    //用于显示 ImageView
    ImageView show;
    //代表从网络下载得到的图片
    Bitmap bitmap;
    Handler handler = new Handler() {
        @Override
        public void handleMessage(Message msg) {
            if (msg.what == 0x123) {//判断该消息是哪个线程发送的
                if(bitmap == null){
                    show.setImageResource(R.drawable.pagefailed_bg);
                }else{
                    //使用 ImageView 显示该图片
                    show.setImageBitmap(bitmap);}}}};
    @Override
    public void onCreate(Bundle savedInstanceState) {
        super.onCreate(savedInstanceState);
        setContentView(R.layout.main);
        show = (ImageView) findViewById(R.id.show);
        new Thread() {
            public void run() {
                try {
                    //定义一个 URL 对象
                    URL url = new URL("http://www.itshixun.com/logo.png");
                    //打开该 URL 对应的资源的输入流
                    InputStream is = url.openStream();
                    //从 InputStream 中解析出图片
                    bitmap = BitmapFactory.decodeStream(is);
                    //发送消息、通知 UI 组件显示该图片
                    handler.sendEmptyMessage(0x123);
                    is.close();
                } catch (Exception e) {
                    String msg = e.getMessage();
                    Log.d("HandlerActivity", msg);
                    //出错也需要进行通知
                    handler.sendEmptyMessage(0x123);}}
        }.start();}
}
```

上述代码中,通过子线程将网络图片下载之后,向 UI 主线程发送一个带有线程标识(读者可自行定义,此处使用 0x123 进行标识)的空消息,然后触发主线程中 handleMessage() 事件处理方法来更新 UI 界面,将图片进行显示。

由于应用程序需要访问网络,所以还需要在 AndroidManifest.xml 文件中加入访问网络的权限,代码如下所示。

【代码5-19】 在 AndroidManifest.xml 中添加访问网络的权限

```
<uses-permission android:name="android.permission.INTERNET"/>
```

运行 HandlerActivity,结果如图 5-10 所示。

· 273 ·

图 5-10　Handler 使用

5.3.2　Handler 的工作机制

上面的实例演示了一个简单的 Handler 的工作过程，其中 Handler 是在主线程中定义的，如果 Handler 在子线程中定义则需更深入地理解其工作原理。

在开发过程中，应尽量避免在 UI 线程中执行耗时操作，否则会导致应用程序长时间无响应。这时可以将主线程中的数据传递给子线程，通过子线程协助完成一些计算量比较大的任务。此种情况下，主线程需要向子线程发送消息，然后在子线程中进行消息处理，因此 Handler 需要定义在子线程中，接收消息并完成相应的消息处理。

下面介绍配合 Handler 工作的其他组件：android.os.Message，用于封装线程之间传递的消息；android.os.MessageQueue，是消息队列，用于负责接收并处理 Handler 发送过来的消息；android.os.Looper，每个线程对应一个 Looper，负责消息队列的管理，将消息从队列中取出交给 Handler 进行处理。

Handler 工作流程如图 5-11 所示。

图 5-11　Handler 工作流程

在创建 Handler 之前，需要先创建 Looper，创建 Looper 的同时会自动创建一个消息队列 MessageQueue。Handler、Looper 和 Message 三者共同实现 Android 系统多线程之间的异步消息处理。

所谓的异步消息处理，是指线程启动后会进入一个无限的循环之中，每循环一次都从消息队列中取出一个消息，然后回调相应的消息处理函数，执行完成一个消息后继续下一次循环。当消息队列为空时，线程则会阻塞等待。其中，Looper 主要负责来创建一个 MessageQueue 队列，然后通过循环体不断地从 MessageQueue 队列中读取消息，而消息的创建者就是一个或多个 Handler。

Looper 类中主要包含 prepare()和 loop()两个静态方法。

- **Looper.prepare()**：在线程中保存一个 Looper 实例，其中保存一个 MessageQueue 对象。Looper.prepare()方法在每个线程中只能调用一次，否则会抛出异常，因此在一个线程中只会存在一个 MessageQueue。
- **Looper.loop()**：当前线程通过无限循环的方式不断从 MessageQueue 队列中读取消息，然后回调 Message.target.dispatchMessage(msg)方法将消息分配给 Handler 对象并进行处理时，其中 Message 的 target 属性即为所关联的 Handler 对象。

在调用 Handler 的构造方法时，需要先获得当前线程中保存的 Looper 实例，进而与 Looper 实例中的 MessageQueue 相关联。通过 Handler 的 sendMessage()方法将 Handler 自身赋给 Message 对象的 target 属性，然后加入 MessageQueue 队列中。

在构造 Handler 实例时，通常会重写 handleMessage()方法，即 Looper.loop()循环中调用 Message.target.dispatchMessage(msg)时最终调用的方法。

5.4 AsyncTask 类

Android 的 Handler 机制为多线程异步消息处理提供了一种完善的处理方式，但是在较为简单的应用下，使用 Handler 会使代码过于烦琐。为了简化操作，Android 1.5 提供了 android.os.AsyncTask 工具类，使得异步任务的处理变得更加简单，不再需要编写任务线程和 Handler 实例也可以完成相同的任务。定义 AsyncTask 的语法格式如下。

【语法】

```
public abstract class AsyncTask < Params, Progress, Result >
```

其中，Params 是启动任务执行的输入参数；Progress 是后台任务执行的进度；Result 是后台计算结果的类型。

在特定场合下，并不是所有的类型都是必需的。如果没有使用某个参数，可以用 java.lang.Void 类型代替。

在执行异步任务时，通常会涉及以下几个步骤。

（1）execute(Params...params)：用于执行一个异步任务，需要在业务代码中调用该方法来触发异步任务的执行。

（2）onPreExecute()：在 execute()方法被调用后立即执行，在后台任务执行之前对 UI 做一些标记。

（3）doInBackground(Params...params)：在 onPreExecute()完成后立即执行，用于执行较为费时的操作，此方法将接收输入参数并返回计算结果。在执行过程中可以调用 publishProgress()来更新进度信息。

（4）onProgressUpdate(Progress...values)：在调用 publishProgress()方法时，自动执行 onProgressUpdate()方法将进度信息直接更新到 UI 组件上。

（5）onPostExecute(Result result)：当后台操作结束时调用该方法，并将计算结果作为输入参数传递到方法中，然后将结果显示到 UI 组件上。

在使用 AsyncTask 工具类时，需要特别注意以下几点。
- 异步任务的实例必须在 UI 线程中创建。
- execute(Params... params)方法必须在 UI 线程中调用。
- 不能手动调用 onPreExecute()、doInBackground()、onProgressUpdate(Progress... values)和 onPostExecute(Result result)等方法。
- 不能在 doInBackground(Params...params)方法中更改 UI 组件的信息。
- 每个任务实例只能执行一次，当执行第二次时将会抛出异常。

下述代码通过一个简单示例来演示使用 AsyncTask 实现异步任务操作，代码如下所示。

【代码 5-20】 async_layout.xml

```xml
<LinearLayout xmlns:android="http://schemas.android.com/apk/res/android" ...>
    <Button
        android:id="@+id/download"
        android:layout_width="fill_parent"
        android:layout_height="wrap_content"
        android:text="Download"/>
    <TextView
        android:id="@+id/tv"
        android:layout_width="fill_parent"
        android:layout_height="wrap_content"
        android:text="当前进度显示"/>
    <ProgressBar
        android:id="@+id/pb"
        android:layout_width="fill_parent"
        android:layout_height="wrap_content"
        style="?android:attr/progressBarStyleHorizontal"/>
</LinearLayout>
```

【代码 5-21】 AsyncTaskActivity.java

```java
public class AsyncTaskActivity extends AppCompatActivity{
    Button download;
    ProgressBar pb;
    TextView tv;
    @Override
    public void onCreate(Bundle savedInstanceState) {
        super.onCreate(savedInstanceState);
        setContentView(R.layout.async_layout);
        pb = (ProgressBar)findViewById(R.id.pb);
        tv = (TextView)findViewById(R.id.tv);
        download = (Button)findViewById(R.id.download);
        download.setOnClickListener(new View.OnClickListener() {
            @Override
            public void onClick(View v) {
                DownloadTask dTask = new DownloadTask();
                dTask.execute(100);});}
    /** 自定义 AsyncTask 子类 */
    private class DownloadTask extends AsyncTask<Integer, Integer, String> {
        @Override
```

```
        protected void onPreExecute() {
            super.onPreExecute();}        //第一个执行方法
    /** doInBackground 方法内部执行后台任务,不可在此方法内修改 UI */
    @Override
        protected String doInBackground(Integer... params) {
            //第二个执行方法,onPreExecute()执行完后执行
            for(int i = 0;i <= 100;i++){
                publishProgress(i);
                try {
                    Thread.sleep(params[0]);
                } catch (InterruptedException e) {
                    e.printStackTrace();}}
            return "执行完毕";}
    /** onProgressUpdate 方法用于更新进度信息 */
    @Override
        protected void onProgressUpdate(Integer... progress) {
            //更新进度条及进度
            pb.setProgress(progress[0]);
            tv.setText(progress[0] + "%");
            super.onProgressUpdate(progress);}
    /** onPostExecute 方法用于在执行完后台任务后更新 UI,显示结果 */
    @Override
        protected void onPostExecute(String result) {
            setTitle(result);       //更新 App 的标题
            super.onPostExecute(result);}}
}
```

上述代码中,自定义了一个 AsyncTask 类的 DownloadTask 子类,该类用于演示 AsyncTask 的几个重要方法的调用顺序,例如 onPreExecute()、doInBackground()等方法。其中,onPreExecute()方法用于在执行后台任务前进行一些 UI 操作,例如初始化 textView 文本等;doInBackground()方法用于处理一些比较耗时的操作,例如下载网络资源等,在该方法中不能做任何有关 UI 的修改,否则会让页面卡死;onProgressUpdate()方法用于更新进度信息;onPostExecute 方法用于在执行完后台任务后更新 UI,显示结果。

运行上述代码,效果如图 5-12 所示。单击 DOWNLOAD 按钮后,界面如图 5-13 所示。当页面完全加载完成后,标题更换为"执行完毕",界面如图 5-14 所示。

    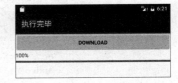

图 5-12 初始界面   图 5-13 进度下载   图 5-14 视图显示

5.5 贯穿任务实现

"GIFT-EMS 礼记"项目为用户提供了日程提醒功能。用户可以为指定的日期添加日程,当到达日期时 App 会以铃声或震动方式提醒用户。App 中支持每个日期添加多个日

程,每个日程添加多个提醒时间。

5.5.1 实现【任务 5-1】

本任务完成用户日程界面,此界面中会显示一个日历,用户可以查看某个日期下的日程记录,还可以为某个日期添加日程记录。编写用户日程界面布局文件,代码如下所示。

【任务 5-1】 user_calendar.xml

```xml
<RelativeLayout xmlns:android="http://schemas.android.com/apk/res/android" …>
    <include
        android:id="@+id/titleBarRelativeLayout"
        layout="@layout/title_bar" />
    <RelativeLayout
        android:id="@+id/bottomRelativeLayout"
        android:layout_width="match_parent"
        android:layout_height="80dp"
        android:layout_alignParentBottom="true"
        android:background="@color/title_bottom"
        android:gravity="center" >
        <Button
            android:id="@+id/addButton"
            style="@style/button"
            android:layout_width="100dp"
            android:layout_height="40dp"
            android:text="添加日程" />
    </RelativeLayout>
    <LinearLayout
        android:layout_width="match_parent"
        android:layout_height="wrap_content"
        android:layout_above="@id/bottomRelativeLayout"
        android:layout_below="@id/titleBarRelativeLayout"
        android:orientation="vertical" >
        <RelativeLayout
            android:layout_width="match_parent"
            android:layout_height="50dp"
            android:layout_marginBottom="10dp"
            android:layout_marginTop="10dp"
            android:background="#EEEEEE" >
            <RelativeLayout
                android:id="@+id/preMonth"
                android:layout_width="70dp"
                android:layout_height="55dp"
                android:layout_centerVertical="true"
                android:layout_toLeftOf="@+id/yearMonthTextView" >
                <TextView
                    android:layout_width="9dp"
                    android:layout_height="14dp"
                    android:layout_alignParentRight="true"
                    android:layout_centerVertical="true"
                    android:layout_marginRight="20dp"
                    android:background="@drawable/bt_calendar_last" />
            </RelativeLayout>
```

```xml
<TextView
    android:id = "@ + id/yearMonthTextView"
    android:layout_width = "110dp"
    android:layout_height = "wrap_content"
    android:layout_centerInParent = "true"
    android:gravity = "center"
    android:textColor = "#aa564b4b"
    android:textSize = "18sp" />
<RelativeLayout
    android:id = "@ + id/nextMonth"
    android:layout_width = "70dp"
    android:layout_height = "55dp"
    android:layout_centerVertical = "true"
    android:layout_toRightOf = "@ + id/yearMonthTextView" >
    <TextView
        android:layout_width = "9dp"
        android:layout_height = "14dp"
        android:layout_alignParentLeft = "true"
        android:layout_centerVertical = "true"
        android:layout_marginLeft = "20dp"
        android:background = "@drawable/bt_calendar_next" />
</RelativeLayout>
</RelativeLayout>
<com.qst.giftems.activity.KCalendar
    android:id = "@ + id/calendarView"
    android:layout_width = "match_parent"
    android:layout_height = "250dp" />
<ListView
    android:id = "@ + id/eventsListView"
    android:layout_width = "match_parent"
    android:layout_height = "wrap_content" />
</LinearLayout>
<include
    android:id = "@ + id/cornerMenuRelativeLayout"
    layout = "@layout/corner_menu" />
</RelativeLayout>
```

上述布局中，主要包含一个日历控件和一个 ListView 列表，日历控件中显示一个月份的日期，单击某个日期时会在 ListView 中显示此日期下的日程。日历控件使用的是一个开源控件 KCalendar，其代码比较复杂，本书不再详细介绍。在日程 ListView 中，每个日程条目的布局如下所示。

【任务 5-1】 user_calendar_item.xml

```xml
<LinearLayout xmlns:android = "http://schemas.android.com/apk/res/android" …
    style = "@style/item"
    android:gravity = "center_vertical"
    android:orientation = "horizontal"
    android:padding = "10dp" >
    <TextView
        android:id = "@ + id/timeTextView"
        style = "@style/text3"
```

```xml
            android:layout_width = "wrap_content"
            android:layout_height = "wrap_content" />
    < TextView
            android:id = "@ + id/titleTextView"
            style = "@style/text1"
            android:layout_width = "0dp"
            android:layout_height = "wrap_content"
            android:layout_marginLeft = "10dp"
            android:layout_weight = "1"
            android:ellipsize = "end"
            android:gravity = "left"
            android:singleLine = "true" />
    < ImageView
            android:id = "@ + id/deleteImageView"
            android:layout_width = "20dp"
            android:layout_height = "20dp"
            android:layout_marginLeft = "10dp"
            android:src = "@drawable/close" />
</LinearLayout >
```

编写日程列表对应的 Activity,代码如下所示。

【任务 5-1】 UserCalendarActivity.java

```java
public class UserCalendarActivity extends BaseActivity {
    static final SimpleDateFormat DATE_FORMAT = new SimpleDateFormat(
            "yyyy - MM - dd", Locale.US);
    static final SimpleDateFormat DATETIME_FORMAT = new SimpleDateFormat(
            "yyyy - MM - dd HH:mm", Locale.US);
    static final SimpleDateFormat TIME_FORMAT = new SimpleDateFormat("HH:mm",
            Locale.US);
    ArrayList < UserCalendar > userCalendars = new ArrayList < UserCalendar >();
    ArrayList < UserCalendar > userCalendarsOfSelectedDate
            = new ArrayList < UserCalendar >();
    Calendar selectedDate = Calendar.getInstance();
    KCalendar calendarView;
    ListView eventsListView;
    Button addButton;
    EventsAdapter eventsAdapter = new EventsAdapter();
    int currentUserCalendarId;
    public UserCalendarActivity() {
        super(R.layout.user_calendar, "日程记录"); }
    @Override
    protected void onCreate(Bundle savedInstanceState) {
        super.onCreate(savedInstanceState);
        calendarView = (KCalendar) findViewById(R.id.calendarView);
        eventsListView = (ListView) findViewById(R.id.eventsListView);
        addButton = (Button) findViewById(R.id.addButton);
        //calendar
        final TextView yearMonthTextView
                = (TextView) findViewById(R.id.yearMonthTextView);
        yearMonthTextView.setText(calendarView.getCalendarYear() + "年"
                + calendarView.getCalendarMonth() + "月");
```

```java
        int year = selectedDate.get(Calendar.YEAR);
        int month = selectedDate.get(Calendar.MARCH) + 1;
        yearMonthTextView.setText(year + "年" + month + "月");
        calendarView.showCalendar(year, month);
        calendarView.setCalendarDayBgColor(selectedDate.getTime(),
                R.drawable.calendar_date_focused);
        calendarView.setOnCalendarClickListener(new OnCalendarClickListener() {
            public void onCalendarClick(int row, int col, String date) {
                int y = selectedDate.get(Calendar.YEAR);
                int m = selectedDate.get(Calendar.MONTH) + 1;
                int d = selectedDate.get(Calendar.DATE);
                try {
                    selectedDate.setTime(DATE_FORMAT.parse(date));
                } catch (ParseException e) {
                    e.printStackTrace();
                    return; }
                int year = selectedDate.get(Calendar.YEAR);
                int month = selectedDate.get(Calendar.MONTH) + 1;
                int day = selectedDate.get(Calendar.DATE);
                if (calendarView.getCalendarMonth() - month == 1
                        || calendarView.getCalendarMonth() - month == -11) {
                    calendarView.lastMonth();
                } else if (month - calendarView.getCalendarMonth() == 1
                        || month - calendarView.getCalendarMonth() == -11) {
                    calendarView.nextMonth();
                } else {
                    calendarView.removeAllBgColor();
                    calendarView.setCalendarDayBgColor(date,
                            R.drawable.calendar_date_focused); }
                showUserCalendarsOfSelectedDate();}});
        calendarView
                .setOnCalendarDateChangedListener(
                    new OnCalendarDateChangedListener() {
                        public void onCalendarDateChanged(int year, int month) {
                            yearMonthTextView.setText(year + "年" + month + "月");}});
        RelativeLayout preMonth = (RelativeLayout) findViewById(R.id.preMonth);
        preMonth.setOnClickListener(new OnClickListener() {
            public void onClick(View v) {
                calendarView.lastMonth();}});
        RelativeLayout nextMonth = (RelativeLayout) findViewById(R.id.nextMonth);
        nextMonth.setOnClickListener(new OnClickListener() {
            public void onClick(View v) {
                calendarView.nextMonth();}});
        eventsListView.setAdapter(eventsAdapter);
        addButton.setOnClickListener(new OnClickListener() {
            @Override
            public void onClick(View v) {
                if (!checkLogin())
                    return;
                Intent intent = new Intent(getApplicationContext(),
                        UserCalendarEventActivity.class);
                intent.putExtra("actionTime",
                        DATETIME_FORMAT.format(selectedDate.getTime()));
```

```java
            startActivity(intent); }});
    //TODO 后续章节改为从服务器获取数据
    SimpleDateFormat sdf = new SimpleDateFormat("yyyy-MM-dd HH:mm:ss",
            Locale.getDefault());
    Date now = new Date();
    int u = 1;
    for (int i = 0; i < 6; i++) {
        //模拟一个随机日期
        Date randomDate = new Date(now.getTime()
                + (int) ((Math.random() - 0.5) * 100 * 24 * 60 * 60 * 1000));
        //此日期下随机创建 1~4 个日程
        int random = (int) (Math.random() / 0.3) + 1;
        for (int j = 0; j < random; j++) {
            UserCalendar uc = new UserCalendar();
            uc.id = u++;
            uc.userId = 1;
            uc.title = "日程" + uc.id;
            uc.actionTime = sdf.format(randomDate);
            userCalendars.add(uc); }}
    ArrayList<String> clist = new ArrayList<String>();
    for (UserCalendar uc : userCalendars) {
        clist.add(uc.actionTime.substring(0, 10));}
    calendarView.addMarks(clist, 0);
    showUserCalendarsOfSelectedDate();}
void showUserCalendarsOfSelectedDate() {
    int year = selectedDate.get(Calendar.YEAR);
    int month = selectedDate.get(Calendar.MONTH) + 1;
    int day = selectedDate.get(Calendar.DATE);
    userCalendarsOfSelectedDate.clear();
    Calendar temp = Calendar.getInstance();
    for (UserCalendar uc : userCalendars) {
        try {
            temp.setTime(DATETIME_FORMAT.parse(uc.actionTime));
            if (temp.get(Calendar.YEAR) == year
                    && temp.get(Calendar.MONTH) + 1 == month
                    && temp.get(Calendar.DATE) == day)
                userCalendarsOfSelectedDate.add(uc);
        } catch (ParseException e) {
            e.printStackTrace();}}
    eventsAdapter.notifyDataSetChanged();}
class EventsAdapter extends BaseAdapter {
    public int getCount() {
        return userCalendarsOfSelectedDate.size();}
    @Override
    public boolean areAllItemsEnabled() {
        return false;}
    public Object getItem(int position) {
        return position;}
    public long getItemId(int position) {
        return position;}
    public View getView(int position, View convertView, ViewGroup parent) {
        final Context context = getApplicationContext();
        final UserCalendar uc = userCalendarsOfSelectedDate.get(position);
```

```java
            if (uc == null)
                return null;
        if (convertView == null) {
            convertView = LayoutInflater.from(context).inflate(
                    R.layout.user_calendar_item, null);
            convertView.setOnClickListener(userCalendarViewOCL);}
        convertView.setTag(uc);
        TextView timeTextView = (TextView) convertView
                .findViewById(R.id.timeTextView);
        TextView titleTextView = (TextView) convertView
                .findViewById(R.id.titleTextView);
        ImageView deleteImageView = (ImageView) convertView
                .findViewById(R.id.deleteImageView);
        timeTextView.setText(uc.actionTime.substring(11));
        titleTextView.setText(uc.title);
        deleteImageView.setTag(uc);
        deleteImageView.setOnClickListener(deleteImageViewOCL);
        return convertView;}
OnClickListener deleteImageViewOCL = new OnClickListener() {
    @Override
    public void onClick(final View v) {
        if (!checkLogin())
            return;
        Dialogs.showSimpleDialog(UserCalendarActivity.this,
                "确定删除这条记录吗?", true, new OnOkListener() {
                    @Override
                    public void onOk() {
                        UserCalendar uc = (UserCalendar) v.getTag();
                        if (uc == null)
                            return;
                        currentUserCalendarId = uc.id;
                        Iterator<UserCalendar> it1
                                = userCalendarsOfSelectedDate
                                    .iterator();
                        while (it1.hasNext()) {
                            UserCalendar u = it1.next();
                            if (u.id == currentUserCalendarId) {
                                it1.remove();
                                break;}}
                        eventsAdapter.notifyDataSetChanged();
                        Iterator<UserCalendar> it2 = userCalendars
                                .iterator();
                        while (it2.hasNext()) {
                            UserCalendar u = it2.next();
                            if (u.id == currentUserCalendarId) {
                                it2.remove();
                                break;}}
                        calendarView.removeAllMarks();
                        ArrayList<String> clist = new ArrayList<String>();
                        for (UserCalendar u : userCalendars) {
                            clist.add(u.actionTime.substring(0, 10));}
                        calendarView.addMarks(clist, 0);
                        //TODO 后续章节加上保存到服务器的功能
```

```
                        }});}};
        OnClickListener userCalendarViewOCL = new OnClickListener() {
            @Override
            public void onClick(View v) {
                if (!checkLogin())
                    return;
                UserCalendar uc = (UserCalendar) v.getTag();
                if (uc == null)
                    return;
                //跳转到日程编辑界面
                Intent intent = new Intent(getApplicationContext(),
                    UserCalendarEventActivity.class);
                intent.putExtra("userCalendarId", uc.id);
                intent.putExtra("title", uc.title);
                intent.putExtra("actionTime", uc.actionTime);
                intent.putExtra("remindTime1", uc.remindTime1);
                intent.putExtra("remindTime2", uc.remindTime2);
                intent.putExtra("remindTime3", uc.remindTime3);
                intent.putExtra("remindTime4", uc.remindTime4);
                intent.putExtra("remindTime5", uc.remindTime5);
                startActivity(intent); }};}
}
```

上述代码中,onCreate()方法模拟生成多条随机日期的日程记录,在日历控件的单击事件中查找对应的日程并显示在下部的 ListView 中。

运行项目,通过右下角菜单进入日程界面,单击某个日期后,底部会显示此日期对应的日程,单击日程右侧的删除图标可以删除该日程,运行结果如图 5-15 所示。

图 5-15 "日程记录"界面

5.5.2 实现【任务 5-2】

本任务完成用户日程编辑界面,首先编写界面布局文件,代码如下所示。

第5章 Intent与BroadcastReceiver

【任务5-2】 user_calendar_event.xml

```xml
<RelativeLayout xmlns:android="http://schemas.android.com/apk/res/android" ...
    android:background="@drawable/background_body"
    android:focusable="true"
    android:focusableInTouchMode="true"
    android:gravity="center">
    <include
        android:id="@+id/titleBarRelativeLayout"
        layout="@layout/title_bar" />
    <RelativeLayout
        android:id="@+id/bottomRelativeLayout"
        android:layout_width="match_parent"
        android:layout_height="80dp"
        android:layout_alignParentBottom="true"
        android:background="@color/title_bottom"
        android:gravity="center">
        <Button
            android:id="@+id/saveButton"
            style="@style/button"
            android:layout_width="100dp"
            android:layout_height="40dp"
            android:text="保存" />
    </RelativeLayout>
    <ScrollView
        android:layout_width="match_parent"
        android:layout_height="match_parent"
        android:layout_below="@id/titleBarRelativeLayout"
        android:layout_above="@id/bottomRelativeLayout">
        <LinearLayout
            android:layout_width="match_parent"
            android:layout_height="wrap_content"
            android:orientation="vertical"
            android:padding="10dp">
            <EditText
                android:id="@+id/titleEditText"
                style="@style/text1"
                android:layout_width="match_parent"
                android:layout_height="80dp"
                android:background="@drawable/background_edit"
                android:gravity="left"
                android:hint="请输入事件内容"
                android:padding="10dp" />
            <LinearLayout
                android:layout_width="match_parent"
                android:layout_height="wrap_content"
                android:layout_marginTop="10dp"
                android:gravity="center_vertical"
                android:orientation="horizontal">
                <TextView
                    style="@style/text1"
                    android:layout_width="wrap_content"
                    android:layout_height="wrap_content"
```

```xml
            android:text = "事件时间: " />
        <Button
            android:id = "@ + id/actionTimeButton"
            style = "@style/button"
            android:layout_width = "match_parent"
            android:layout_height = "40dp"
            android:text = "9999 - 99 - 99 99:99"
            android:textColor = "@color/selected"
            android:textSize = "16sp" />
</LinearLayout>
<View
    android:layout_width = "match_parent"
    android:layout_height = "1dp"
    android:layout_marginTop = "10dp"
    android:background = "@color/title_bottom" />
<LinearLayout
    android:id = "@ + id/remind1"
    android:layout_width = "match_parent"
    android:layout_height = "wrap_content"
    android:layout_marginTop = "10dp"
    android:gravity = "center_vertical"
    android:orientation = "horizontal" >
    <Spinner
        android:id = "@ + id/remindTime1Spinner"
        android:layout_width = "0dp"
        android:layout_height = "wrap_content"
        android:layout_weight = "1" />
    <ImageView
        android:id = "@ + id/deleteRemind1ImageView"
        android:layout_width = "20dp"
        android:layout_height = "20dp"
        android:layout_marginLeft = "10dp"
        android:src = "@drawable/close" />
</LinearLayout>
<LinearLayout
    android:id = "@ + id/remind2"
    android:layout_width = "match_parent"
    android:layout_height = "wrap_content"
    android:gravity = "center_vertical"
    android:orientation = "horizontal" >
    <Spinner
        android:id = "@ + id/remindTime2Spinner"
        android:layout_width = "0dp"
        android:layout_height = "wrap_content"
        android:layout_weight = "1" />
    <ImageView
        android:id = "@ + id/deleteRemind2ImageView"
        android:layout_width = "20dp"
        android:layout_height = "20dp"
        android:layout_marginLeft = "10dp"
        android:src = "@drawable/close" />
</LinearLayout>
<LinearLayout
```

```xml
        android:id = "@ + id/remind3"
        android:layout_width = "match_parent"
        android:layout_height = "wrap_content"
        android:gravity = "center_vertical"
        android:orientation = "horizontal" >
        <Spinner
            android:id = "@ + id/remindTime3Spinner"
            android:layout_width = "0dp"
            android:layout_height = "wrap_content"
            android:layout_weight = "1" />
        <ImageView
            android:id = "@ + id/deleteRemind3ImageView"
            android:layout_width = "20dp"
            android:layout_height = "20dp"
            android:layout_marginLeft = "10dp"
            android:src = "@drawable/close" />
    </LinearLayout>
    <LinearLayout
        android:id = "@ + id/remind4"
        android:layout_width = "match_parent"
        android:layout_height = "wrap_content"
        android:gravity = "center_vertical"
        android:orientation = "horizontal" >
        <Spinner
            android:id = "@ + id/remindTime4Spinner"
            android:layout_width = "0dp"
            android:layout_height = "wrap_content"
            android:layout_weight = "1" />
        <ImageView
            android:id = "@ + id/deleteRemind4ImageView"
            android:layout_width = "20dp"
            android:layout_height = "20dp"
            android:layout_marginLeft = "10dp"
            android:src = "@drawable/close" />
    </LinearLayout>
    <LinearLayout
        android:id = "@ + id/remind5"
        android:layout_width = "match_parent"
        android:layout_height = "wrap_content"
        android:gravity = "center_vertical"
        android:orientation = "horizontal" >
        <Spinner
            android:id = "@ + id/remindTime5Spinner"
            android:layout_width = "0dp"
            android:layout_height = "wrap_content"
            android:layout_weight = "1" />
        <ImageView
            android:id = "@ + id/deleteRemind5ImageView"
            android:layout_width = "20dp"
            android:layout_height = "20dp"
            android:layout_gravity = "center_vertical"
            android:layout_marginLeft = "10dp"
            android:src = "@drawable/close" />
```

```xml
            </LinearLayout>
            <Button
                android:id = "@ + id/addRemindButton"
                style = "@style/button"
                android:layout_width = "100dp"
                android:layout_height = "40dp"
                android:layout_gravity = "right"
                android:layout_marginTop = "10dp"
                android:text = "添加提醒" />
        </LinearLayout>
    </ScrollView>
    <include
        android:id = "@ + id/cornerMenuRelativeLayout"
        layout = "@layout/corner_menu" />
</RelativeLayout>
```

上述布局中主要包括三部分：上部为日程事件的内容输入框，中间为日程的日期选择，下部为日程的提醒列表。一个日程事件中最多允许添加 5 个提醒，每个提醒可以指定提醒的时间。接下来编写用户日程编辑界面对应的 Activity，代码如下所示。

【任务 5-2】 UserCalendarEventActivity.java

```java
public class UserCalendarEventActivity extends BaseActivity {
    UserCalendar userCalendar = new UserCalendar();
    EditText titleEditText;
    Button actionTimeButton;
    LinearLayout remind1, remind2, remind3, remind4, remind5;
    Spinner remindTime1Spinner, remindTime2Spinner, remindTime3Spinner,
            remindTime4Spinner, remindTime5Spinner;
    ImageView deleteRemind1ImageView, deleteRemind2ImageView,
            deleteRemind3ImageView, deleteRemind4ImageView,
            deleteRemind5ImageView;
    Button addRemindButton;
    Button saveButton;
    public UserCalendarEventActivity() {
        super(R.layout.user_calendar_event, "用户日程");}
    @Override
    protected void onCreate(Bundle savedInstanceState) {
        super.onCreate(savedInstanceState);
        userCalendar.id = getIntent().getIntExtra("userCalendarId", 0);
        userCalendar.actionTime = getIntent().getStringExtra("actionTime");
        if (userCalendar.id == 0) {
            titleTextView.setText("添加日程");
        } else {
            userCalendar.title = getIntent().getStringExtra("title");
            userCalendar.remindTime1 = getIntent()
                    .getIntExtra("remindTime1", 0);
            userCalendar.remindTime2 = getIntent()
                    .getIntExtra("remindTime2", 0);
            userCalendar.remindTime3 = getIntent()
                    .getIntExtra("remindTime3", 0);
            userCalendar.remindTime4 = getIntent()
                    .getIntExtra("remindTime4", 0);
```

```java
            userCalendar.remindTime5 = getIntent()
                    .getIntExtra("remindTime5", 0);}
            …//省略获取各组件的findViewById()方法
ArrayAdapter<RemindTime> aa = new ArrayAdapter<RemindTime>(
        getApplicationContext(), R.layout.spinner_item, REMIND_TIMES);
remindTime1Spinner.setAdapter(aa);
remindTime2Spinner.setAdapter(aa);
remindTime3Spinner.setAdapter(aa);
remindTime4Spinner.setAdapter(aa);
remindTime5Spinner.setAdapter(aa);
actionTimeButton.setText(userCalendar.actionTime);
titleEditText.setText(userCalendar.title);
showReminds();
final OnTimeSetListener onTimeSetListener = new OnTimeSetListener() {
    @Override
    public void onTimeSet(TimePicker view, int hourOfDay, int minute) {
        String date = actionTimeButton.getText().toString()
                .substring(0, 10);
        actionTimeButton.setText(date + " "
                + (hourOfDay < 10 ? "0" + hourOfDay : "" + hourOfDay)
                + ":" + (minute < 10 ? "0" + minute : "" + minute));}};
actionTimeButton.setOnClickListener(new OnClickListener() {
    @Override
    public void onClick(View v) {
        String time = actionTimeButton.getText().toString();
        int hour = 12;
        int minute = 0;
        try {
            hour = Integer.parseInt(time.substring(11, 13));
            minute = Integer.parseInt(time.substring(14, 16));
        } catch (Exception e) {
            e.printStackTrace();}
        final TimePickerDialog dialog = new TimePickerDialog(
                UserCalendarEventActivity.this, onTimeSetListener,
                hour, minute, true);
        dialog.show();}});
remindTime1Spinner
        .setOnItemSelectedListener(new OnItemSelectedListener() {
            @Override
            public void onItemSelected(AdapterView<?> parent,
                    View view, int position, long id) {
                userCalendar.remindTime1 = ((RemindTime) remindTime1Spinner
                        .getSelectedItem()).minutes;}
            @Override
            public void onNothingSelected(AdapterView<?> parent) {
            }});
remindTime2Spinner
        .setOnItemSelectedListener(new OnItemSelectedListener() {
            @Override
            public void onItemSelected(AdapterView<?> parent,
```

```java
                        View view, int position, long id) {
                    userCalendar.remindTime2 = ((RemindTime) remindTime2Spinner
                            .getSelectedItem()).minutes;}
                @Override
                public void onNothingSelected(AdapterView<?> parent) {
                }});
        remindTime3Spinner
                .setOnItemSelectedListener(new OnItemSelectedListener() {
                    @Override
                    public void onItemSelected(AdapterView<?> parent,
                            View view, int position, long id) {
                        userCalendar.remindTime3 = ((RemindTime) remindTime3Spinner
                                .getSelectedItem()).minutes;}
                    @Override
                    public void onNothingSelected(AdapterView<?> parent) {
                    }});
        remindTime4Spinner
                .setOnItemSelectedListener(new OnItemSelectedListener() {
                    @Override
                    public void onItemSelected(AdapterView<?> parent,
                            View view, int position, long id) {
                        userCalendar.remindTime4 = ((RemindTime) remindTime4Spinner
                                .getSelectedItem()).minutes;}
                    @Override
                    public void onNothingSelected(AdapterView<?> parent) {
                    }});
        remindTime5Spinner
                .setOnItemSelectedListener(new OnItemSelectedListener() {
                    @Override
                    public void onItemSelected(AdapterView<?> parent,
                            View view, int position, long id) {
                        userCalendar.remindTime5 = ((RemindTime) remindTime5Spinner
                                .getSelectedItem()).minutes;}
                    @Override
                    public void onNothingSelected(AdapterView<?> parent) {
                    }});
        deleteRemind1ImageView.setOnClickListener(new OnClickListener() {
            @Override
            public void onClick(View v) {
                userCalendar.remindTime1 = userCalendar.remindTime2;
                userCalendar.remindTime2 = userCalendar.remindTime3;
                userCalendar.remindTime3 = userCalendar.remindTime4;
                userCalendar.remindTime4 = userCalendar.remindTime5;
                userCalendar.remindTime5 = 0;
                showReminds();}});
        deleteRemind2ImageView.setOnClickListener(new OnClickListener() {
            @Override
            public void onClick(View v) {
                userCalendar.remindTime2 = userCalendar.remindTime3;
                userCalendar.remindTime3 = userCalendar.remindTime4;
                userCalendar.remindTime4 = userCalendar.remindTime5;
                userCalendar.remindTime5 = 0;
                showReminds();}});
```

```java
        deleteRemind3ImageView.setOnClickListener(new OnClickListener() {
            @Override
            public void onClick(View v) {
                userCalendar.remindTime3 = userCalendar.remindTime4;
                userCalendar.remindTime4 = userCalendar.remindTime5;
                userCalendar.remindTime5 = 0;
                showReminds();}});
        deleteRemind4ImageView.setOnClickListener(new OnClickListener() {
            @Override
            public void onClick(View v) {
                userCalendar.remindTime4 = userCalendar.remindTime5;
                userCalendar.remindTime5 = 0;
                showReminds();}});
        deleteRemind5ImageView.setOnClickListener(new OnClickListener() {
            @Override
            public void onClick(View v) {
                userCalendar.remindTime5 = 0;
                showReminds();}});
        addRemindButton.setOnClickListener(new OnClickListener() {
            @Override
            public void onClick(View v) {
                if (userCalendar.remindTime5 != 0)
                    return;
                addRemindButton.setText("添加提醒");
                if (userCalendar.remindTime4 != 0) {
                    remind5.setVisibility(View.VISIBLE);
                    userCalendar.remindTime5 = ((RemindTime) remindTime5Spinner
                            .getSelectedItem()).minutes;
                    addRemindButton.setVisibility(View.GONE);
                    return; }
                if (userCalendar.remindTime3 != 0) {
                    remind4.setVisibility(View.VISIBLE);
                    userCalendar.remindTime4 = ((RemindTime) remindTime4Spinner
                            .getSelectedItem()).minutes;
                    return; }
                if (userCalendar.remindTime2 != 0) {
                    remind3.setVisibility(View.VISIBLE);
                    userCalendar.remindTime3 = ((RemindTime) remindTime3Spinner
                            .getSelectedItem()).minutes;
                    return;}
                if (userCalendar.remindTime1 != 0) {
                    remind2.setVisibility(View.VISIBLE);
                    userCalendar.remindTime2 = ((RemindTime) remindTime2Spinner
                            .getSelectedItem()).minutes;
                    return; }
                remind1.setVisibility(View.VISIBLE);
                userCalendar.remindTime1 = ((RemindTime) remindTime1Spinner
                        .getSelectedItem()).minutes; } });
        saveButton.setOnClickListener(new OnClickListener() {
            @Override
            public void onClick(View v) {
                userCalendar.title = titleEditText.getText().toString().trim();
                userCalendar.actionTime = actionTimeButton.getText().toString();
```

```java
            if (StringUtils.isEmpty(userCalendar.title)) {
                showMessage("请输入日程内容");
                return; }
            //TODO 后续加上保存功能
        }}); }
void showReminds() {
    remind1.setVisibility(View.GONE);
    remind2.setVisibility(View.GONE);
    remind3.setVisibility(View.GONE);
    remind4.setVisibility(View.GONE);
    remind5.setVisibility(View.GONE);
    remindTime1Spinner.setSelection(0);
    remindTime2Spinner.setSelection(0);
    remindTime3Spinner.setSelection(0);
    remindTime4Spinner.setSelection(0);
    remindTime5Spinner.setSelection(0);
    int remindCount = 0;
    if (userCalendar.remindTime1 != 0)
        remindCount++;
    if (userCalendar.remindTime2 != 0)
        remindCount++;
    if (userCalendar.remindTime3 != 0)
        remindCount++;
    if (userCalendar.remindTime4 != 0)
        remindCount++;
    if (userCalendar.remindTime5 != 0)
        remindCount++;
    if (remindCount == 0) {
        addRemindButton.setText("设置提醒");
        addRemindButton.setVisibility(View.VISIBLE);
    } else if (remindCount > 0 && remindCount < 5) {
        addRemindButton.setText("添加提醒");
        addRemindButton.setVisibility(View.VISIBLE);
    } else if (remindCount == 5) {
        addRemindButton.setVisibility(View.GONE);
    }
    if (userCalendar.remindTime1 != 0) {
        remind1.setVisibility(View.VISIBLE);
        RemindTime.select(remindTime1Spinner, userCalendar.remindTime1);
        if (userCalendar.remindTime2 != 0) {
            remind2.setVisibility(View.VISIBLE);
            RemindTime.select(remindTime2Spinner, userCalendar.remindTime2);
            if (userCalendar.remindTime3 != 0) {
                remind3.setVisibility(View.VISIBLE);
                RemindTime.select(remindTime3Spinner,
                    userCalendar.remindTime3);
                if (userCalendar.remindTime4 != 0) {
                    remind4.setVisibility(View.VISIBLE);
                    RemindTime.select(remindTime4Spinner,
                        userCalendar.remindTime4);
                    if (userCalendar.remindTime5 != 0) {
                        remind5.setVisibility(View.VISIBLE);
                        RemindTime.select(remindTime5Spinner,
```

```
                            userCalendar.remindTime5);
                    addRemindButton.setVisibility(View.GONE);}}}}}}
    public static class RemindTime {
        public int minutes;
        public String name;
        public RemindTime(int minutes, String name) {
            super();
            this.minutes = minutes;
            this.name = name; }
        @Override
        public String toString() {
            return name; }
        static void select(Spinner spinner, int minutes) {
            for (int i = 0; i < spinner.getCount(); i++) {
                RemindTime rt = (RemindTime) spinner.getItemAtPosition(i);
                if (rt.minutes == minutes) {
                    spinner.setSelection(i);
                    break; }}}
        static final RemindTime[] REMIND_TIMES = {
            new RemindTime(10, "提前 10 分钟提醒"),
            new RemindTime(60, "提前 1 小时提醒"),
            new RemindTime(360, "提前 6 小时提醒"),
            new RemindTime(720, "提前 12 小时提醒"),
            new RemindTime(1440, "提前 1 天提醒"),
            new RemindTime(4320, "提前 3 天提醒"),
            new RemindTime(10080, "提前 1 星期提醒") };}
```

运行项目,进入日程列表界面,单击某个日程或下部的"添加日程"按钮,进入用户"日历事件"编辑界面。在"日历事件"编辑界面中,可以设置"事件时间",单击"设置提醒"按钮完成提醒的添加。运行结果如图 5-16 所示。

图 5-16 用户日程编辑界面

5.5.3 实现【任务 5-3】

本任务完成用户日程提醒功能,当到达用户所设置的日程提醒时间时,需要弹出提醒界面,并按照用户设置的提醒方式,播放提醒声音或者发出震动。

定时提醒功能可以使用 Android 提供的 AlermManager 来实现,在指定的时间间隔之后执行一个 PendingIntent,通过 PendingIntent 可以发送一个广播,然后在广播接收器中打开提醒界面。

首先编写提醒界面的布局文件,代码如下所示。

【任务 5-3】 user_calendar_remind.xml

```xml
<RelativeLayout xmlns:android="http://schemas.android.com/apk/res/android"
    xmlns:tools="http://schemas.android.com/tools"
    android:id="@+id/container"
    android:layout_width="match_parent"
    android:layout_height="match_parent"
    android:background="#80000000"
    android:gravity="center">
    <ImageView
        android:layout_width="match_parent"
        android:layout_height="match_parent"
        android:layout_marginLeft="20dp"
        android:layout_marginRight="20dp"
        android:alpha="0.2"
        android:src="@drawable/logo1" />
    <LinearLayout
        android:layout_width="match_parent"
        android:layout_height="match_parent"
        android:gravity="center"
        android:orientation="vertical"
        android:padding="30dp">
        <TextView
            style="@style/text3"
            android:layout_width="match_parent"
            android:layout_height="wrap_content"
            android:text="GiftEMS\n日程提醒"
            android:textSize="26sp" />
        <TextView
            android:id="@+id/actionTimeTextView"
            style="@style/text1"
            android:layout_width="match_parent"
            android:layout_height="wrap_content"
            android:layout_marginTop="20dp"
            android:gravity="left"
            android:textColor="#FFFFFF"
            android:textSize="16sp" />
        <TextView
            android:id="@+id/titleTextView"
            style="@style/text1"
            android:layout_width="match_parent"
            android:layout_height="wrap_content"
            android:layout_marginTop="20dp"
            android:textColor="#FFFFFF"
            android:gravity="left"
            android:textSize="16sp" />
```

```xml
<TextView
    style = "@style/text1"
    android:layout_width = "match_parent"
    android:layout_height = "wrap_content"
    android:layout_marginTop = "40dp"
    android:text = "单击屏幕停止提醒"
    android:textColor = "#FFFFFF"
    android:textSize = "14sp" />
    </LinearLayout>
</RelativeLayout>
```

上述布局中,使用 TextView 显示提醒信息。接下来编写提醒界面对应的 Activity,代码如下。

【任务 5-3】 UserCalendarRemindActivity.java

```java
public class UserCalendarRemindActivity extends Activity {
    PowerManager.WakeLock wakeLock;    //唤醒锁屏
    MediaPlayer player;                //铃声播放器
    RelativeLayout container;
    TextView titleTextView;
    TextView actionTimeTextView;
    @Override
    protected void onCreate(Bundle savedInstanceState) {
        super.onCreate(savedInstanceState);
        setContentView(R.layout.user_calendar_remind);
        String title = getIntent().getStringExtra("title");
        String actionTime = getIntent().getStringExtra("actionTime");
        if (StringUtils.isEmpty(title) || StringUtils.isEmpty(actionTime)) {
            finish();
            return;
        }
        container = (RelativeLayout) findViewById(R.id.container);
        titleTextView = (TextView) findViewById(R.id.titleTextView);
        actionTimeTextView = (TextView) findViewById(R.id.actionTimeTextView);
        titleTextView.setText(title);
        actionTimeTextView.setText(actionTime);
        container.setOnClickListener(new OnClickListener() {
            @Override
            public void onClick(View v) {
                if (player != null && player.isPlaying()) {
                    player.stop();
                } else {
                    Vibrator vibrator
                        = (Vibrator) getSystemService(Context.VIBRATOR_SERVICE);
                    vibrator.cancel();}}});
        //锁屏后也会显示,会点亮屏幕并保持屏幕常亮
        getWindow().addFlags(
                WindowManager.LayoutParams.FLAG_SHOW_WHEN_LOCKED
                        | WindowManager.LayoutParams.FLAG_DISMISS_KEYGUARD);
        getWindow().addFlags(
                WindowManager.LayoutParams.FLAG_KEEP_SCREEN_ON
                        | WindowManager.LayoutParams.FLAG_TURN_SCREEN_ON);}
```

```java
    @Override
    protected void onResume() {
        super.onResume();
        String remindType = getIntent().getStringExtra("remindType");
        if (remindType != null) {
            PowerManager pm
                = (PowerManager) getSystemService(Context.POWER_SERVICE);
            wakeLock = pm.newWakeLock(PowerManager.ACQUIRE_CAUSES_WAKEUP
                | PowerManager.SCREEN_DIM_WAKE_LOCK, "SimpleTimer");
            wakeLock.acquire();
            if (remindType.equals("alerm")) {
                Uri uri = RingtoneManager.getActualDefaultRingtoneUri(this,
                    RingtoneManager.TYPE_RINGTONE);
                player = MediaPlayer.create(this, uri);
                player.setLooping(true);
                player.start();
            } else if (remindType.equals("vibrator")) {
                Vibrator vibrator
                    = (Vibrator) getSystemService(Context.VIBRATOR_SERVICE);
                vibrator.vibrate(new long[] { 0, 500, 1000, 1500, 2000, 2500 }, -1);}}}
    @Override
    protected void onPause() {
        super.onPause();
        if (wakeLock != null)
            wakeLock.release();
        wakeLock = null;
        if (player != null)
            player.release();
        player = null;}
    public static class RemindBroadcastReceiver extends BroadcastReceiver {
        @Override
        public void onReceive(Context context, Intent intent) {
            boolean remind = true;              //TODO 后续改为从已存储文件中取值
            if (!remind)
                return;
            String title = intent.getStringExtra("title");
            String actionTime = intent.getStringExtra("actionTime");
            Intent itt = new Intent(context, UserCalendarRemindActivity.class);
            itt.putExtra("title", title);
            itt.putExtra("actionTime", actionTime);
            itt.putExtra("remindType", "alerm"); //TODO 后续改为从已存储文件中取值
            itt.setFlags(Intent.FLAG_ACTIVITY_NEW_TASK);
            context.startActivity(itt);}}
}
```

在 UserCalendarRemindActivity 代码中，各方法功能分别如下：

（1）在 onCreate()方法中，从 Intent 中获取提醒的标题和日程时间并显示在界面上，通过设定 Activity 的显示参数实现锁屏后也能显示提醒，并在提醒时保持屏幕常亮。

（2）在 onResume()方法中，首先获取电源管理器 PowerManager 对象，并通过其 newWakeLock()方法获取 WakeLock 对象，从而阻止 CPU 休眠。根据提醒的类型决定播放铃声或者发出震动。

（3）在 onPause()方法中，释放 WakeLock 和播放器。

（4）静态内部类 RemindBroadcastReceiver 是一个接收器，用于接收 AlermManager 发出的广播，其中 onReceive()方法用于启动一个 UserCalendarRemindActivity 并向其传递 Intent 数据。注意，RemindBroadcastReceiver 需要注册才能生效。

此处，使用内部类来编写 RemindBroadcastReceiver 仅仅是因为这个广播接收器只在 UserCalendarRemindActivity 中使用一次，其他位置并不会使用。

除此之外，还需要为项目添加 android.permission.WAKE_LOCK 和 android.permission.VIBRATE 权限，分别使用 WakeLock 和震动的权限。

为测试日程提醒，修改 UserCalendarEventActivity，在单击"保存"按钮时，根据日程提醒的时间，设置 AlermManager 执行 PendingIntent。修改后的 UserCalendarEventActivity 代码如下所示。

【任务 5-3】 UserCalendarEventActivity.java

```java
/** 日历事件 */
public class UserCalendarEventActivity extends BaseActivity {
    ...//省略其他部分代码
    @Override
    protected void onCreate(Bundle savedInstanceState) {
        ...//省略
        saveButton.setOnClickListener(new OnClickListener() {
            @Override
            public void onClick(View v) {
                userCalendar.title = titleEditText.getText().toString().trim();
                userCalendar.actionTime = actionTimeButton.getText().toString();
                if (StringUtils.isEmpty(userCalendar.title)) {
                    showMessage("请输入日程内容");
                    return; }
                //TODO 后续加上保存到服务器功能
                setUserCalendarRemind();}});}
    public void setUserCalendarRemind() {
        SimpleDateFormat sdf = new SimpleDateFormat("yyyy-MM-dd HH:mm",
                Locale.US);
        Date now = new Date();
        try {
            Date actionTime = sdf.parse(userCalendar.actionTime);
            if (!actionTime.after(now))
                return;
            Calendar calendar = Calendar.getInstance();
            ArrayList<Long> remindTimes = new ArrayList<Long>();
            calendar.setTime(actionTime);
            remindTimes.add(calendar.getTimeInMillis());
            if (userCalendar.remindTime1 > 0) {
                calendar.setTime(actionTime);
                calendar.add(Calendar.MINUTE, -userCalendar.remindTime1);
                if (calendar.getTime().after(now))
                    remindTimes.add(calendar.getTimeInMillis());}
            if (userCalendar.remindTime2 > 0) {
                calendar.setTime(actionTime);
                calendar.add(Calendar.MINUTE, -userCalendar.remindTime2);
```

```
                if (calendar.getTime().after(now))
                    remindTimes.add(calendar.getTimeInMillis());}
            if (userCalendar.remindTime3 > 0) {
                calendar.setTime(actionTime);
                calendar.add(Calendar.MINUTE, - userCalendar.remindTime3);
                if (calendar.getTime().after(now))
                    remindTimes.add(calendar.getTimeInMillis());}
            if (userCalendar.remindTime4 > 0) {
                calendar.setTime(actionTime);
                calendar.add(Calendar.MINUTE, - userCalendar.remindTime4);
                if (calendar.getTime().after(now))
                    remindTimes.add(calendar.getTimeInMillis());}
            if (userCalendar.remindTime5 > 0) {
                calendar.setTime(actionTime);
                calendar.add(Calendar.MINUTE, - userCalendar.remindTime5);
                if (calendar.getTime().after(now))
                    remindTimes.add(calendar.getTimeInMillis());}
            AlarmManager alarmManager
                = (AlarmManager) getSystemService(ALARM_SERVICE);
            for (Long time : remindTimes) {
                Intent intent = new Intent(
                    this,
                    UserCalendarRemindActivity
                        .RemindBroadcastReceiver.class);
                intent.putExtra("title", userCalendar.title);
                intent.putExtra("actionTime", userCalendar.actionTime);
                PendingIntent pendingIntent = PendingIntent.getBroadcast(this,
                        userCalendar.id, intent, Intent.FLAG_ACTIVITY_NEW_TASK);
                alarmManager.set(AlarmManager.RTC_WAKEUP, time, pendingIntent);}
        } catch (Exception e) {
            e.printStackTrace();}}
    }
```

上述代码的 setUserCalendarRemind() 方法中，首先将日程的 5 个提醒时间分别与当前时间比较，计算所有需要提醒的时间。然后获取 AlermManager 对象，遍历每个提醒时间，通过 RemindBroadcastReceiver 的 Intent 对象传入日程的标题和时间。最后创建 PendingIntent 对象并调用 AlermManager 的 set() 方法向其注册提醒。

运行应用程序，进入日程编辑界面，修改某个日程的时间并添加提醒，然后保存这个日程。当提醒时间到达时，启动提醒界面，如图 5-17 所示。

需要注意，上述提醒功能是在保存日程时

图 5-17　用户日程编辑界面

注册到 AlermManager 的,而实际上应该是在用户登录 App 后自动将其所有日程的提醒时间注册到 AlermManager,从而保证不会遗漏提醒。此处仅仅演示通过 AlermManager、PendingIntent 等组件完成提醒功能,"GIFT-EMS"系统的完整业务实现在后续章节中将继续完善。

注意

> 限于篇幅,本书不再对 AlermManager 和 PowerManager 详细介绍,读者可以参阅其他资料。

本章总结

小结

- Intent 是 Android 中的一个消息传递机制,提供了一种通用的消息系统,可以激活 Android 应用的三个核心组件：Activity、Service 和 BroadcastReceiver。
- Intent 对象是一个信息的捆绑包,包含了接收该 Intent 的组件所需要的信息。
- 广播是一种广泛运用的、在应用程序之间传输信息的机制,而 BroadcastReceiver 是对发送出来的广播进行过滤接收并响应的一类组件。
- Intent 类提供了许多 Action,包括标准的 Activity 动作、标准的 Broadcast 动作、标准的类别动作和标准的附加数据动作。
- Handler 类的作用包括：在新启动的线程中发送消息和在主线程中获取和处理消息。
- AsyncTask 使得创建异步任务变得更加简单,不再需要编写任务线程和 Handler 实例也可完成相同的任务。

Q&A

问题：简述 Handler 的机制。

回答：Android 提供了 Handler 和 Looper 来满足线程间的通信要求。Handler 遵守先进先出原则,Looper 类用来管理特定线程内对象之间的消息交换(MessageExchange)。

（1）**Looper**：一个线程可以产生一个 Looper 对象,来管理此线程中的 Message Queue（消息队列）。

（2）**Handler**：通过构造 Handler 对象来与 Looper 沟通,以便 push 新消息传入 Message Queue 中,或者接收 Looper 从 Message Queue 队列取出并发送来的消息。

（3）**Message Queue**(消息队列)：用来存放线程放入的消息。

（4）**线程**：UI Thread 通常就是 MainThread,而 Android 启动程序时会替该线程建立一个 Message Queue。

章节练习

习题

1. 下面在 AndroidManifest.xml 文件中注册 BroadcastReceiver 方式正确的是_____。

   A.
```
< receiver android:name = "NewBroad">
    < intent - filter >
        < action android:name = "android.provider.action.NewBroad"/> </action >
    </ intent - filter >
</receiver >
```

   B.
```
< receiver android:name = "NewBroad">
    < intent - filter >
        < action android:name = "android.provider.action.NewBroad"/>
    </ intent - filter >
</receiver >
```

   C.
```
< receiver android:name = "NewBroad">
    < action android:name = "android.provider.action.NewBroad"/>
    </action >
</receiver >
```

   D.
```
< intent - filter >
    < receiver android:name = "NewBroad">
        < action android:name = "android.provider.action.NewBroad"/>
    </receiver >
</ intent - filter >
```

2. 在 Android 中下列属于 Intent 的作用的是_____。

   A. 实现应用程序间的数据共享
   B. 是一段长的生命周期,没有用户界面的程序,可以保持应用在后台运行,而不会因为切换页面而消失
   C. 可以实现界面间的切换,可以包含动作和动作数据,连接四大组件的纽带
   D. 处理一个应用程序整体性的工作

上机

训练目标：广播、Handler 的综合使用。

第5章 Intent与BroadcastReceiver

培养能力	熟练运用广播机制和 Handler 机制		
掌握程度	★★★★★	难度	中
代码行数	400	实施方式	重复编码
结束条件	熟练运用广播机制和 Handler 机制		

参考训练内容
编写一个广播接收器对象,在单击按钮时,发送广播给该接收器对象,并在对象中调用 Handler 来弹出一个提示窗口

第6章 数据存储

本章任务是完成"GIFT-EMS 礼记"中数据保存相关功能：
- 【任务 6-1】 完成保存用户登录信息功能。
- 【任务 6-2】 完成设置信息保存功能。
- 【任务 6-3】 完成购物袋功能。

知 识 点	Listen(听)	Know(懂)	Do(做)	Revise(复习)	Master(精通)
Android 数据存储方式	★	★			
文件存储	★	★	★	★	★
SharedPreferences 存储	★	★	★	★	★
SQLite 数据库	★	★	★	★	★

第6章 数据存储

6.1 数据存储简介

应用程序必然涉及数据的输入和输出，Android 应用程序也不例外，需要将数据存储到硬件设备中。存放数据需要使用数据存储机制，Android 提供了以下几种数据存储方式。

- **文件存储**：Android 应用是使用 Java 语言来开发的，因此 Java 中关于文件的 I/O 操作大部分可以移植到 Android 应用开发上。如果只有少量数据需要保存，且数据格式无须结构化，则使用普通的文件进行数据存储即可。
- **SharedPreferences 存储**：数据是以 key-value 键值对的方式进行组织和管理，并保存到 XML 文件中。如果要存储的数据格式很简单，都是普通的字符串、数值等，例如小游戏的玩家积分、音效、配置信息等，可以采用 SharedPreferences 存储。相对于其他方式，SharedPreferences 是一个轻量级的存储机制，该方式实现比较简单，适合存储少量且数据结构简单的数据。
- **SQLite 数据库存储**：Android 系统内置了 SQLite 数据库，SQLite 是一个轻量级数据库，没有后台进程，整个数据库对应一个文件，便于在不同设备之间进行移植。Android 为访问 SQLite 数据库提供了大量的 API，可以非常便捷地进行添加、删除和更新等操作。相比 SharedPreferences 和文件存储，使用 SQLite 较为复杂，该方式通常应用于数据量较多且需要进行结构化存储的情况。

6.2 文件存储

文件存储方式不受类型限制，可以将一些数据直接以文件的形式保存在设备中，例如文本文件、PDF、音频、图片等。存储类型复杂的数据时，通常采用文件存储。Java 提供一套完整的 I/O 流体系，通过 I/O 流可以非常方便地访问磁盘中的文件，同样 Android 也支持 I/O 流方式来访问手机等移动设备中的存储文件。

6.2.1 I/O 流操作文件

进行 I/O 流操作文件时，需要先获得文件的输入流和输出流。在 Android 应用程序中，可以通过上下文环境 Context 对象提供的 openFileInput() 和 openFileOuput() 两个方法分别来获得文件的输入流和输出流，这两个方法的具体介绍如下。

- **FileInputStream openFileInput(String name)**：用于获取应用程序的数据文件夹下指定 name 文件名的标准文件输入流，以便读取设备中的文件。
- **FileOutputStream openFileOuput(String name, int mode)**：用于获取应用程序的数据文件夹下指定 name 文件名的标准文件输出流，以便将数据写入设备的文件中。

其中 openFileOutput() 方法的第二个参数 mode 用于指定输出流的模式，即打开文件进行操作的模式。Context 类中提供了 4 个静态常量用于表示不同的输出模式，如表 6-1 所示。

表 6-1　4 种文件读写模式

模　式	功　能　描　述
Context.MODE_PRIVATE	私有模式,该模式所创建的文件都是私有文件,只能被应用本身所访问。因此,该模式下所写入的内容会覆盖原来文件的内容
Context.MODE_APPEND	附加模式,该模式首先会检查文件是否存在,若文件不存在,则创建新文件;若文件存在,则在原文件的末尾追加内容
Context.MODE_WORLD_READABLE	可读模式,该模式的文件可以被其他应用程序读取
Context.MODE_WORLD_WRITABLE	可写模式,该模式的文件可以被其他应用程序读写

从 Android 4.2 开始,不推荐使用 Context.MODE_WORLD_WRITABLE 可读模式和 Context.MODE_WORLD_READABLE 可写模式,由于这两种模式允许其他应用程序操作本应用程序所创建的文件数据,很容易引起安全漏洞,因此在高版本的 Android 系统中尽量不要采用这两种模式。如果应用程序需要暴露自己的数据以便其他应用程序进行访问,可以使用第 7 章所介绍的 ContentProvider 来实现。

除此之外,Context 上下文对象还提供了一些方法来访问应用程序的数据文件夹,如表 6-2 所示。

表 6-2　访问数据文件夹的方法

方　法	功　能　描　述
File getDir(String name,int mode)	在应用程序的数据文件夹下获取或创建 name 对应的子目录
File getFilesDir()	获取应用程序的数据文件夹的绝对路径
String[] fileList()	返回应用程序的数据文件夹下的所有文件
boolean deleteFile(String name)	删除应用程序的数据文件夹下的指定文件

下述代码演示如何通过文件输入流读取文件,代码如下所示。

【示例】　获取文件输入流读取文件

```
String file = "qst.txt";                                          //定义文件名
FileInputStream fileInputStream = openFileInput(file);            //获取指定文件的文件输入流
byte[] buffer = new byte[fileInputStream.available()];            //定义一个字节缓存数组
fileInputStream.read(buffer);                                     //将数据读到缓存区
fileInputStream.close();                                          //关闭文件输入流
```

下述代码演示如何通过文件输出流写入文件,即将数据保存到文件中。

【示例】　获取文件输出流写文件

```
//获取文件输出流,操作模式是私有
FileOutputStream fileOutputStream = openFileOutput(file,Context.MODE_PRIVATE);
String strContent = "QST 青软实训";
fileOutputStream.write(strContent.getBytes());        //将内容写入文件
fileOutputStream.close();
```

下述代码演示如何使用 I/O 流对文件进行读写操作,其中 XML 布局文件的代码如下所示。

【代码 6-1】 activity_file_io.xml

```xml
<LinearLayout xmlns:android = "http://schemas.android.com/apk/res/android" …>
    <EditText
        android:id = "@+id/editFileOut"
        android:layout_width = "match_parent"
        android:layout_height = "wrap_content"
        android:lines = "4" />
    <Button
        android:id = "@+id/btnWrite"
        android:layout_width = "wrap_content"
        android:layout_height = "wrap_content"
        android:text = "保存文件" />
    <EditText
        android:id = "@+id/editFileIn"
        android:layout_width = "match_parent"
        android:layout_height = "wrap_content"
        android:cursorVisible = "false"
        android:editable = "false"
        android:lines = "4" />
    <Button
        android:id = "@+id/btnRead"
        android:layout_width = "wrap_content"
        android:layout_height = "wrap_content"
        android:text = "读取文件" />
</LinearLayout>
```

上述界面布局比较简单,只包含两个文本框和两个按钮,分别用于保存文件和读取文件两种操作。接着编写 Activity 程序部分,代码如下所示。

【代码 6-2】 FileIOActivity.java

```java
public class FileIOActivity extends AppCompatActivity {
    private EditText editFileIn, editFileOut;          //声明两个文本框
    private Button btnRead, btnWrite;                  //声明两个按钮
    final String FILE_NAME = "qstIO.txt";              //指定文件名
    @Override
    public void onCreate(Bundle savedInstanceState) {
        super.onCreate(savedInstanceState);
        setContentView(R.layout.activity_file_io);
        Log.d("FileIO","FileIOActivity");
        //获取两个文本框
        editFileIn = (EditText) findViewById(R.id.editFileIn);
        editFileOut = (EditText) findViewById(R.id.editFileOut);
        //获取两个按钮
        Button btnRead = (Button) findViewById(R.id.btnRead);
        Button btnWrite = (Button) findViewById(R.id.btnWrite);
        //绑定 btnRead 按钮的事件监听器
```

```
            btnRead.setOnClickListener(new OnClickListener() {
                @Override
                public void onClick(View v) {
                    //读取指定文件中的内容,并在 editFileIn 文本框中显示出来
                    editFileIn.setText(read());}});
            //绑定 btnWrite 按钮的事件监听器
            btnWrite.setOnClickListener(new OnClickListener() {
                @Override
                public void onClick(View source) {
                    write(editFileOut.getText().toString());      //将 edit1 中的内容写入文件中
                    editFileOut.setText("");}});}                 //清空 editFileOut 文本框中的内容
    private String read() {
        try {
            FileInputStream fis = openFileInput(FILE_NAME);   //打开文件输入流
            byte[] buff = new byte[1024];
            int hasRead = 0;
            StringBuilder sb = new StringBuilder("");
            //读取文件内容
            while ((hasRead = fis.read(buff)) > 0) {
                sb.append(new String(buff, 0, hasRead)); }
            fis.close();                                      //关闭文件输入流
            return sb.toString();
        } catch (Exception e) {
            e.printStackTrace();}
        return null; }
    private void write(String content) {
        try {
            FileOutputStream fos = openFileOutput(FILE_NAME,
                Context.MODE_APPEND);                         //以追加模式打开文件输出流
            PrintStream ps = new PrintStream(fos);            //将 FileOutputStream 包装成 PrintStream
            ps.println(content);                              //输出文件内容
            ps.close();                                       //关闭文件输出流
            Toast.makeText(FileIOActivity.this, "保存成功", Toast.LENGTH_LONG)
                .show();                                      //使用 Toast 显示保存成功
        } catch (Exception e) {
            e.printStackTrace();}
    }
}
```

上述代码的核心操作就是文件的保存和读取,其中 read()和 write()两个方法分别用于读文件和写文件操作;代码中分别对 btnRead 和 btnWrite 按钮设置了事件监听器,并在事件处理方法中调用相应的 read()或 write()方法实现文件的读取或保存。

运行上述程序,先在第一个文本框中输入信息,并单击"保存文件"按钮,将用户输入的信息保存到 qstIO.txt 文件中;单击"读取文件"按钮,从 qstIO.txt 文件中读取内容,并在第二个文本框中显示出来,显示结果如图 6-1 所示。

Android 应用程序的数据文件默认保存在/data/data/包名/files 目录下。在 Android Device Monitor 的 File Explorer 选项卡中,展开/data/data/com.qst.chapter06/files 目录,在该目录下可以看到保存的 qstIO.txt 数据文件,如图 6-2 所示。

图 6-1　保存和读取文件　　　　图 6-2　通过 Android Device Monitor 查看文件

6.2.2　读写 SD 卡文件

在 Android 应用程序中，通过 Context 的 openFileInput()和 openFileOuput()方法操作文件时，所操作的文件都存放在应用程序的数据文件夹中，由于手机内置的存储空间有限，所以此类操作文件的大小会受限制。为了更好地存取一些大文件，Android 应用程序往往需要读写 SD 卡中的文件。

注意

> SD 卡（Secure Digital Memory Card）是一种基于半导体快闪记忆器的多功能存储卡，具有容量大、性能高、安全性好等多种特点，被广泛地用于便携式移动设备，例如手机、数码相机、PDA 等。SD 卡极大地扩充了手机的存储能力。

读写 SD 卡文件的步骤如下。

（1）使用 Environment.getExternalStorageState()方法判断是否插入 SD 卡，且应用程序具有读写 SD 卡的权限；

（2）使用 Environment.getExternalStorageDirectory()方法获取 SD 卡的目录；

（3）使用文件输入流（FileInputStream、FileReader）或输出流（FileOutputStream、FileWriter）来读写 SD 卡中的文件。

【示例】　读 SD 卡上的文件

```
//1.如果手机插入了SD卡,而且应用程序具有访问SD的权限
if (Environment.getExternalStorageState().equals(Environment.MEDIA_MOUNTED)){
    //2.获取SD卡对应的存储目录
    File sdCardDir = Environment.getExternalStorageDirectory();
    Log.d("FileIO","" + sdCardDir);
    //3.获取指定文件对应的输入流
    FileInputStream fis = new FileInputStream(sdCardDir.getCanonicalPath()
        + FILE_NAME);
    ...//读文件
}
```

Android 应用程序读写 SD 卡中的文件时,需要注意以下两点:①确保已插入 SD 卡,对于模拟器而言,在创建 AVD 时需要指明 SD 卡大小,也可以通过 mksdcard 命令来创建 SD 虚拟存储卡;②在 AndroidManifest.xml 程序清单文件中配置 SD 卡的读写权限。代码如下所示。

```xml
<!-- 在 SD 卡中创建与删除文件权限 -->
<uses-permission android:name="android.permission.MOUNT_UNMOUNT_FILESYSTEMS"/>
<!-- 向 SD 卡写入数据权限 -->
<uses-permission android:name="android.permission.WRITE_EXTERNAL_STORAGE"/>
```

下述代码演示如何读写 SD 卡中的文件。Activity 所使用 XML 布局文件与 FileIOActivity 应用完全相同,此处不再重复 XML 布局文件的代码,而在 Activity 中的操作是基于 SD 卡文件操作,具体代码如下所示。

【代码 6-3】 SDActivity.java

```java
public class SDActivity extends AppCompatActivity; {
    private EditText editFileIn, editFileOut;         //声明两个文本框
    private Button btnRead, btnWrite;                 //声明两个按钮
    final String FILE_NAME = "/qstSD.txt";            //指定文件名
    @Override
    public void onCreate(Bundle savedInstanceState) {
        super.onCreate(savedInstanceState);
        setContentView(R.layout.activity_file_io);
        Log.d("FileIO","FileIOActivity");
        //获取两个文本框
        editFileIn = (EditText) findViewById(R.id.editFileIn);
        editFileOut = (EditText) findViewById(R.id.editFileOut);
        //获取两个按钮
        Button btnRead = (Button) findViewById(R.id.btnRead);
        Button btnWrite = (Button) findViewById(R.id.btnWrite);
        //绑定 btnRead 按钮的事件监听器
        btnRead.setOnClickListener(new OnClickListener() {
            @Override
            public void onClick(View v) {
                //读取指定文件中的内容,并在 editFileIn 文本框章显示出来
                editFileIn.setText(read());}});
        //绑定 btnWrite 按钮的事件监听器
        btnWrite.setOnClickListener(new OnClickListener() {
            @Override
            public void onClick(View source) {
                write(editFileOut.getText().toString());//将 edit1 中的内容写入文件中
                editFileOut.setText("");}});}          //清空 editFileOut 文本框中的内容
    private String read() {
        try {
            //如果手机插入了 SD 卡,而且应用程序具有访问 SD 卡的权限
            if (Environment.getExternalStorageState().equals(
                Environment.MEDIA_MOUNTED)) {
                //获取 SD 卡对应的存储目录
                File sdCardDir = Environment.getExternalStorageDirectory();
                Log.d("FileIO","FileIOActivity");
                //获取指定文件对应的输入流
```

```
                FileInputStream fis = new FileInputStream(
                        sdCardDir.getCanonicalPath() + FILE_NAME);
                //将指定输入流包装成 BufferedReader
                BufferedReader br = new BufferedReader(new InputStreamReader(
                        fis));
                StringBuilder sb = new StringBuilder("");
                String line = null;
                while ((line = br.readLine()) != null) {
                    sb.append(line); }      //循环读取文件内容
                br.close();         //关闭资源
                return sb.toString();}
        } catch (Exception e) {
            e.printStackTrace();}
        return null; }
    private void write(String content) {
        try {
            //如果手机插入了 SD 卡,而且应用程序具有访问 SD 卡的权限
            if (Environment.getExternalStorageState().equals(
                    Environment.MEDIA_MOUNTED)) {
                //获取 SD 卡的目录
                File sdCardDir = Environment.getExternalStorageDirectory();
                File targetFile = new File(sdCardDir.getCanonicalPath()
                        + FILE_NAME);
                //以指定文件创建 RandomAccessFile 对象
                RandomAccessFile raf = new RandomAccessFile(targetFile, "rw");
                //将文件记录指针移动到最后
                raf.seek(targetFile.length());
                //输出文件内容
                raf.write(content.getBytes());
                raf.close();}                  //关闭 RandomAccessFile
            Toast.makeText(SDActivity.this, "保存成功",
                    Toast.LENGTH_LONG).show();           //使用 Toast 显示保存成功
        } catch (Exception e) {
            e.printStackTrace();}}
}
```

上述代码在 write()方法中,使用 RandomAccessFile 向指定的文件追加内容。

对于模拟器而言,其 SD 卡对应的路径为/storage/emulated/0 目录。运行上述代码,在 Android Device Monitor 的 File Explorer 选项卡中,展开 SD 卡所对应的目录,在该目录下可以看到保存的 qstSD.txt 数据文件,如图 6-3 所示。

图 6-3 通过 Android Device Monitor 查看 SD 卡文件

> **注意**
>
> 　　除了使用 Environment.getExternalStorageDirectory()方法来获取 SD 卡的路径外，还可以直接判断 SD 卡所对应的路径是否存在，这样也可以知道手机是否插入了 SD 卡。

6.2.3 文件浏览器

本节将使用 File 类开发一个文件浏览器，用于查看 SD 卡中的文件信息。文件浏览器的 XML 布局文件使用 ListView 组件来显示指定目录中的全部文件和文件夹，代码如下所示。

【代码 6-4】 activity_file_browser.xml

```xml
<RelativeLayout xmlns:android="http://schemas.android.com/apk/res/android" …>
    <!-- 显示当前路径的文本框 -->
    <TextView
        android:id="@+id/path"
        android:layout_width="match_parent"
        android:layout_height="wrap_content"
        android:layout_gravity="center_horizontal"
        android:layout_alignParentTop="true"/>
    <!-- 列出当前路径下所有文件的 ListView -->
    <ListView
        android:id="@+id/list"
        android:layout_width="wrap_content"
        android:layout_height="wrap_content"
        android:divider="#000"
        android:dividerHeight="1px"
        android:layout_below="@id/path"/>
    <!-- 返回上一级目录的按钮 -->
    <Button android:id="@+id/parent"
        android:layout_width="38dp"
        android:layout_height="34dp"
        android:background="@drawable/home"
        android:layout_centerHorizontal="true"
        android:layout_alignParentBottom="true"/>
</RelativeLayout>
```

【代码 6-5】 liner.xml

```xml
<LinearLayout xmlns:android="http://schemas.android.com/apk/res/android" …>
    <!-- 定义一个 ImageView,用于作为列表项的一部分. -->
    <ImageView
        android:id="@+id/icon"
        android:layout_width="40dp"
        android:layout_height="40dp"
        android:paddingLeft="10dp" />
```

```xml
<!-- 定义一个TextView,用于作为列表项的一部分. -->
<TextView
    android:id = "@ + id/file_name"
    android:layout_width = "wrap_content"
    android:layout_height = "wrap_content"
    android:gravity = "center_vertical"
    android:paddingBottom = "10dp"
    android:paddingLeft = "10dp"
    android:paddingTop = "10dp"
    android:textSize = "16sp" />
</LinearLayout>
```

【代码 6-6】 FileBrowserActivity.java

```java
public class FileBrowserActivity extends AppCompatActivity {
    ListView listView;
    TextView textView;
    File currentParent;      //记录当前的父文件夹
    File[] currentFiles;     //记录当前路径下的所有文件的文件数组
    @Override
    public void onCreate(Bundle savedInstanceState) {
        super.onCreate(savedInstanceState);
        setContentView(R.layout.activity_file_browser);
        //获取列出全部文件的ListView
        listView = (ListView) findViewById(R.id.list);
        textView = (TextView) findViewById(R.id.path);
        //获取系统的SD卡的目录
        File root = new File("/storage/emulated/0");
        //如果SD卡存在
        if (root.exists()) {
            currentParent = root;
            currentFiles = root.listFiles();
            //使用当前目录下的全部文件、文件夹来填充ListView
            inflateListView(currentFiles); }
        //为ListView的列表项的单击事件绑定监听器
        listView.setOnItemClickListener(new OnItemClickListener() {
            @Override
            public void onItemClick(AdapterView<?> parent, View view,
                int position, long id) {
                //用户单击了文件,直接返回,不做任何处理
                if (currentFiles[position].isFile())
                    return;
                //获取用户单击的文件夹下的所有文件
                File[] tmp = currentFiles[position].listFiles();
                if (tmp == null || tmp.length == 0) {
                    Toast.makeText(FileBrowserActivity.this,
                        "当前路径不可访问或该路径下没有文件",
                        Toast.LENGTH_SHORT).show();
                } else {
```

```java
                    //获取用户单击的列表项对应的文件夹,设为当前的父文件夹
                    currentParent = currentFiles[position]; //
                    //保存当前的父文件夹内的全部文件和文件夹
                    currentFiles = tmp;
                    //再次更新 ListView
                    inflateListView(currentFiles); }}});
    //获取上一级目录的按钮
    Button parent = (Button) findViewById(R.id.parent);
    parent.setOnClickListener(new OnClickListener() {
        @Override
        public void onClick(View source) {
            try {
                if (!currentParent.getCanonicalPath()
                        .equals("/storage/emulated/0")){
                    //获取上一级目录
                    currentParent = currentParent.getParentFile();
                    //列出当前目录下所有文件
                    currentFiles = currentParent.listFiles();
                    //再次更新 ListView
                    inflateListView(currentFiles); }
            } catch (IOException e) {
                e.printStackTrace();}}});}
    private void inflateListView(File[] files) { //①
        //创建一个 List 集合,List 集合的元素是 Map
        List<Map<String, Object>> listItems
                = new ArrayList<Map<String, Object>>();
        for (int i = 0; i < files.length; i++) {
            Map<String, Object> listItem = new HashMap<String, Object>();
            //如果当前 File 是文件夹,使用 folder 图标,否则使用 file 图标
            if (files[i].isDirectory()) {
                listItem.put("icon", R.drawable.folder);
            } else {
                listItem.put("icon", R.drawable.file);}
            listItem.put("fileName", files[i].getName());
            listItems.add(listItem);}                  //添加 List 项
        SimpleAdapter simpleAdapter = new SimpleAdapter(this, listItems,
                R.layout.liner, new String[] { "icon", "fileName" }, new int[] {
                        R.id.icon, R.id.file_name }); //创建一个 SimpleAdapter
        listView.setAdapter(simpleAdapter);             //为 ListView 设置 Adapter
        try {
            textView.setText("当前路径为: " + currentParent.getCanonicalPath());
        } catch (IOException e) {
            e.printStackTrace();}}
}
```

上述代码中,主要使用 File 的 listFiles()方法来获取指定目录中的所有文件及文件夹,①处所定义的 inflateListView()方法实现使用 File[]数组来填充 ListView 组件,填充时程序会根据文件的类型设置相应的图标。

运行结果如图 6-4 所示。

第6章 数据存储

图 6-4　文件浏览器

6.3　使用 SharedPreferences

SharedPreferences 能够保存简单格式的数据，主要用于类似配置信息格式的数据，这些数据都以 key-value 键值对形式存储在 XML 文件中。

6.3.1　SharedPreferences 和 SharedPreferences.Editor 接口

使用 SharedPreferences 方式存储数据时，需要用到 SharedPreferences 和 SharedPreferences.Editor 接口，这两个接口位于 android.content 包中。其中 SharedPreferences 接口提供了获取数据的方法，其常用的方法及功能如表 6-3 所示。

表 6-3　SharedPreferences 常用方法

方　　法	功　能　描　述
boolean contains(String key)	判断 SharedPreferences 是否包含指定 key 的数据
SharedPreferences.Editor edit()	返回 SharedPreferences.Editor 编辑对象
Map<String,?> getAll()	获取 SharedPreferences 中所有 key-value 对，返回值的类型为 Map 类型
xxx getXxx(String key, xxx defValue)	返回 SharedPreferences 中指定 key 的数据值，如果 key 不存在，则返回指定的默认 defValue 值。xxx 是数据类型，可以是 String、boolean、int、long、float

SharedPreferences 接口本身没有提供写入数据的能力，需要使用 SharedPreferences.Editor 内部接口来实现。调用 SharedPreferences 的 edit() 方法即可获得所对应的 Editor 编辑对象。SharedPreferences.Editor 编辑接口中常用的方法如表 6-4 所示。

表 6-4　SharedPreferences.Editor 接口常用方法

方　　法	功　能　描　述
SharedPreferences.Editor clear()	清除 SharedPreferences 中所有数据
SharedPreferences.Editor putXxx(String key, xxx value)	将指定 key 所对应的数据保存到 SharedPreferences 中。xxx 是数据类型，可以是 String、boolean、int、long、float
SharedPreferences.Editor remove(String key)	删除 SharedPreferences 中指定 key 所对应的数据
boolean commit()	但 Editor 编辑完成后，使用该方法提交内容，以便数据保存到 SharedPreferences 中

当使用 SharedPreferences 中的 getXxx()方法获取数据,以及使用 SharedPreferences. Editor 的 putXxx()方法保存数据时,需要根据数据的类型调用相应的方法,例如,获取一个整型数据时,使用 getInt()方法;而保存一个整型数据时,则使用 putInt()方法。

> **注意**
>
> SharedPreferences 和 SharedPreferences. Editor 需要组合使用,SharedPreferences 负责读取数据,而 SharedPreferences. Editor 负责保存数据。

SharedPreferences 本身只是一个接口,不能直接实例化,只能通过 Context 上下文对象所提供的 getSharedPreferences()方法来获取 SharedPreferences 实例对象。关于 getSharedPreferences(String name,int mode)方法的参数说明如下。

- 参数 name 用于指定存储数据的 XML 文件名,该文件名无需后缀(.xml),系统会自动添加.xml 后缀,并在/data/data/包名/shared_prefs/目录中创建该文件。
- 参数 mode 用于设定文件的操作模式,取值可以是 Context. MODE_WORLD_READABLE(可读)、Context. MODE_WORLD_WRITEABLE(可写)和 Context. MODE_PRIVATE(私有)三种。

> **注意**
>
> 从 Android 4.2 开始不再推荐 MODE_WORLD_READABLE(可读)和 MODE_WORLD_WRITEABLE(可写)这两种模式。

6.3.2 SharedPreferences 操作步骤

使用 SharedPreferences 进行数据操作时步骤如下:
(1)使用 getSharedPreferences()方法获取一个 SharedPreferences 实例对象;
(2)使用 SharedPreferences 实例对象的 edit()方法,获取 SharedPreferences. Editor 编辑对象;
(3)使用 SharedPreferences. Editor 编辑对象的 putXxx()方法来保存数据;
(4)使用 SharedPreferences. Editor 编辑对象的 commit()方法将数据提交到 XML 文件中;
(5)使用 SharedPreferences 对象的 getXxx()方法来读取数据。
下述代码演示如何使用 SharedPreferences 进行数据操作。

【代码 6-7】 activity_shared_preferences. xml

```
<?xml version = "1.0" encoding = "utf - 8"?>
<LinearLayout xmlns:android = "http://schemas.android.com/apk/res/android"
    android:layout_width = "match_parent"
    android:layout_height = "match_parent"
    android:gravity = "center">
    <Button
        android:layout_width = "wrap_content"
```

```
            android:layout_height = "wrap_content"
            android:text = "写入"
            android:id = "@ + id/btnWriter"
            android:layout_gravity = "center"
            android:layout_marginRight = "10dp" />
    <Button
            android:layout_width = "wrap_content"
            android:layout_height = "wrap_content"
            android:text = "读取"
            android:id = "@ + id/btnReader"
            android:layout_gravity = "center" />
</LinearLayout>
```

【代码 6-8】 SharedPreferencesActivity.java

```java
public class SharedPreferencesActivity extends AppCompatActivity {
    SharedPreferences preferences;
    SharedPreferences.Editor editor;
    @Override
    public void onCreate(Bundle savedInstanceState) {
        super.onCreate(savedInstanceState);
        setContentView(R.layout.activity_shared_preferences);
        //获取 SharedPreferences 对象
        preferences = getSharedPreferences("qstPreferences",
            Context.MODE_PRIVATE);
        editor = preferences.edit();
        Button read = (Button) findViewById(R.id.btnWriter);
        Button write = (Button) findViewById(R.id.btnReader);
        read.setOnClickListener(new OnClickListener() {
            @Override
            public void onClick(View arg0) {
                //读取字符串数据
                String time = preferences.getString("time", null);
                //读取 int 类型的数据
                int randNum = preferences.getInt("random", 0);
                String result = time == null ? "您暂时还未写入数据" :
                    "写入时间为:" + "\n" + time + "\n上次生成的随机数为：" + randNum;
                Toast.makeText(SharedPreferencesActivity.this, result,
                    Toast.LENGTH_SHORT).show();}});        //使用 Toast 提示信息
        write.setOnClickListener(new OnClickListener() {
            @Override
            public void onClick(View arg0) {
                SimpleDateFormat sdf = new SimpleDateFormat("yyyy 年 MM 月 dd 日 "
                    + "hh:mm:ss");
                editor.putString("time", sdf.format(new Date()));        //存入当前时间
                editor.putInt("random", (int) (Math.random() * 100));    //存入一个随机数
                editor.commit();}});}                                    //提交所有存入的数据
}
```

上述代码中,在保存 Date 日期时,由于 SharedPreferences 不能直接保存 Date 类型的数据,因此需要使用 SimpleDateFormat 将日期转换成一个字符串后,再调用 putString()方法进行保存。运行结果如图 6-5 所示。

图 6-5 使用 SharedPreferences 进行数据的存储和读取

打开 Android Device Monitor 的 File Explorer 选项卡,展开/data/data/com.qst.chapter06/shared_prefs 目录,查看 qstPreferences.xml 文件,如图 6-6 所示。

图 6-6 通过 Android Device Monitor 查看 qstPreferences.xml 文件

使用导出文件按钮 将 qstPreferences.xml 文件导出到计算机中,在计算机中查看其内容,文件中的数据如下所示。

【代码 6-9】 qstPreferences.xml

```
<?xml version = '1.0' encoding = 'utf-8' standalone = 'yes' ?>
< map >
    < string name = "time">2016 年 10 月 08 日 07:20:29</string>
    < int name = "random" value = "39" />
</map>
```

6.4 SQLite 数据库

SQLite 是一种免费、开源的轻量级数据库，Android 系统中已集成了 SQLite。SQLite 只是一个嵌入式的数据库引擎，其底层就是一个数据库文件，不需占用系统太多资源，在内存中不到 1MB 的内存空间就可运行，因此被广泛地应用在资源有限的小型移动设备（如手机、PDA 等）上来存取适量数据。

6.4.1 SQLite 简介

SQLite 数据库支持绝大部分的 SQL 92 语法，并为数据的增、删、改、查等操作提供了高效的方法，允许使用 SQL 语句操作数据库中的数据，使用非常方便。

SQLite 数据库具有以下特征。

（1）轻量级。大多数数据库的读写模型是基于 C/S 架构设计的，该架构下的数据库分为客户端和服务器端。C/S 架构数据库是重量型的数据库，系统功能复杂且尺寸较大。SQLite 和 C/S 模式的数据库软件不同，不使用分布式架构作为数据引擎。SQLite 数据库功能简单且尺寸较小，一般只需要带上 DDL，就可使用 SQLite 数据库。

（2）独立性好。SQLite 与底层操作系统无关，其核心引擎既不需要安装，也不依赖任何第三方软件，SQLite 几乎能在所有的操作系统上运行，具有较好的独立性。

（3）操作简单。提供了基本数据库、表以及记录的操作，包括数据库的创建、数据库的删除、表的创建、表的删除、记录的插入、记录的删除、记录的更新、记录的查询。

（4）便于管理和维护。SQLite 数据库具有较强的数据隔离性。SQLite 的一个文件包含了数据库的所有信息（例如表、视图、触发器），有利于数据的管理和维护。

（5）可移植性好。SQLite 数据库应用可快速无缝移植到大部分操作系统，如 Android、Windows Mobile、Symbian、Palm 等。

（6）语言无关。SQLite 数据库与语言无关，支持很多语言，例如 Python、.Net、C/C++、Java、Ruby、Perl 等。

（7）事务性。SQLite 数据库采用独立事务处理机制，SQLite 遵守 ACID（Atomicity、Consistency、Isolation、Durability）原则，使用数据库的独占性和共享锁处理事务。此种方式规定必须获得该共享锁后才能执行写操作。因而，SQLite 既允许数据库被多个进程并发读取，又能保证最多只有一个进程写数据。这种方式可有效地防止读脏数据、不可重复读、丢失修改等异常。

6.4.2 SQLiteDatabase 类

Android 提供了创建和使用 SQLite 数据库的 API。其中，SQLiteDatabase 代表一个数据库对象，提供了操作数据库的一些方法，通过以下几个静态方法可以打开数据库。

- **openDatabase(String path，SQLiteDatabase.CursorFactory factory，int flags)**：打开 path 所指定的 SQLite 数据库；
- **openOrCreateDatabase(String path，SQLiteDatabase.CursorFactory factory)**：打开或

创建（如果文件不存在）path 所指定的 SQLite 数据库；
- **openOrCreateDatabase**（File file，SQLiteDatabase.CursorFactory factory）：打开或创建（如果文件不存在）file 所指定的 SQLite 数据库。

获取 SQLiteDatabase 对象后，可以调用相应的方法对 SQLite 数据库进行操作。SQLiteDatabase 类常用的操作方法如表 6-5 所示。

表 6-5 SQLiteDatabase 常用操作方法

方 法	功 能 描 述
insert(String table, String nullColumnHack, ContentValues values)	插入一条记录
delete(String table, String whereClause, String[] whereArgs)	删除一条记录
query (boolean distinct, String table, String[] columns, String selection, String[] selectionArgs, String groupBy, String having, String orderBy, String limit)	查询记录
update(String table, ContentValues value, String whereClause, String[] whereArgs)	修改记录
execSQL(String sql)	执行一条 SQL 语句
rawQuery(String sql, String[] selectionArgs)	执行带占位符的 SQL 查询
beginTransaction()	开始事务
endTransaction()	结束事务
close()	关闭数据库

6.4.3 SQLite 数据库的创建和删除

1. 创建或打开数据库

使用 openDatabase()方法打开指定的数据库时，需要以下三个参数：
- path 用于指定数据库的路径，若指定的数据库不存在，则抛出 FileNotFoundException 异常。
- factory 用于构造查询时的游标，若 factory 为 null，则表示使用默认的 factory 构造游标。
- flags 指定了数据库打开的模式，SQLite 定义了 4 种数据库打开模式，分别是 OPEN_READONLY（只读）、OPEN_READWRITE（可读可写）、CREATE_IF_NECESSARY（若数据库不存在则先创建数据库）、NO_LOCALIZED_COLLATORS（不按照本地化语言对数据进行排序）。数据库打开模式可以同时指定多个，中间使用"|"进行分隔即可。

【示例】 使用 openDatabase()方法打开指定的数据库

```
SQLiteDatabase sqliteDatabase = SQLiteDatabase
        .openDatabase("qst_Student.db", null, NO_LOCALIZED_COLLATORS);
```

使用 openOrCreateDatabase()方法打开或创建数据库时，数据库默认不按照本地化语言对数据进行排序，其作用同 openDatabase(path, factory, CREATE_IF_NECESSARY)一样。因为创建 SQLite 数据库的过程就是在文件系统中创建一个 SQLite 数据库的文件，所以应用程序必须对创建数据库的目录具有可写的权限，否则会抛出 SQLiteException 异常。

【示例】 使用 openOrCreateDatabase()方法打开或创建指定的数据库

```
SQLiteDatabase sqliteDatabase = SQLiteDatabase
        .openOrCreateDatabase ("qst_Student.db", null);
```

2．删除数据库

Context 上下文环境提供 deleteDatabase() 方法来删除指定的数据库。例如，在 Activity 中可使用下面示例代码来删除指定的数据库。

【示例】 使用 deleteDatabase()方法删除数据库

```
deleteDatabase("qst_Student.db");    //删除数据库 qst_Student.db
```

3．关闭数据库

调用 SQLiteDatabase 实例对象的 close() 方法可以关闭数据库，代码如下所示。

【示例】 使用 close()方法关闭数据库

```
sqliteDatabase.close(); //关闭数据库,sqliteDatabase 是一个实例对象
```

6.4.4 表的创建和删除

1．创建表

数据库包含多个表，每个表可存储多条记录。SQLite 没有专门提供方法来创建表，但是可以通过 execSQL()方法来执行创建表的 SQL 语句，代码如下所示。

【示例】 使用 execSQL()方法创建表

```
//创建表的 SQL 语句
String sql = "CREATE TABLE student(ID INTEGER PRIMARY KEY, age INTEGER,name TEXT)";
//执行该 SQL 语句创建表
sqliteDatabase.execSQL(sql);
```

2．删除表

SQLite 没有专门提供方法来删除表，也可以通过 execSQL()方法来执行删除表的 SQL 语句，代码如下所示。

【示例】 使用 execSQL()方法删除表

```
String sql = "DROP TABLE student";        //删除表的 SQL 语句
sqliteDatabase.execSQL(sql);              //删除 student 表
```

6.4.5 记录的插入、修改和删除

1．插入记录

向表中插入记录有两种实现方式：insert()方法和 execSQL()方法。

使用 SQLiteDatabase 的 insert()方法向 SQLite 数据库的表中插入数据,语法格式如下。

【语法】

```
insert(String table, String nullColumnHack, ContentValues values)
```

第 1 个参数 table 是需要插入数据的表名称;第 2 个参数 nullColumnHack 是空列的默认值;第 3 个参数 values 是 ContentValues 类型的对象,用于封装列名和列值的 Map 集合,代表一条记录信息。

【示例】 使用 insert()方法插入记录

```
ContentValues contentValues = new ContentValues();     //创建 ContentValues 对象
//将 ID、age 和 name 放入 contentValues
contentValues.put("ID", 1);
contentValues.put("age", 26);
contentValues.put("name", "StudentA");
//调用 insert()方法将 contentValues 对象封装的数据插入到 student 表中
sqliteDatabase.insert("student", null, contentValues);
```

ContentValues 可以对数据进行封装,在使用的时候能够更加便捷,因此 Android 推荐使用 ContentValues 来代替 SQL 语句进行数据操作。ContentValues 存储的值只能是基本类型,不能是对象类型。

使用 execSQL()方法向数据库表中插入数据时,需要先编写插入数据的 SQL 语句,然后使用 execSQL()方法执行该语句完成数据的插入,示例代码如下所示。

【示例】 使用 execSQL()方法插入记录

```
//定义插入 SQL 语句
String sql = "INSERT INTO student (ID,age,name) values (1, 26, 'StudentA')";
//调用 execSQL()方法执行 SQL 语句,将数据插入 student 表中
sqliteDatabase.execSQL(sql);
```

2. 修改记录

与插入记录类似,更新记录也有两种实现方式:update()方法和 execSQL()方法。

使用 SQLiteDatabasede 的 update()方法可以对数据库表中的数据进行更新,语法格式如下。

【语法】

```
update(String table, ContentValues value, String whereClause, String[] whereArgs)
```

第 1 个参数 table 是需要更新数据的表的名称;第 2 个参数 value 是更新的记录信息,ContentValues 对象类型的数据;第 3 个参数 whereClause 是更新条件(where 子句);第 4 个参数 whereArgs 是更新条件所需的参数数组。

【示例】 使用 update()方法修改记录

```
ContentValues contentValues = new ContentValues();     //创建 ContentValues 对象
contentValues.put("ID", 1);
```

```
contentValues.put("age", 25);                          //更新 age 为 25
contentValues.put("name", "StudentA");
//调用 update()方法更新 student 表中名为 StudentA 的数据
sqliteDatabase.update("student", contentValues, "name = StudentA", null);
```

使用 execSQL()方法更新数据时,需先编写更新数据的 SQL 语句,然后使用 execSQL()方法执行该语句实现数据的更新,示例代码如下所示。

【示例】 使用 execSQL()方法修改记录

```
String sql = "UPDATE student SET age = 25 where name = 'StudentA'";   //定义更新 SQL 语句
sqliteDatabase.execSQL(sql);      //调用 execSQL()方法执行 SQL 语句更新 student 表中的记录
```

3. 删除记录

删除记录也有以两种实现方式:delete()方法和 execSQL()方法。

使用 SQLiteDatabasede 的 delete()方法可以删除数据库表中的表数据,语法格式如下。

【语法】

```
delete(String table,String whereClause,String[] whereArgs)
```

第 1 个参数 table 是需要删除数据的表的名称;第 2 个参数 whereClause 是删除条件;第 3 个参数 whereArgs 是删除条件所需的参数数组。

【示例】 使用 delete()方法删除记录

```
sqliteDatabase.delete("student","name = ?",new String[]{"StudentA"});
```

使用 execSQL()方法删除表数据需要先编写删除记录的 SQL 语句,然后使用 execSQL()方法执行该语句实现数据的删除,示例代码如下所示。

【示例】 使用 execSQL()方法删除记录

```
Stringsql = "DELETE FORM student where name = 'StudentA'";   //定义更新 SQL 语句
sqliteDatabase.execSQL(sql);      //调用 execSQL()方法执行 SQL 语句更新 student 表中的记录
```

6.4.6 数据查询与 Cursor 接口

使用 SQLiteDatabase 的 query()方法可以查询记录。SQLiteDatabase 中提供了 6 种 query()方法用于不同方式的查询,常用的 query()方法的语法格式如下。

【语法】

```
public Cursor query (boolean distinct, String table, String[] columns,
    String selection, String[] selectionArgs, String groupBy, String having,
    String orderBy, String limit);
```

distinct 是一个可选的布尔值,用来说明返回的值是否只包含唯一的值;table 是表名称;columns 是由列名称构成的数组;selection 是条件 where 子句,可以包含"?"通配符,在子句中用作占位符;selectionArgs 是参数数组,替换 where 子句中的"?"占位符;groupBy

表示分组列;having 是分组条件;orderBy 是排序列;limit 是一个可选的字符串,用来对返回的行数进行限制。

query()方法返回一个 Cursor 游标对象,相当于 JDBC 中的结果集 ResultSet。游标提供了一种对表检索操作的灵活方式,其实质是一种能够从检索的结果集中每次提取一条记录的机制。游标是由结果集(可以是 0 条、1 条或由相关的选择语句检索出的多条记录)和结果集中指向特定记录的游标位置组成。当决定对结果集进行处理时,必须声明一个指向该结果集的游标。Cursor 游标常用的方法如表 6-6 所示。

表 6-6 Cursor 游标常用方法

方 法	功 能 描 述
move(int offset)	以当前的位置为基准,将 Cursor 移动到偏移量为 offset 的位置。移动成功则返回 true,失败则返回 false。当 offset 为正值时,游标向前移动,负值时向后移动
moveToPosition(int position)	将 Cursor 移动到绝对位置 position 处。移动成功则返回 true,失败则返回 false。需要注意的是:moveToPosition 移动到一个绝对位置,而 move 以当前位置为基准移动
moveToNext()	将 Cursor 向前移动一个位置,成功则返回 true,失败返回 false,其功能等同于 move(1)
moveToLast()	将 Cursor 移动到最后一条记录,成功则返回 true,失败则返回 false。若当前记录数为 count,则其功能等同于 moveToPosition(count)
moveToFisrt()	将 Cursor 移动到第一条记录,成功则返回 true,失败则返回 false,其功能等同于 moveToPosition(1)
isBeforeFirst()	判断 Cursor 是否指向第一项数据之前。若指向第一项数据之前,则返回 true,否则返回 false
isAfterLast()	判断 Cursor 是否指向最后一项数据之后。若指向最后一项数据之后,则返回 true,否则返回 false
isClosed()	判断 Cursor 是否关闭。若 Cursor 关闭则返回 true,否则返回 false
isFirst()	判断 Cursor 是否指向第一项记录
isLast()	判断 Cursor 是否指向最后一条记录
isNull(int columnIndex)	判断指定的位置 columnIndex 的记录是否存在
getCount()	获取当前表的行数(即记录总数)
getInt(int columnIndex)	获取指定列索引的 int 类型值
getString(int columnIndex)	获取指定列索引的 String 类型值

【示例】 使用 query()方法查询记录

```
Cursor cursor = sqliteDatabase.query(true, "student", null, "name = StudentA", null,
        null, null,null,null);                  //查询获得游标
//将游标移动到第一条记录,并判断
if(cursor.moveToFirst()){
    //获得列信息
    int id = cursor.getInt(0);
    int age = cursor.getInt(1);
    String name = cursor.getString(3);
    Log.debug("id + ":" + age + ":" + name ");      //输出
}
```

6.4.7 事务处理

SQLiteDatabase 提供以下几个方法来控制事务：
- beginTransaction()方法用于开始事务。
- endTransaction()方法用于结束事务。
- inTransaction()方法用于判断当前上下文是否处于事务环境中。如果当前上下文处于事务中，则返回 true；否则返回 false。
- setTransactionSuccessful()方法用来设置事务成功标志。如果程序在事务执行过程中调用该方法设置事务成功，则可以提交事务；否则程序将对事务进行回滚。

下述示例代码演示事务处理过程。

【示例】 事务处理过程

```
sqliteDatabase.beginTransaction();           //开始事务
try{
    ...//执行 DML 语句
    //调用 setTransactionSuccessful()方法设置事务成功，
    //否则 endTransaction()方法将回滚事务
    sqliteDatabase.setTransactionSuccessful();
}finally{
    sqliteDatabase.endTransaction();         //由事务标志决定是提交事务，还是回滚事务
}
```

6.4.8 SQLiteOpenHelper 类

SQLiteOpenHelper 是 SQLiteDatabase 的一个帮助类，用来管理数据库的创建和版本更新。通过继承 SQLiteOpenHelper 类可以隐藏开发过程中不需要直接调用的方法。通常需要定义一个类来继承 SQLiteOpenHelper，并重写 onCreate()和 onUpgrade()两个方法。SQLiteOpenHelper 类的常用方法如表 6-7 所示。

表 6-7 SQLiteOpenHelper 常用方法

方　法	功　能　描　述
SQLiteOpenHelper(Context context, String name, SQLiteDatabase.CursorFactory, int version)	构造函数，第二个参数是数据库名称
onCreate(SQLiteDatabase db)	创建数据库时调用
onUpgrade(SQLiteDatabase db, int oldVersion, int newVersion)	版本更新时调用
getReadableDatabase()	创建或打开一个只读数据库
getWritableDatabase()	创建或打开一个读写数据库

下面使用 SQLiteOpenHelper 来实现音乐播放列表的添加、删除和浏览功能，具体步骤如下。

(1) 创建一个数据库工具类 DBHelper，该类继承 SQLiteOpenHelper，并重写 onCreate()和 onUpgrade()方法，然后添加 insert()、delete()和 query()方法，分别实现数据的添加、删除和查询功能。

【代码6-10】 DBHelper.java

```java
public class DBHelper extends SQLiteOpenHelper {
    private static final String DB_NAME = "music.db";    //数据库名称
    private static final String TBL_NAME = "MusicTbl";   //表名
    private SQLiteDatabase db;                            //声明SQLiteDatabase对象
    //构造函数
    DBHelper(Context c) {
        super(c, DB_NAME, null, 2); }
    @Override
    public void onCreate(SQLiteDatabase db) {
        this.db = db;                                     //获取SQLiteDatabase对象
        String CREATE_TBL = "create table MusicTbl(_id integer primary key
            autoincrement,name text,singer text) ";       //创建表
        db.execSQL(CREATE_TBL); }
    //插入
    public void insert(ContentValues values) {
        SQLiteDatabase db = getWritableDatabase();
        db.insert(TBL_NAME, null, values);
        db.close();}
    //查询
    public Cursor query() {
        SQLiteDatabase db = getReadableDatabase();;
        Cursor cursor = db.query(TBL_NAME, null, null, null, null, null, null);
        return cursor; }
    //删除
    public void del(int id) {
        if (db == null)
            db = getWritableDatabase();
        db.delete(TBL_NAME, "_id = ?", new String[] { String.valueOf(id) });}
    //关闭数据库
    public void close() {
        if (db != null)
            db.close();}
    @Override
    public void onUpgrade(SQLiteDatabase db, int oldVersion, int newVersion) {
    }
}
```

（2）创建添加音乐的 AddMusicActivity 及对应的 XML 布局文件。在布局文件中提供两个文本框和一个按钮，文本框分别用于输入音乐名和歌手名，当单击"添加"按钮时，将数据插入到表中，代码如下所示。

【代码6-11】 add.xml

```xml
<LinearLayout xmlns:android = "http://schemas.android.com/apk/res/android" ...>
</LinearLayout>
```

【代码6-12】 AddMusicActivity.java

```java
public class AddActivity extends AppCompatActivity {
    private EditText et1, et2;
    private Button b1;
```

```
@Override
public void onCreate(Bundle savedInstanceState) {
    super.onCreate(savedInstanceState);
    setContentView(R.layout.add);
    this.setTitle("添加收藏信息");
    et1 = (EditText) findViewById(R.id.EditTextName);
    et2 = (EditText) findViewById(R.id.EditTextSinger);
    b1 = (Button) findViewById(R.id.ButtonAdd);
    b1.setOnClickListener(new OnClickListener() {
        public void onClick(View v) {
            String name = et1.getText().toString();   //获取用户输入的文本信息
            String singer = et2.getText().toString();
            ContentValues values = new ContentValues();
                                                //创建 ContentValues 对象,封装记录信息
            values.put("name", name);
            values.put("singer", singer);
            //创建数据库工具类 DBHelper
            DBHelper helper = new DBHelper(getApplicationContext());
            helper.insert(values);                   //调用 insert()方法插入数据
            //跳转到 QueryActivity,显示音乐列表
            Intent intent = new Intent(AddMusicActivity.this,
                QueryActivity.class);
            startActivity(intent);}});}
}
```

运行上述代码,当单击"添加"按钮时,先将用户输入的音乐名和歌手信息封装到 ContentValues 对象中,再调用 DBHelper 的 insert()方法将数据插入数据库中,最后跳转到 QueryActivity 显示音乐列表。

(3) 创建显示音乐列表的 QueryActivity。

【代码 6-13】 row.xml

```
<LinearLayout xmlns:android = "http://schemas.android.com/apk/res/android" …>
    <LinearLayout
        android:orientation = "horizontal"
        android:layout_width = "match_parent"
        android:layout_height = "wrap_content">
    <TextView
        android:layout_width = "wrap_content"
        android:layout_height = "wrap_content"
        android:text = "Text0"
        android:textColor = "#000"
        android:textSize = "20sp"
        android:id = "@ + id/text0"
        android:layout_weight = "1"
        android:gravity = "left " />
    <TextView
        android:layout_width = "wrap_content"
        android:layout_height = "wrap_content"
        android:text = "Text1"
        android:textColor = "#000"
        android:textSize = "20sp"
```

```xml
            android:id = "@ + id/text1"
            android:layout_weight = "1"
            android:gravity = "left" />
        <TextView
            android:layout_width = "wrap_content"
            android:layout_height = "wrap_content"
            android:text = "Text2"
            android:textColor = "#000"
            android:textSize = "20sp"
            android:id = "@ + id/text2"
            android:layout_weight = "1"
            android:gravity = "left" />
    </LinearLayout>
</LinearLayout>
```

【代码 6-14】 music_list.xml

```xml
<LinearLayout xmlns:android = "http://schemas.android.com/apk/res/android" ...>
    <ListView
        android:layout_width = "fill_parent"
        android:layout_height = "wrap_content"
        android:id = "@ + id/music_listview"
        android:layout_weight = "1" />
</LinearLayout>
```

【代码 6-15】 QueryActivity.java

```java
public class QueryActivity extends AppCompatActivity {
    ListView listView;
    DBHelper helpter
    @Override
    public void onCreate(Bundle savedInstanceState) {
        super.onCreate(savedInstanceState);
        setContentView(R.layout.music_list);
        listView = (ListView)findViewById(R.id.music_listview);
        this.setTitle("浏览音乐列表信息");
        helpter = new DBHelper(this);
        use_cursor();                                                   //查询数据,获取游标
        final AlertDialog.Builder builder = new AlertDialog.Builder(this);    //提示对话框
        //设置 ListView 单击监听器
        listView.setOnItemClickListener(new OnItemClickListener() {
            @Override
            public void onItemClick(AdapterView<?> arg0, View arg1, int arg2,
                long arg3) {
                final long temp = arg3;
                builder.setMessage("真的要删除该记录吗?")
                    .setPositiveButton("是",
                        new DialogInterface.OnClickListener() {
                            public void onClick(DialogInterface dialog,
                                int which) {
                                dbHelper.del((int) temp);    //删除数据
                                use_cursor();}               //重新查询数据
                    }).setNegativeButton("否",
```

```
                    new DialogInterface.OnClickListener() {
                        public void onClick(DialogInterface dialog,
                            int which) {
                        }});
                    AlertDialog dialog = builder.create();
                    dialog.show();}});
        dbHelper.close();}
    private void use_cursor(){
        Cursor cursor = dbHelper.query();                        //查询数据,获取游标
        String[] from = {"_id","name","singer"};                 //列表项数据
        int[] to = {R.id.text0,R.id.text1,R.id.text2};           //列表项 ID
        SimpleCursorAdapter adapter = new SimpleCursorAdapter(getApplicationContext(),
    R.layout.row,cursor,from,to);                                //适配器
        listView.setAdapter(adapter);}                           //为列表视图添加适配器
}
```

上述代码中,调用 DBHelper 的 query()方法查询数据库并返回一个 Cursor 游标,然后使用 SimpleCursorAdapter 适配器将数据绑定到 ListView 控件上;接下来在 ListView 控件上注册单击监听,当单击某条记录时,显示一个警告对话框提示是否删除,单击"是"按钮,则调用 DBHelper 的 del()方法删除指定记录的信息。

运行程序,输入音乐名称和歌手信息后,单击"添加"按钮添加一条音乐信息,如图 6-7 所示。在音乐列表页面中单击某条记录,弹出警告对话框提示删除一条记录,如图 6-8 所示。单击"是"则删除该记录,单击"否"则取消删除操作。

图 6-7　添加音乐记录

图 6-8　删除音乐记录

6.4.9　使用 ListView 滑动分页

当数据较多的情况下,在一个页面中不能完全显示时,可以使用 ListView 实现滑动分页效果。ListView 滑动分页是经常用到的,用于分页加载数据。

下述代码演示使用 ListView 实现滑动分页。

【代码 6-16】　listview.xml

```
<LinearLayout xmlns:android = "http://schemas.android.com/apk/res/android" …>
    <ListView
```

```xml
android:id = "@ + id/listView1"
android:layout_width = "match_parent"
android:layout_height = "wrap_content" >
    </ListView>
</LinearLayout>
```

【代码 6-17】 listview_item.xml

```xml
< LinearLayout xmlns:android = "http://schemas.android.com/apk/res/android" ··· >
    < TextView
        android:id = "@ + id/list_item_text"
        android:layout_width = "fill_parent"
        android:layout_height = "fill_parent"
        android:gravity = "center"
        android:textSize = "20sp"
        android:paddingTop = "10dp"
        android:paddingBottom = "10dp"/>
</LinearLayout>
```

【代码 6-18】 load_more.xml

```xml
< LinearLayout xmlns:android = "http://schemas.android.com/apk/res/android" ··· >
    < Button
        android:id = "@ + id/loadMoreButton"
        android:layout_width = "fill_parent"
        android:layout_height = "wrap_content"
        android:onClick = "loadMore"
        android:text = "加载更多" />
</LinearLayout>
```

【代码 6-19】 ListViewAdapter.java

```java
public class ListViewAdapter extends BaseAdapter {
    private static Map< Integer, View > m = new HashMap< Integer, View >();
    private List< String > items;
    private LayoutInflater inflater;
    public ListViewAdapter(List< String > items, Context context) {
        super();
        this.items = items;
        this.inflater = (LayoutInflater) context
            .getSystemService(Context.LAYOUT_INFLATER_SERVICE);}
    @Override
    public int getCount() {
        //TODO Auto - generated method stub
        return items.size();}
    @Override
    public Object getItem(int position) {
        //TODO Auto - generated method stub
        return items.get(position);}
    @Override
    public long getItemId(int position) {
        //TODO Auto - generated method stub
        return position;}
```

```java
        @Override
        public View getView(int position, View contentView, ViewGroup arg2) {
            //TODO Auto-generated method stub
            contentView = m.get(position);
            if(contentView == null){
                contentView = inflater.inflate(R.layout.listview_item, null);
                TextView text = (TextView) contentView
                    .findViewById(R.id.list_item_text);
                text.setText(items.get(position));}
            m.put(position, contentView);
            return contentView;}
        public void addItem(String item) {
            items.add(item);   }
    }
```

【代码 6-20】 ListViewActivity.java

```java
public class ListViewActivity extends Activity implements OnScrollListener    {
    List<String> items = new ArrayList<String>();
    private ListView listView;
    private int visibleLastIndex = 0;                    //最后的可视项索引
    private int visibleItemCount;                        //当前窗口可见项总数
    private ListViewAdapter adapter;
    private View loadMoreView;
    private Button loadMoreButton;
    private Handler handler = new Handler();
    @Override
        public void onCreate(Bundle savedInstanceState) {
            super.onCreate(savedInstanceState);
            setContentView(R.layout.listview);
            loadMoreView = getLayoutInflater()
                    .inflate(R.layout.load_more, null);
            loadMoreButton = (Button) loadMoreView
                    .findViewById(R.id.loadMoreButton);
            loadMoreButton.setOnClickListener(new OnClickListener() {
                @Override
                public void onClick(View v) {
                    //TODO Auto-generated method stub
                    loadMoreButton.setText("正在加载...");    //设置按钮文字 loading
                    handler.postDelayed(new Runnable() {
                        @Override
                        public void run() {
                        loadData();
                        adapter.notifyDataSetChanged();         //数据集变化后,通知 adapter
                        //设置选中项
                        listView
                            .setSelection(visibleLastIndex - visibleItemCount + 1);
                        loadMoreButton.setText("加载更多");//恢复按钮文字
                        }
                    }, 1000);
                }});
            listView = (ListView) this.findViewById(R.id.listView1);
            listView.addFooterView(loadMoreView);               //设置列表底部视图
            //listView.addHeaderView(v)                         //设置列表顶部视图
            initAdapter();
```

```java
        listView.setAdapter(adapter);              //自动为 ID 是 list 的 ListView 设置适配器
        listView.setOnScrollListener(this);                    //添加滑动监听
        listView.setOnItemClickListener(new OnItemClickListener() {
            @Override
            public void onItemClick(AdapterView<?> arg0, View view,
                int position, long arg3) {
                //TODO Auto-generated method stub
                Toast.makeText(getApplicationContext(),
                    items.get(position),Toast.LENGTH_SHORT).show();
            }});}
    /**初始化适配器*/
    private void initAdapter(){
        for (int i = 0; i < 12; i++){
            items.add("用户编号: " + String.valueOf(i + 1)); }
            adapter = new ListViewAdapter(items,this);  }
    /**滑动时被调用*/
    @Override
    public void onScroll(AbsListView view, int firstVisibleItem, int
            visibleItemCount, int totalItemCount){
        this.visibleItemCount = visibleItemCount;
        visibleLastIndex = firstVisibleItem + visibleItemCount - 1;}
    /**滑动状态改变时被调用*/
    @Override
    public void onScrollStateChanged(AbsListView view, int scrollState) {
        int itemsLastIndex = adapter.getCount() - 1;//数据集最后一项的索引
        int lastIndex = itemsLastIndex + 1;         //加上底部的 loadMoreView 项
        if(scrollState == OnScrollListener.SCROLL_STATE_IDLE
            && visibleLastIndex == lastIndex) {
            //如果是自动加载,则可以在这里放置异步加载数据的代码
            Log.i("LOADMORE", "loading...");}}
    /**模拟加载数据*/
    private void loadData() {
        int count = adapter.getCount();
        for (int i = count; i < count + 20; i++) {
            adapter.addItem("用户编号: " + String.valueOf(i + 1)); }  }
}
```

运行上述代码,结果如图 6-9 所示。

图 6-9 ListView 向下滑动分页

第 6 章　数据存储

本章总结

小结

- Android 提供了多种数据存储方式，包括文件存储、SharedPreferences 和 SQLite。
- 文件存储方式不受类型限制，可以将一些数据直接以文件的形式保存在设备中。
- Android 支持使用 I/O 流方式来访问手机等移动设备上存储的文件。
- 在 Android 应用程序中，可以通过 Context 上下文环境提供的 openFileInput()和 openFileOuput()方法分别获得文件的输入流和输出流。
- SD 卡是一种基于半导体快闪记忆器的多功能存储卡，扩充了手机的存储能力。
- SharedPreferences 保存的数据都以 key-value 键值对的方式存储在 XML 文件中。
- 使用 SharedPreferences 中的 getXxx()方法获取数据，使用 SharedPreferences.Editor 的 putXxx()方法保存数据。
- SQLite 是一种免费开源、支持多种语言的数据库。
- SQLiteDatabase 代表一个数据库对象，提供了操作数据库的一些方法。
- SQLiteDatabase 的 query()方法的返回值是一个 Cursor 游标对象，可以查询记录。
- SQLiteOpenHelper 是 SQLiteDatabase 的一个帮助类，用来管理数据库的创建和版本更新。
- 使用 ListView 控件可以实现滑动分页。

Q&A

问题：简述 Android 的几种数据存储方式及各自的特点。

回答：Android 提供的数据存储方式有文件存储、SharedPreferences 和 SQLite。其中，文件存储用于保存少量数据，且数据格式无须结构化；SharedPreferences 保存的数据都以 key-value 键值对的方式存储在 XML 文件中，用于少量且数据结构简单的数据；SQLite 是一个轻量级数据库，提供了大量的 API，可以非常便捷地进行添加、删除、更新等操作，适合数据量较多且需要进行结构化存储的情况。

章节练习

习题

1. 在 Android 中，以 XML 文件来存储的方式是_____。
   A. 文件　　　　　　　　　　　B. SharedPreferences
   C. SQLite　　　　　　　　　　D. 网络
2. 在 Android 中，用于存储较多且结构化数据的方式是_____。
   A. 文件　　　　　　　　　　　B. SharedPreferences

C. SQLite   D. 网络
3. 下面说法不正确的是_____。
   A. 文件适合存储无须结构化的数据
   B. SharedPreferences 适合小数据量的存储
   C. SQLite 适合嵌入式设备进行数据存储
   D. Android 应用程序中无法使用 Java 标准的 I/O 机制
4. 下面关于 SQLite 的说法,不正确的是_____。
   A. SQLite 支持事务   B. SQLite 只能用于 Android 系统
   C. SQLite 不支持完整的 SQL 规范   D. SQLite 支持很多语言

上机

1. 训练目标：SQLite 应用。

培养能力	熟练使用 SQLite 保存数据		
掌握程度	★★★★★	难度	中
代码行数	200	实施方式	重复编码
结束条件	保存数据		
参考训练内容 编写代码,读取所有联系人的信息,并存储在自定义的 SQLite 表中			

2. 训练目标：SQLite 应用。

培养能力	熟练使用 SQLite 实现数据的增、删、改、查		
掌握程度	★★★★★	难度	难
代码行数	500	实施方式	重复编码
结束条件	数据的增、删、改、查		
参考训练内容 使用 SQLite 实现图书信息管理系统,图书信息包括书名、书号、价格以及出版日期			

第7章 ContentProvider数据共享

本章任务是完成"GIFT-EMS 礼记"项目中的购买功能：
- 【任务 7-1】 完成购买下单功能，可以从通讯录中获取联系人。
- 【任务 7-2】 完成订单列表和订单回收站功能。

知 识 点	Listen(听)	Know(懂)	Do(做)	Revise(复习)	Master(精通)
ContentProvider 简介	★	★			
开发 ContentProvider 程序	★	★	★	★	★
操作系统的 ContentProvider	★	★	★	★	

7.1 ContentProvider 简介

在某些情况下，Android 应用程序需要对外暴露自己的数据，以便其他应用程序进行访问，从而完成系统中不同 Android 应用程序之间的数据共享，这就需要使用 ContentProvider。ContentProvider 是不同应用程序之间进行数据交换的标准 API，也是所有应用程序之间数据存储和检索的一个桥梁，其作用是使各个应用程序之间实现数据共享。一个应用程序可以通过使用 ContentProvider 将自己的数据共享给其他应用程序，其他应用程序再通过 ContentResolver 来访问共享的数据。

7.1.1 ContentProvider 类

ContentProvider 是 Android 应用的四大组件之一，与 Activity、Service、BroadcastReceiver 类似，需要在 AndroidManifest.xml 配置文件中进行配置。android.content.ContentProvider 类主要功能用于存储、检索数据，并向应用程序提供访问数据的接口，其常用的方法如表 7-1 所示。

表 7-1 ContentProvider 类常用方法

方　　法	功　能　描　述
public abstract boolean onCreate()	创建 ContentProviderh 后会被调用
public abstract Uri insert(Uri uri, ContentValues values)	根据 Uri 插入 values 对应的数据
public abstract int delete(Uri uri, String selection, String[] selectionArgs)	根据 Uri 删除 selection 条件所匹配的全部记录
public abstract int update(Uri uri, ContentValues values, String selection, String[] selectionArgs)	根据 Uri 修改 selection 条件所匹配的全部记录
public abstract Cursor query(Uri uri, String[] projection, String selection, String[] selectionArgs, String sortOrder)	根据 Uri 查询 selection 条件所匹配的全部记录，其中 projection 是一个列名列表，表明只选出指定的数据列
public abstract String getType(Uri uri)	获得当前 Uri 所代表的 MIME 数据类型
public final Context getContext()	获得 Context 对象

ContentProvider 类中的 insert()、delete()、update()、query() 和 getType() 等方法都是抽象方法，因此需要通过实现这些抽象方法来实现对数据进行增、删、改、查操作。

在 ContentProvider 的增、删、改、查操作方法中，都用到类型为 Uri 的参数，Uri 是 ContentProvider 对外提供一个自身数据集的唯一标识。当一个 ContentProvider 管理多个数据集时，该 ContentProvider 将会为每个数据集分配一个独立且唯一的 Uri。Uri 的语法格式如下：

【语法】

content://数据路径/标识 ID(可选)

"content://"是 ContentProvider 规定的协议，用来标识 ContentProvider 所管理的 schema，所有的 Uri 都以"content://"开头；"数据路径"用于查找所要操作的 ContentProvider；"标识 ID"是可选的，标识不同数据资源，当访问不同资源时，该 ID 是动态改变的。

【示例】 返回设备中存储的所有图片的 Uri

content://media/internal/images

【示例】 返回 ID 为 5 的联系人信息的 Uri

content://contacts/people/5

Android 提供了 Uri 工具类来定义 Uri，该工具类的静态方法 parse() 可以将一个字符串转换成 Uri 对象。

【示例】 Uri.parse()静态方法

```
Uri uri = Uri.parse("content://media/internal/images");
Uri uri = Uri.parse("content://contacts/people/5");
```

Android 系统提供了 UriMatcher 工具类对 Uri 进行匹配判断，该工具类提供了以下两个常用的方法：①void addURI(String authority, String path, int code)，用于注册 Uri，其中参数 authority 和 path 组合成一个 Uri，而参数 code 代表 Uri 对应的标识码；②int match(Uri uri)，根据前面注册的 Uri 判断指定的 Uri 对应的标识码，如果找不到匹配的标识码，则返回-1。

【示例】 UriMatcher 工具类的使用

```
UriMatcher matcher = new UriMatcher(UriMatcher.NO_MATCH);
matcher.addURI("contacts","people/#",1);
matcher.match(Uri.parse("content://contacts/people/5");
```

除了 UriMatcher，Android 还提供了 ContentUris 工具类，该工具类用于操作 Uri 字符串，其提供的两个方法如下：①withAppendedId(uri,id)，用于为 Uri 路径加上 ID 部分；②parseId(uri)，用于从指定的 Uri 中解析出所包含的 ID 值。

【示例】 ContentUris 工具类的使用

```
Uri uri = Uri.parse("content://qdu.edu/student");
Uri resultUri = ContentUris.withAppendedId(uri, 3);
//生成后的 Uri 为 content://qdu.edu/student/3
Uri uri = Uri.parse("content://qdu.edu/student/3");
long personid = ContentUris.parseId(uri);          //获取的结果为 3
```

7.1.2 ContentResolver 类

ContentProvider 中共享的数据不能被 Android 应用程序直接访问，而是通过操作 ContentResolver 来间接操作 ContentProvider 中的数据。ContentResolver 是内容解析器，提供了对 ContentProvider 数据进行查询、插入、修改和删除等操作的方法。通常情况下，ContentProvider 是单实例模式的，当多个应用程序通过 ContentResolver 来操作 ContentProvider 中的数据时，ContentResolver 操作将会委托给同一个 ContentProvider 进行处理。

每个应用程序的上下文都有一个默认的 ContentResolver 实例对象，通过 Context 的 getContentResolver()方法来获取 ContentResolver 实例对象，示例代码如下所示。

【示例】 获取默认的 ContentResolver 实例对象

```
ContentResolver cr = getContentResolver();    //Activity中获得默认的 ContentResolver 对象
```

ContentResolver 类常用的方法如表 7-2 所示。

表 7-2 ContentResolver 类常用方法

方 法	功 能 描 述
insert(Uri uri, ContentValues values)	向 Uri 对应的 ContentProvide 中插入 values 对应的数据
delete(Uri uri, String where, String[] selectionArgs)	删除 Uri 对应的 ContentProvide 中 where 匹配的数据
update(Uri uri, ContentValues values, String where, String[] selectionArgs)	更新 Uri 对应的 ContentProvide 中 where 匹配的数据
query(Uri uri, String[] projection, String selection, String[] selectionArgs, String sortOder)	查询 Uri 对应的 ContentProvide 中 where 匹配的数据

下述代码使用 ContentResolver 的 query()方法来查询数据并返回一个指向结果集的游标 Cursor。

【示例】 查询

```
ContentResolver resolver = getContentResolver();    //获取 ContentResolver 对象
Cursor cursor = resolver.query(Contacts.CONTENT_URI, null, null, null, null);
```

其中，常量 CONTENT_URI 用来标识某个特定的 ContentProvider 和数据集。

ContentResolver.insert()方法用于向 ContentProvider 中插入一个新的记录，并返回一个 Uri，该 Uri 的内容是由 ContentProvider 的 Uri 加上新记录的 ID 扩展得到的。下述代码演示 insert()方法的用法。

【示例】 向 ContentProvider 插入数据

```
ContentValues contentValues = new ContentValues();
values.put(Contacts._ID, 1);                                //联系人 ID
contentValues.put(Contacts.DISPLAY_NAME, "zhangsan");       //联系人名
ContentResolver resolver = getContentResolver();            //获取 ContentResolver 对象
Uri uri = resolver.insert(Contacts.CONTENT_URI, contentValues); //插入
```

使用 ContentResolver.insert()方法向 ContentProvider 中增加记录时，需要先将数据封装到 ContentValues 对象中，然后调用 ContentResolver.insert()方法保存数据。

下述代码使用 ContentResolver.update()方法实现记录的更新操作。

【示例】 更新 ContentProvider 中的数据

```
//创建一个新值
ContentValues contentValues = new ContentValues();
contentValues.put(Contacts.DISPLAY_NAME, "zhangsan");
ContentResolver resolver = getContentResolver();       //获取 ContentResolver 对象
resolver.update(Contacts.CONTENT_URI, contentValues, "_id = 5", null);  //更新
```

下述代码使用 ContentResolver.delete()方法删除记录。

【示例】 删除 ContentProvider 中的数据

```
ContentResolver resolver = getContentResolver();       //获取 ContentResolver 对象
//删除单个记录
resolver.delete(Uri.withAppendedPath(Contacts.CONTENT_URI, 41), null, null);
//删除前 5 行记录
resolver.delete(Contacts.CONTENT_URI, "_id < 5", null);
```

第 7 章 ContentProvider 数据共享

如果要删除单个记录,可以调用 ContentResolver.delete()方法,通过给该方法传递一个特定行的 Uri 对象来实现删除操作。如果要对多行记录执行删除操作,就需要给 delete()方法传递被删除的记录类型的 Uri 和 where 条件子句。

7.2 开发 ContentProvider 程序

开发 ContentProvider 程序的步骤如下:
(1) 创建一个 ContentProvider 子类,并实现 query()、insert()、update()和 delete()等方法。
(2) 在 AndroidManifest.xml 配置文件中注册 ContentProvider,并指定 android:authorities 属性。
(3) 使用 ContentProvider。Activity 和 Service 等组件都可以获取 ContentProvider 对象,并调用该对象相应的方法进行操作。

7.2.1 编写 ContentProvider 子类

下述代码创建一个 ContentProvider 子类,并实现 query()、insert()、update()和 delete()方法。

【代码 7-1】 FirstProvide.java

```java
public class FirstProvider extends ContentProvider {
    //第一次创建该 CoontentProvide 时调用该方法
    @Override
    public boolean onCreate() {
        Log.i("FirstProvider"," === onCreate 方法被调用 === ");
        return true;}
    //实现查询方法,该方法返回查询得到的 Cursor
    @Override
    public Cursor query(Uri uri, String[] projection, String where,
              String[] whereArgs, String sortOrder) {
        Log.i("FirstProvider"," === query 方法被调用 === ");
        Log.i("FirstProvider","uri 参数为: " + uri + "where 参数为: " + where);
        return null;}
    //该方法的返回值代表了该 ContentProvider 所提供数据的 MIME 类型
    @Override
    public String getType(Uri uri) {
        return null;}
    //实现插入的方法,该方法应该返回新插入的记录的 Uri
    @Override
    public Uri insert(Uri uri, ContentValues values) {
    Log.i("FirstProvider"," === insert 方法被调用 === ");
    Log.i("FirstProvider","values 参数为: " + values);
    return null;}
    //实现删除方法,该方法应该返回被删除的记录条数
    @Override
    public int delete(Uri uri, String where, String[] whereArgs) {
        Log.i("FirstProvider"," === delete 方法被调用 === ");
```

```
            Log.i("FirstProvider","where 参数为: " + where);
            return 0;}
    //实现更新方法,该方法应该返回被更新的记录条数
    @Override
    public int update(Uri uri, ContentValues values, String where,
                String[] whereArgs) {
        Log.i("FirstProvider"," === update 方法被调用 === ");
        Log.i("FirstProvider","where 参数为: " + where + ",values 参数为: " + values);
        return 0;
    }
```

7.2.2 注册 ContentProvider

在 AndroidManifest.xml 配置文件中注册 ContentProvider,只需在< application >元素中添加< provider >子元素即可,其示例代码如下。

【代码 7-2】 使用< provider >子元素注册 **ContentProvider**

```xml
<!-- 注册一个 ContentProvider -->
<provider
    android:name = ".FirstProvider"
    android:authorities = "com.qst.chapter07.Firstprovider"
    android:exported = "true">
</provider>
```

name 属性用于指定 ContentProvider 的实现类；authorities 属性用于指定该 ContentProvider 对应的 Uri,相当于给 ContentProvider 分配一个域名；exported 属性用于指定该 ContentProvider 是否允许其他应用调用。

7.2.3 使用 ContentProvider

下述代码通过一个 Activity 对 ContentProvider 进行使用。首先创建相应的 XML 布局文件,代码如下所示。

【代码 7-3】 activity_main.xml

```xml
<?xml version = "1.0" encoding = "utf-8"?>
<RelativeLayout xmlns:android = "http://schemas.android.com/apk/res/android"
    xmlns:tools = "http://schemas.android.com/tools"
    android:layout_width = "match_parent"
    android:layout_height = "match_parent"
    android:paddingBottom = "@dimen/activity_vertical_margin"
    android:paddingLeft = "@dimen/activity_horizontal_margin"
    android:paddingRight = "@dimen/activity_horizontal_margin"
    android:paddingTop = "@dimen/activity_vertical_margin"
    tools:context = "com.qst.chapter07.MainActivity">
    <Button
        android:layout_width = "wrap_content"
        android:layout_height = "wrap_content"
        android:text = "新增"
        android:id = "@ + id/insert"
```

```xml
            android:layout_alignParentTop = "true"
            android:layout_alignParentLeft = "true"
            android:layout_alignParentStart = "true" />
    <Button
        android:layout_width = "wrap_content"
        android:layout_height = "wrap_content"
        android:text = "更改"
        android:id = "@ + id/update"
        android:layout_alignBottom = "@ + id/insert"
        android:layout_alignParentRight = "true"
        android:layout_alignParentEnd = "true" />
    <Button
        android:layout_width = "wrap_content"
        android:layout_height = "wrap_content"
        android:text = "删除"
        android:id = "@ + id/delete"
        android:layout_below = "@ + id/insert"
        android:layout_alignParentLeft = "true"
        android:layout_alignParentStart = "true"
        android:layout_marginTop = "50dp" />
    <Button
        android:layout_width = "wrap_content"
        android:layout_height = "wrap_content"
        android:text = "查询"
        android:id = "@ + id/find"
        android:layout_alignBottom = "@ + id/delete"
        android:layout_alignParentRight = "true"
        android:layout_alignParentEnd = "true" />
</RelativeLayout>
```

【代码 7-4】 FirstProvideActivity.java

```java
public class FirstProvideActivity extends AppCompatActivity{
    ContentResolver contentResolver;
    Uri uri = Uri.parse("content://com.qst.chapter07.providers.firstprovider/");
    @Override
    public void onCreate(Bundle savedInstanceState)    {
        super.onCreate(savedInstanceState);
        setContentView(R.layout.activity_main);
        contentResolver = getContentResolver();}       //获取系统的 ContentResolver 对象
    public void query(View source)    {
        //调用 ContentResolver 的 query()方法
        //实际返回的是该 Uri 对应的 ContentProvider 的 query()的返回值
        Cursor c = contentResolver.query(uri, null,
            "query_where", null, null);
        Toast.makeText(this, "远程 ContentProvide 返回的 Cursor 为: " + c,
            Toast.LENGTH_SHORT).show(); }
    public void insert(View source) {
        ContentValues values = new ContentValues();
        values.put("name", "fkjava");
        //调用 ContentResolver 的 insert()方法
        //实际返回的是该 Uri 对应的 ContentProvider 的 insert()的返回值
        Uri newUri = contentResolver.insert(uri, values);
```

```
            Toast.makeText(this, "远程 ContentProvide 新插入记录的 Uri 为: "
                + newUri, Toast.LENGTH_SHORT).show();}
    public void update(View source) {
        ContentValues values = new ContentValues();
        values.put("name", "fkjava");
        //调用 ContentResolver 的 update()方法
        //实际返回的是该 Uri 对应的 ContentProvider 的 update()的返回值
        int count = contentResolver.update(uri, values,
                "update_where", null);
        Toast.makeText(this, "远程 ContentProvide 更新记录数为: "
                + count, Toast.LENGTH_SHORT).show();}
    public void delete(View source) {
        //调用 ContentResolver 的 delete()方法
        //实际返回的是该 Uri 对应的 ContentProvider 的 delete()的返回值
        int count = contentResolver.delete(uri,
                "delete_where", null);
        Toast.makeText(this, "远程 ContentProvide 删除记录数为: "
                + count, Toast.LENGTH_SHORT).show();}
}
```

上述代码中，通过 getContentResolver()方法获取系统的 contentResolver 对象，在按钮单击时实现 Uri 相对应的 ContentProvider 的增删改查功能，在 logcat 中看到如图 7-1 所示的日志信息。

图 7-1　ContentProvider 的使用

7.3　操作系统的 ContentProvider

Android 系统本身提供了大量的 ContentProvider，例如联系人信息、系统的多媒体信息等，程序员自己开发 Android 应用程序时可以通过 ContentResolver 来调用系统 ContentProvider 所提供的 query()、insert()、update()和 delete()方法，如此即可对 Android 内部数据进行操作。

7.3.1　管理联系人

Android 系统用于管理联系人的 ContentProvider 的几个 Uri 如下。
- **ContactsContract. Contacts. CONTENT_URI**：管理联系人的 Uri；
- **ContactsContract. CommonDataKinds. Phone. CONTENT_URI**：管理联系人的电话 Uri；
- **ContactsContract. CommonDataKinds. Email. CONTENT_URI**：管理联系人的 E-mail

的 Uri。

下述程序代码使用 ContentProvider 对联系人进行管理与维护。

【代码 7-5】 contacts.xml

```xml
<LinearLayout xmlns:android="http://schemas.android.com/apk/res/android" …>
    <LinearLayout
        android:orientation="vertical"
        android:layout_width="match_parent"
        android:layout_height="match_parent"
        android:layout_weight="1">
        <EditText
            android:layout_width="match_parent"
            android:layout_height="wrap_content"
            android:id="@+id/name"
            android:hint="姓名" />
        <EditText
            android:layout_width="match_parent"
            android:layout_height="wrap_content"
            android:id="@+id/phone"
            android:hint="电话" />
        <EditText
            android:layout_width="match_parent"
            android:layout_height="wrap_content"
            android:id="@+id/email"
            android:hint="邮箱" />
    </LinearLayout>
    <LinearLayout
        android:orientation="horizontal"
        android:layout_width="match_parent"
        android:layout_height="wrap_content"
        android:gravity="bottom">
        <Button
            android:layout_width="wrap_content"
            android:layout_height="wrap_content"
            android:text="添加"
            android:id="@+id/add"
            android:layout_weight="1" />
        <Button
            android:layout_width="wrap_content"
            android:layout_height="wrap_content"
            android:text="查找"
            android:id="@+id/search"
            android:layout_weight="1"
            android:layout_marginLeft="5dp" />
    </LinearLayout>
</LinearLayout>
```

【代码 7-6】 result.xml

```xml
<LinearLayout xmlns:android="http://schemas.android.com/apk/res/android" …>
    <ExpandableListView
        android:layout_width="match_parent"
        android:layout_height="match_parent"
```

```
            android:id = "@ + id/list"
            android:layout_weight = "1" />
</LinearLayout>
```

【代码 7-7】 ContactsActivity.java

```java
public class ContactsActivity extends AppCompatActivity{
    Button search;
    Button add;
    @Override
    public void onCreate(Bundle savedInstanceState) {
        super.onCreate(savedInstanceState);
        setContentView(R.layout.contacts);
        //获取系统界面中查找、添加两个按钮
        search = (Button) findViewById(R.id.search);
        add = (Button) findViewById(R.id.add);
        search.setOnClickListener(new OnClickListener() {
            @Override
            public void onClick(View source) {
                //使用 List 来封装系统的联系人信息、指定联系人的电话号码、E-mail 等详情
                final ArrayList<String> names = new ArrayList<>();
                final ArrayList<ArrayList<String>> details = new ArrayList<>();
                //使用 ContentResolver 查找联系人数据
                Cursor cursor = getContentResolver().query(
                    ContactsContract.Contacts.CONTENT_URI, null, null, null, null);
                //遍历查询结果,获取系统中所有联系人
                while (cursor.moveToNext()) {
                    //获取联系人 ID
                    String contactId = cursor.getString(cursor
                        .getColumnIndex(ContactsContract.Contacts._ID));
                    //获取联系人的名字
                    String name = cursor.getString(cursor.getColumnIndex(
                        ContactsContract.Contacts.DISPLAY_NAME));
                    names.add(name);
                    //使用 ContentResolver 查找联系人的电话号码
                    Cursor phones = getContentResolver().query(
                        ContactsContract.CommonDataKinds.Phone.CONTENT_URI,
                        null, ContactsContract.CommonDataKinds.Phone.CONTACT_ID
                            + " = " + contactId, null, null);
                    ArrayList<String> detail = new ArrayList<>();
                    //遍历查询结果,获取该联系人的多个电话号码
                    while (phones.moveToNext()) {
                        //获取查询结果中电话号码列中数据
                        String phoneNumber = phones.getString(phones
                            .getColumnIndex(ContactsContract
                                .CommonDataKinds.Phone.NUMBER));
                        detail.add("电话号码: " + phoneNumber);}
                    phones.close();
                    //使用 ContentResolver 查找联系人的 E-mail 地址
                    Cursor emails = getContentResolver().query(
                        ContactsContract.CommonDataKinds.Email.CONTENT_URI,
                        null, ContactsContract.CommonDataKinds.Email
                            .CONTACT_ID + " = " + contactId, null, null);
```

```java
            //遍历查询结果,获取该联系人的多个 E-mail 地址
            while (emails.moveToNext()) {
                //获取查询结果中 E-mail 地址列中数据
                String emailAddress = emails.getString(emails
                    .getColumnIndex(ContactsContract
                        .CommonDataKinds.Email.DATA));
                detail.add("邮件地址: " + emailAddress); }
            emails.close();
            details.add(detail);}
    cursor.close();
    //加载 result.xml 界面布局代表的视图
    View resultDialog = getLayoutInflater().inflate(
        R.layout.result, null);
    //获取 resultDialog 中 ID 为 list 的 ExpandableListView
    ExpandableListView list = (ExpandableListView) resultDialog
        .findViewById(R.id.list);
    //创建一个 ExpandableListAdapter 对象
    ExpandableListAdapter adapter =
        new BaseExpandableListAdapter() {
            //获取指定组位置、指定子列表项处的子列表项数据
            @Override
            public Object getChild(int groupPosition,
                            int childPosition) {
                return details.get(groupPosition).get(
                    childPosition); }
            @Override
            public long getChildId(int groupPosition,
                            int childPosition) {
                return childPosition;}
            @Override
            public int getChildrenCount(int groupPosition) {
                return details.get(groupPosition).size();}
            private TextView getTextView() {
                AbsListView.LayoutParams lp = new AbsListView
                    .LayoutParams(ViewGroup.LayoutParams.MATCH_PARENT,
                        64);
                TextView textView = new TextView(
                    ContactsActivity.this);
                textView.setLayoutParams(lp);
                textView.setGravity(Gravity.CENTER_VERTICAL
                    | Gravity.LEFT);
                textView.setPadding(36, 0, 0, 0);
                textView.setTextSize(20);
                return textView;}
            //该方法决定每个子选项的外观
            @Override
            public View getChildView(int groupPosition,
                int childPosition, boolean isLastChild,
                View convertView, ViewGroup parent) {
                TextView textView = getTextView();
                textView.setText(getChild(groupPosition,
                    childPosition).toString());
                return textView;}
```

```java
                    //获取指定组位置处的组数据
                    @Override
                    public Object getGroup(int groupPosition) {
                        return names.get(groupPosition);}
                    @Override
                    public int getGroupCount() {
                        return names.size();}
                    @Override
                    public long getGroupId(int groupPosition) {
                        return groupPosition;}
                    //该方法决定每个组选项的外观
                    @Override
                    public View getGroupView(int groupPosition,
                        boolean isExpanded, View convertView,
                        ViewGroup parent) {
                        TextView textView = getTextView();
                        textView.setText(getGroup(groupPosition)
                            .toString());
                        return textView;}
                    @Override
                    public boolean isChildSelectable(int groupPosition,
                        int childPosition) {
                        return true;}
                    @Override
                    public boolean hasStableIds() {
                        return true;}};
            //为 ExpandableListView 设置 Adapter 对象
            list.setAdapter(adapter);
            //使用对话框来显示查询结果
            new AlertDialog.Builder(ContactsActivity.this)
                .setView(resultDialog).setPositiveButton("确定", null)
                .show();}});
    //为 add 按钮的单击事件绑定监听器
    add.setOnClickListener(new OnClickListener() {
        @Override
        public void onClick(View v) {
            //获取程序界面中的三个文本框的内容
            String name = ((EditText) findViewById(R.id.name))
                .getText().toString();
            String phone = ((EditText) findViewById(R.id.phone))
                .getText().toString();
            String email = ((EditText) findViewById(R.id.email))
                .getText().toString();
            //创建一个空的 ContentValues
            ContentValues values = new ContentValues();
            //向 RawContacts.CONTENT_URI 执行一个空值插入
            //目的是获取系统返回的 rawContactId
            Uri rawContactUri = getContentResolver().insert(
                ContactsContract.RawContacts.CONTENT_URI, values);
            long rawContactId = ContentUris.parseId(rawContactUri);
            values.clear();
            values.put(Data.RAW_CONTACT_ID, rawContactId);
            //设置内容类型
            values.put(Data.MIMETYPE, StructuredName.CONTENT_ITEM_TYPE);
            values.put(StructuredName.GIVEN_NAME, name);   //设置联系人名字
```

```
        getContentResolver().insert(android.provider.ContactsContract
            .Data.CONTENT_URI, values);        //向联系人 Uri 添加联系人名字
        values.clear();
        values.put(Data.RAW_CONTACT_ID, rawContactId);
        values.put(Data.MIMETYPE, Phone.CONTENT_ITEM_TYPE);
        values.put(Phone.NUMBER, phone);        //设置联系人的电话号码
        values.put(Phone.TYPE, Phone.TYPE_MOBILE);   //设置电话类型
        getContentResolver().insert(android.provider.ContactsContract
            .Data.CONTENT_URI, values);        //向联系人电话号码 Uri 添加电话号码
        values.clear();
        values.put(Data.RAW_CONTACT_ID, rawContactId);
        values.put(Data.MIMETYPE, Email.CONTENT_ITEM_TYPE);
        values.put(Email.DATA, email);          //设置联系人的 E-mail 地址
        values.put(Email.TYPE, Email.TYPE_WORK);   //设置该电子邮件的类型
        getContentResolver().insert(android.provider.ContactsContract
            .Data.CONTENT_URI, values);      //向联系人 E-mail Uri 添加 E-mail 数据
        Toast.makeText(ContactsActivity.this, "联系人数据添加成功",
            Toast.LENGTH_SHORT).show(); }});}
}
```

上述代码要读取、添加联系人信息,因此要在 AndroidManifest.xml 文件中进行授权,让应用程序能够读取 Contacts 信息。

【代码 7-8】 在 AndroidManifest.xml 授予读写联系人信息的权限

```
<!-- 授予读联系人 ContentProvider 的权限 -->
<uses-permission android:name="android.permission.READ_CONTACTS"/>
<!-- 授予写联系人 ContentProvider 的权限 -->
<uses-permission android:name="android.permission.WRITE_CONTACTS"/>
```

运行上述代码,结果如图 7-2 所示。输入姓名、电话、邮箱,单击"添加"按钮后,出现相应的提示信息,如图 7-3 所示,表示该联系人添加成功。

图 7-2 管理联系人首页

图 7-3 添加联系人成功

单击"查找"按钮,弹出信息提示对话框,如图 7-4 所示。单击对话框中的 QST QingRuanShiXun,显示该联系人的电话和邮箱的详细信息,如图 7-5 所示。

图 7-4　查找联系人　　　　图 7-5　联系人详细信息

7.3.2 管理多媒体

Android 提供了 Camera API 来支持拍照、拍摄视频，用户所拍摄的照片、视频等多媒体内容都存放在固定位置，其他应用程序可以通过 ContentProvider 进行访问。

Android 系统为多媒体提供了相应 ContentProvider 的 Uri，具体如下：MediaStore.Audio.Media.EXTERNAL_CONTENT_URI 存储在外部 SD 存储卡中的音频文件的 Uri；MediaStore.Audio.Media.INTERNAL_CONTENT_URI 存储在手机内存中音频文件的 Uri；MediaStore.Images.Media.EXTERNAL_CONTENT_URI 存储在外部 SD 存储卡中图片文件的 Uri；MediaStore.Images.Media.INTERNAL_CONTENT_URI 存储在手机内存中图片文件的 Uri；MediaStore.Video.Media.EXTERNAL_CONTENT_URI 存储在外部 SD 存储卡中视频文件的 Uri；MediaStore.Video.Media.INTERNAL_CONTENT_URI 存储在手机内存中视频文件的 Uri。

下述程序代码使用 ContentProvider 管理多媒体内容。

【代码 7-9】　media.xml

```xml
<LinearLayout xmlns:android = "http://schemas.android.com/apk/res/android" …>
    <LinearLayout
        android:orientation = "horizontal"
        android:layout_width = "match_parent"
        android:layout_height = "wrap_content"
        android:gravity = "center">
        <Button
            android:layout_width = "wrap_content"
            android:layout_height = "wrap_content"
            android:text = "添加"
            android:id = "@ + id/add"
            android:layout_weight = "1" />
        <Button
            android:layout_width = "wrap_content"
            android:layout_height = "wrap_content"
```

```xml
            android:text = "查看"
            android:id = "@ + id/view"
            android:layout_weight = "1" />
    </LinearLayout>
    <LinearLayout
        android:orientation = "horizontal"
        android:layout_width = "match_parent"
        android:layout_height = "match_parent"
        android:paddingTop = "5dp">
        <ListView
            android:layout_width = "match_parent"
            android:layout_height = "match_parent"
            android:id = "@ + id/show"
            android:layout_weight = "1" />
    </LinearLayout>
</LinearLayout>
```

【代码 7-10】 line.xml

```xml
<LinearLayout xmlns:android = "http://schemas.android.com/apk/res/android" …>
    <LinearLayout
        android:orientation = "horizontal"
        android:layout_width = "match_parent"
        android:layout_height = "match_parent"
        android:layout_weight = "1">
        <TextView
            android:layout_width = "wrap_content"
            android:layout_height = "wrap_content"
            android:textAppearance = "?android:attr/textAppearanceMedium"
            android:text = "Medium Text"
            android:id = "@ + id/pic_id"
            android:layout_weight = "1"
            android:gravity = "center_horizontal" />
        <TextView
            android:layout_width = "wrap_content"
            android:layout_height = "wrap_content"
            android:textAppearance = "?android:attr/textAppearanceMedium"
            android:text = "Medium Text"
            android:id = "@ + id/name"
            android:layout_weight = "1"
            android:gravity = "center_horizontal" />
        <TextView
            android:layout_width = "wrap_content"
            android:layout_height = "wrap_content"
            android:textAppearance = "?android:attr/textAppearanceMedium"
            android:text = "Medium Text"
            android:id = "@ + id/title"
            android:layout_weight = "1"
            android:gravity = "center_horizontal" />
    </LinearLayout>
</LinearLayout>
```

【代码 7-11】 view.xml

```xml
<LinearLayout xmlns:android = "http://schemas.android.com/apk/res/android" …>
    <ImageView
        android:layout_width = "match_parent"
        android:layout_height = "match_parent"
        android:id = "@ + id/image"
        android:layout_gravity = "center_vertical" />
</LinearLayout>
```

【代码 7-12】 MediaStoreActivity.java

```java
public class MediaStoreActivity extends AppCompatActivity {
    Button add;
    Button view;
    ListView show;
    ArrayList<String> ids = new ArrayList<>();
    ArrayList<String> names = new ArrayList<>();
    ArrayList<String> fileNames = new ArrayList<>();
    ArrayList<String> filePaths = new ArrayList<>();
    @Override
    protected void onCreate(Bundle savedInstanceState) {
        super.onCreate(savedInstanceState);
        setContentView(R.layout.media);
        add = (Button) findViewById(R.id.add);
        view = (Button) findViewById(R.id.view);
        show = (ListView) findViewById(R.id.show);
        //为 view 按钮的单击事件绑定监听器
        view.setOnClickListener(new View.OnClickListener() {
            @Override
            public void onClick(View view) {
                //清空 ids、names、fileName 集合里原有的数据
                ids.clear();
                names.clear();
                fileNames.clear();
                filePaths.clear();
                //通过 ContentResolver 查询所有图片信息
                Cursor cursor = getContentResolver()
                        .query(MediaStore.Images.Media.EXTERNAL_CONTENT_URI,
                                null, null, null, null);
                while (cursor.moveToNext()) {
                //获取图片的 ID
                String id = cursor.getString(
                        cursor.getColumnIndex(MediaStore.Images.Media._ID));
                //获取图片的 DISPLAY_NAME
                String name = cursor.getString(cursor.getColumnIndex(
                                MediaStore.Images.Media.DISPLAY_NAME));
                //获取图片的 TITLE
                String title = cursor.getString(cursor.getColumnIndex(
                                MediaStore.Images.Media.TITLE));
                //获取图片的保存位置的数据
                byte[] data = cursor.getBlob(cursor.getColumnIndex(
                                MediaStore.Images.Media.DATA));
                ids.add(id);               //将图片名添加到 ids 集合中
                names.add(name);           //将图片 DISPLAY_NAME 添加到 names 集合中
                fileNames.add(title);      //将图片 TITLE 添加到 flieNames 集合中
```

```java
        //将图片保存路径添加到filePaths集合中
        filePaths.add(new String(data, 0, data.length - 1)); }
    //创建一个List集合,List集合的元是Map
    List<Map<String, Object>> listItems = new ArrayList<>();
    //将ids、names、fileNames三个集合对象的数据转换到Map集合中
    for (int i = 0; i < names.size(); i++) {
        Map<String, Object> listItem = new HashMap<>();
        listItem.put("id", ids.get(i));
        listItem.put("name", names.get(i));
        listItem.put("title",fileNames.get(i) + ".jpg");
        listItems.add(listItem); }
    //创建一个SimpleAdapter
        SimpleAdapter simpleAdapter = new SimpleAdapter
                (MediaStoreActivity.this, listItems, R.layout.line,
                    new String[]{"id", "name","title"},
                    new int[]{R.id.pic_id, R.id.name,R.id.title});
        //为show ListView组件设置Adapter
        show.setAdapter(simpleAdapter); }});
//为show ListView的列表项单击事件添加监听器
show.setOnItemClickListener(new MyOnItemClickListener());
//为add按钮的单击事件绑定监听器
add.setOnClickListener(new View.OnClickListener() {
    @Override
    public void onClick(View view) {
        //创建ContentValues对象,准备插入数据
        ContentValues values = new ContentValues();
        values.put(MediaStore.Images.Media.DISPLAY_NAME,"金字塔");
        //设置多媒体类型为image/jpeg
        values.put(MediaStore.Images.Media.MIME_TYPE,"image/jpeg");
        //插入数据,返回所插入数据对应的Uri
        Uri uri = getContentResolver()
            .insert(MediaStore.Images.Media.EXTERNAL_CONTENT_URI,values);
        //加载应用程序下的jinzita图片
        Bitmap bitmap = BitmapFactory.decodeResource(
        MediaStoreActivity.this.getResources(),R.drawable.jinzita);
        OutputStream os = null;
    try{
        //获取刚插入的数据的Uri对应的输出流
        os = getContentResolver().openOutputStream(uri);
        //将Bitmap图片保存到Uri对应的数据节点中
        bitmap.compress(Bitmap.CompressFormat.JPEG,100,os);
        os.close();
        }catch (Exception e){
            e.printStackTrace();}}});}
private class MyOnItemClickListener implements AdapterView
                                            .OnItemClickListener {
    @Override
    public void onItemClick(AdapterView<?> parent,
        View source, int position, long id) {
        //加载view.xml界面布局代表的视图
        View viewDialog = getLayoutInflater().inflate(R.layout.view,null);
        //获取viewDialog中ID为image的组件
        ImageView image = (ImageView) viewDialog.findViewById(R.id.image);
        //设置image显示指定图片
        image.setImageBitmap(BitmapFactory.
                            decodeFile(filePaths.get(position)));
        //使用对话框显示用户单击的图片
```

```
            new AlertDialog.Builder(MediaStoreActivity.this)
                .setView(viewDialog).setPositiveButton("确定",null).show();}}
}
```

上述代码需要读写外部存储设备中的多媒体文件,因此必须为该应用授予读写外部存储设备的权限,即在 AndroidManifest.xml 文件中进行授权,其配置信息如下所示。

【代码 7-13】 为应用程序授予读写外部存储设备的权限

```
<!-- 授予读取外部存储设备的访问权限 -->
< uses - permission android:name = "android.permission.READ_EXTERNAL_STORAGE"/>
<!-- 授予写入外部存储设备的访问权限 -->
< uses - permission android:name = "android.permission.WRITE_EXTERNAL_STORAGE"/>
```

启动应用后,单击"添加"按钮,将应用程序中 drawable 目录下名为 jinzita.jpg 的图片保存到手机 MediaStore/Images,然后单击"查看"按钮后,将相册中的图片列表信息显示出来,结果如图 7-6 所示。

单击"添加"按钮,图片的路径、DISPLAY_NAME、MINE_TYPE、TITLE 等信息会保存到 data/data/com.android.providers.media/databases/目录下的 external.db 文件中,如图 7-7 所示。

图 7-6 管理多媒体应用程序主界面　　　　　　　图 7-7 信息存储路径

将该文件保存到本地,使用 SQLite 可视化工查看 external.db 数据文件,如图 7-8 所示。
单击"金字塔",在弹出的对话框中会展示所添加的金字塔图片,如图 7-9 所示。

图 7-8 插入图片所对应的数据文件　　　　　　　图 7-9 查看图片

本章总结

小结

- ContentProvider 是 Android 应用的四大组件之一。
- ContentProvider 类提供了 insert()、delete()、update()、query() 和 getType() 等操作数据的抽象方法。
- Uri 是每一个 ContentProvider 都对外提供一个自身数据集的唯一标识。
- 在开发过程中通过 ContentResolver 来间接操作 ContentProvider 所提供的数据。
- 每个应用程序的上下文都有一个默认的 ContentResolver 实例对象,可以调用 getContentResolver() 方法获取 ContentResolver 实例对象。

Q&A

问题:简述开发 ContentProvider 程序的步骤。

回答:先编写一个 ContentProvider 子类,该子类需要实现 query()、insert()、update() 和 delete() 等方法;然后在 AndroidManifest.xml 配置文件中注册 ContentProvider,并指定 android:authorities 属性;最后使用 ContentProvider,Activity 和 Service 等组件都可以获取 ContentProvider 对象,并调用相应的方法进行操作。

章节练习

习题

1. Android 使用_____实现应用程序之间的数据共享。
   A. 文件   B. SharedPreferences
   C. SQLite   D. ContentProvider
2. 下面对 ContentProvider 描述错误的是_____。
   A. 使用 ContentProvider 能够实现 Android 应用程序之间的数据共享
   B. ContentProvider 与 Activity、Service、BroadcastReceiver 并称为 Android 应用的四大组件
   C. 可以直接使用 ContentProvider 类中提供的 insert()、delete()、update()、query() 和 getType() 方法对数据进行操作
   D. ContentProvider 是通过 Uri 对外提供一个自身数据集的唯一标识
3. 下面_____类可以对 Uri 进行匹配判断。
   A. ContentProvider   B. ContentResolver
   C. Uri   D. UriMatcher
4. 外界的程序可以通过_____访问 ContentProvider 提供的数据。

上机

训练目标：ContentProvider 数据共享。

培养能力	熟练使用 ContentProvider 实现数据共享		
掌握程度	★★★★★	难度	中
代码行数	450	实施方式	重复编码
结束条件	使用 ContentProvider 实现数据共享		

参考训练内容
(1) 使用 ContentProvider 管理系统联系人，并以列表形式显示出来；
(2) 列表有三列：序号、姓名和电话

第 8 章

Service 服务

本章任务是完成"GIFT-EMS 礼记"项目中的赠礼留言相关功能和用户日程提醒 Service：
- 【任务 8-1】 完成录制赠礼留言功能。
- 【任务 8-2】 完成扫描二维码功能。
- 【任务 8-3】 完成播放赠礼留言功能。
- 【任务 8-4】 完成日程提醒的 Service。

知 识 点	Listen（听）	Know（懂）	Do（做）	Revise（复习）	Master（精通）
Service 分类	★	★			
编写 Service	★	★	★	★	★
Service 生命周期	★	★	★	★	★
远程 Service	★	★	★		
系统 Service	★	★	★	★	

8.1 Service 简介

Android 中 Service 组件表示一种服务，专门用于执行一些持续性的、耗时长并且无须与用户界面交互的操作。Activity 可以显示用户界面，完成用户和应用程序的交互，而

Service 不同，Service 的运行是不可见的，通常用于执行一些无须用户交互，并需要持续运行的任务，例如从网络上搜索内容、更新 ContentProvider、激活 Notification、播放音乐等。

Service 拥有独立的生命周期，其启动、停止以及运行期的控制可以由其他组件完成，包括 Activity、BroadcastReceiver 或者其他的 Service，这些使用 Service 的组件可以称为 Service 的客户端。

一个处于运行状态的 Service 拥有的优先级要比暂停和停止状态的 Activity 级别更高，因此，当系统资源匮乏时，Service 被 Android 终止的可能性更小。当 Android 系统需要为运行前台资源释放更多的内存时，可能会终止正在运行的 Service，这是 Service 被系统提前终止的唯一可能情况。如果系统终止了一个运行的 Service，当系统发现有资源可用时，可以自动重新启动这个 Service。Service 优先级可以提升到与前台 Activity 相同的级别，此时 Service 将更不容易被系统终止，但是由于 Service 通常具有更长的运行期，因此过多优先级较高的 Service 势必会降低系统的性能。

Service 没有界面（最多只能显示一个通知），当 Service 所对应的应用程序界面不可见时，Service 仍运行于应用程序主线程中，因此，如果在 Service 中需要执行耗时操作，必须新开线程运行，否则会阻塞主线程，从而造成界面卡顿。

Android 系统中提供了大量可以直接调用的系统 Service，例如播放音乐、震动、闹钟、通知栏消息等，通过向这些 Service 传递特定的数据，可以方便地运行系统服务。

8.1.1 Service 分类

Service 作为 Android 的基本组件之一，也具有相应的生命周期及回调函数，按照运行形式和使用方式的不同，可以对 Service 进行归类。

按照运行的进程不同，可以将 Service 分为本地（Local）Service 和远程（Remote）Service。

- 本地 Service 运行于其客户端的应用程序进程中，当客户端终止后，本地 Service 也会被终止。
- 远程 Service 运行于独立的进程中，与其客户端之间需要进行跨进程的通信，Android 中的进程间通信依赖于 Android 接口定义语言（Android Interface Definition Language，AIDL），当客户端终止后，这种远程 Service 会继续运行。

按照运行的形式分为前台 Service 和后台 Service。

- 前台 Service 在运行时，会在状态栏显示一个 ONGOING 状态的 Notification，用以提示用户服务正在运行，当前台服务终止后，Notification 会消失。
- 后台 Service 在运行时，没有状态栏通知。

按照使用 Service 的方式可以分为启动（Start）方式 Service、绑定（Bind）方式 Service 和混合方式 Service。

- 启动方式 Service 通过调用 Context.startService()启动 Service，运行过程中与客户端不进行通信，如果不调用停止方法，其会一直运行。
- 绑定方式 Service 通过调用 Context.bindService()启动 Service，在运行时可以与绑定的客户端通信，当客户端终止时 Service 也将终止。
- 混合方式 Service 将启动方式和绑定方式混合使用，即以 Start 和 Bind 两种方式来启动 Service。

8.1.2 Service 基本示例

当应用程序需要一种无界面交互并且可持续运行的组件时,Service 是一种最合适的选择。当使用 Service 时,首先需要创建 Service,并重写在 Service 生命周期的各个阶段需要执行的操作,最后启动 Service 即可。当使用已有的 Service 组件时,则可以直接启动即可。

创建一个 Service 组件只需要两步,而启动 Service 可以使用 Start 和 Bind 两种方式。创建 Service 的步骤如下:通过继承 Service 的方式来定义一个 Service 的子类;在应用程序的 AndroidManifest.xml 中配置 Service 组件。

1. 编写 Service 类

Android 提供了 android.app.Service 抽象类,作为所有 Service 的父类,其包含一个抽象方法,语法格式如下。

【语法】

```
public abstract IBinder onBind(Intent intent);
```

在编写 Service 时,需要继承 android.app.Service 抽象类并实现 onBind()方法即可,代码如下所示。

【代码 8-1】 MyService1.java

```
//一个空的 Service 示例
public class MyService1 extends Service {
    @Override
    public IBinder onBind(Intent intent) {
        return null; }
}
```

上述示例代码中,MyService1 类继承了 android.app.Service 类,并实现了 onBind()方法,因此 MyService1 类就可以配置为一个 Service 组件。

MyService1 类中目前还没有添加任何的业务操作,在本章后续小节中将逐渐丰富其内容。

与 Activity 类似,Service 也是由 Android 系统构造并管理的一种组件,因此 Service 类必须提供一个 public 的无参数构造方法,以保证系统能够构造 Service 的实例。

2. 配置 Service

编写完成 Service 类后,还需要在应用程序的 AndroidManifest.xml 中配置 Service 组件。配置完成后,Android 才能在 APK 应用安装时解析出 Service 组件的信息,从而允许其他组件启动这个 Service。在 AndroidManifest.xml 中,每个 Service 组件都需要在<application>元素的一个<service>子元素中进行配置,下列代码演示对 MyService1 类的配置。

【代码 8-2】 AndroidManifest.xml

```xml
<?xml version = "1.0" encoding = "utf-8"?>
< manifest xmlns:android = "http://schemas.android.com/apk/res/android"
    package = "com.qst.chapter08">
    < application
        android:allowBackup = "true"
        android:icon = "@drawable/ic_launcher"
        android:label = "@string/app_name"
        android:theme = "@style/AppTheme" >
        < activity
            android:name = ".MainActivity"
            android:label = "@string/app_name" >
            < intent - filter >
                < action android:name = "android.intent.action.MAIN" />
                < category android:name = "android.intent.category.LAUNCHER" />
            </ intent - filter >
        </ activity >
        < service android:name = "com.qst.chapter08.MyService1" />
    </ application >
</ manifest >
```

上述配置文件中，在< application >元素中添加了一个< service >子元素，并指定其 name 属性值为 com.qst.chapter08.MyService1，从而完成了 Service 组件 MyService1 的配置。

3. 启动 Service

在完成 Service 组件的编写和配置后，即可以在其他组件中启动这个 Service 了，启动 Service 有 Start 和 Bind 两种方式，本节只介绍 Start 启动方式。下列代码演示了如何以 Start 方式启动 Service。

【代码 8-3】 MainActivity.java

```java
public class MainActivity extends AppCompatActivity {
    @Override
    protected void onCreate(Bundle savedInstanceState) {
        super.onCreate(savedInstanceState);
        setContentView(R.layout.activity_main);
        Intent intent = new Intent(this, MyService1.class);
        startService(intent);
    }
}
```

上述 MainActivity 的 onCreate()方法中，首先构造了一个 Intent 对象，并传入 MyService1.class 参数来指定所要启动的组件类型，然后调用 Context 对象的 startService()方法启动 Service 组件。

注意

本节只是简单介绍了 Service 的基本使用，在 8.2 节中将结合 Service 生命周期对 Service 的运行过程进行详细介绍。

8.2 Service 详解

Service 组件需要通过 Context 对象进行启动,有两种启动方式:Start 和 Bind 方式,分别对应于 Context 的 startService()和 bindService()方法。通过 Start 和 Bind 方式启动 Service 时,生命周期有所不同,在运行过程中会调用相应的生命周期方法。

与 Service 生命周期相关的回调方法如表 8-1 所示。

表 8-1 与 Service 生命周期相关的回调方法

方 法	功 能 描 述
onCreate()	用于创建 Service 组件
onStartCommand(Intent intent, int flags, int started)	通过 Start 方式启动 Service 时调用
onBind(Intent intent)	通过 Bind 方式启动 Service
onUnbind(Intent intent)	通过 Bind 方式取消 Service 绑定
onRebind(Intent intent)	通过 Bind 方式重新绑定 Service
onDestroy()	用于销毁 Service

无论使用 Start 还是 Bind 方式来启动 Service,都会经历 onCreate()和 onDestroy()方法。如果采用 Start 方式,在启动时会调用 Service 的 onStartCommand()方法;如果采用 Bind 方式,在启动时则会调用 Service 的 onBind()方法,当取消绑定时会调用 onUnbind()方法,重新绑定时会调用 onRebind()方法。

注意

> 从 Android 2.0 开始,Service 的 onStart()方法已经不再推荐使用,由 onStartCommand()方法取代。

8.2.1 Start 方式启动 Service

Start 方式是通过调用 Context.startService()方法来启动 Service 的,Service 将自行管理生命周期,并会一直运行下去,直到 Service 调用自身的 stopSelf()方法或其他组件调用该 Service 的 stopService()方法时为止。当然在系统资源不足的情况下,Android 也会结束 Service。需要注意的是:一个组件通过 startService()方法启动 Service 后,该 Service 和启动这个 Service 的组件之间并没有关联,即使组件被销毁,也不影响该 Service 的运行。

Start 方式启动 Service 的生命周期如图 8-1 所示。

当使用 Start 方式启动 Service 时,会自动调用 onStartCommand()方法。如果该 Service 是第一次启动,则会先调用 onCreate()方法,然后再调用 onStartCommand()方法,否则直接调用 onStartCommand()方法。

图 8-1 Start 方式启动的 Service 的生命周期

与 Activity 类似，在系统资源缺乏时，Service 也有可能被 Android 强行终止，根据 Service 的 onStartCommand() 方法返回值也不同，Service 可能会自动重新启动，而 onStartCommand() 方法的参数则与 Service 被系统重新启动时的状态有关。关于 onStartCommand() 方法的语法结构如下所示。

【语法】

```
public int onStartCommand(Intent intent, int flags, int startId)
```

其中：
- 参数 intent 为在启动 Service 时所传入的 Intent 对象。
- 参数 flags 取值范围是 0、Service.START_FLAG_REDELIVERY 和 Service.START_FLAG_RETRY。flags 参数的值与 onStartCommand() 方法返回值有一定的关系。当 flags 为 0 时代表正常启动 Service。而在 Service 因某种情况下被系统异常终止后，如果调用该 Service 的 onStartCommand() 方法返回值为 Service.START_STICKY 或 Service.START_REDELIVER_INTENT 时，系统会在资源可用时自动重新启动该 Service。根据方法的返回值不同，此时系统在调用 onStartCommand() 方法时向 flags 参数传入的数据也不同。如果 onStartCommand() 方法返回值为 Service.START_STICKY，则 flags 参数会传入 Service.START_FLAG_RETRY；如果 onStartCommand() 的返回值为 Service.START_REDELIVER_INTENT，则 flags 参数会传入 Service.START_FLAG_REDELIVERY 值。
- 参数 startId 为启动请求的 ID，用于唯一标识一次启动请求，在调用 stopSelfResult() 方法停止 Service 时，可以传入特定的 startId，用于对停止 Service 的操作附加条件。

onStartCommand() 方法的返回值可以是以下三种情况。

(1) Service.START_NOT_STICKY：如果 Service 进程被终止，系统将保留 Service 的状态为开始状态，但不会自动重启该 Service，直到 startService(Intent intent) 方法再次被调用。

(2) Service.START_STICKY：如果 Service 进程被终止，系统将保留 Service 的状态为开始状态，但不保留原来的 Intent 对象。随后系统会尝试重新创建 Service，由于服务状态为开始状态，所以创建服务后一定会调用 onStartCommand() 方法。如果在此期间没有任何启动命令被传递到 Service，那么参数 Intent 将为 null。

(3) Service.START_REDELIVER_INTENT：如果 Service 进程被终止，系统会自动重启该服务，并将 Service 被终止前接收到的最后一个 Intent 对象传入 onStartCommand() 方法。

下列代码演示使用 Start 方式来启动 Service。首先自定义一个 Service 组件 MyService2，代码如下所示。

【代码 8-4】 MyService2.java

```
public class MyService2 extends Service {
    @Override
    public void onCreate() {
        Log.i("MyService2", "onCreate");}
    @Override
```

```
    public int onStartCommand(Intent intent, int flags, int startId) {
        Log.i("MyService2", "onStartCommand");
        final String message = intent.getStringExtra("message");
        Log.i("MyService2", "intent:" + message + ",flags:" + flags
                + ",startId:" + startId);
        return super.onStartCommand(intent, flags, startId);}
    @Override
    public void onDestroy() {
        Log.i("MyService2", "onDestroy");}
    @Override
    public IBinder onBind(Intent intent) {
        return null;}
}
```

上述代码中,重写了 onCreate()、onStartCommand()和 onDestroy()方法,每个方法中输出相应的 Log 信息,在 onStartCommand()方法中还输出 Intent 参数信息以及 flags 和 startId 参数值。

在 MyService2 的 onStartCommand()方法中最后返回 super.onStartCommand()方法的返回值,即调用父类的默认返回值。在父类 Service 中,onStartCommand()方法返回值为 Service.START_STICKY,即 Service 被异常终止后,系统再次启动 Service 并调用 onStartCommand()方法时,Intent 参数将传入 null。

在 AndroidManifest.xml 中配置 MyService2,代码如下所示。

【代码 8-5】 AndroidManifest.xml 中配置 MyService2

```
< service android:name = "com.qst.chapter08.MyService2" />
```

然后编写 Activity,实现 MyService2 的启动和停止,代码如下所示。

【代码 8-6】 MainActivity.java

```
public class MainActivity extends AppCompatActivity {
    private Button startButton;
    private Button stopButton;
    @Override
    protected void onCreate(Bundle savedInstanceState) {
        super.onCreate(savedInstanceState);
        setContentView(R.layout.activity_main);
        startButton = (Button) findViewById(R.id.startButton);
        stopButton = (Button) findViewById(R.id.stopButton);
        startButton.setOnClickListener(new OnClickListener() {
            @Override
            public void onClick(View v) {
                Intent intent = new Intent(MainActivity.this, MyService2.class);
                intent.putExtra("message", "hello!");
                startService(intent);
            }
        });
        stopButton.setOnClickListener(new OnClickListener() {
            @Override
            public void onClick(View v) {
```

```
                    Intent intent = new Intent(MainActivity.this, MyService2.class);
                    stopService(intent);
                }
            });
        }
    }
```

上述 MainActivity 代码中,为 startButton 和 stopButton 添加了单击事件:

(1) 在 startButton 事件处理方法中,创建了一个 Intent 对象,指定组件的类型为 MyService2.class,并传入 message 附加数据,然后调用 startService()方法启动了 MyService2。

(2) 在 stopButton 处理方法中,也构造了同样的 Intent 对象,然后调用 stopServie()方法停止了 MyService2。

MainActivity 的界面结构非常简单,不再列出其布局 XML 代码。运行程序,结果如图 8-2 所示。

图 8-2 MainActivity

单击"启动 MyService2"按钮,在 Logcat 中输出如下:

```
11-30 04:46:36.065: I/MyService2(2648): onCreate
11-30 04:46:36.065: I/MyService2(2648): onStartCommand
11-30 04:46:36.065: I/MyService2(2648):intent:hello!,flags:0,startId:1
```

从输出结果可以看出,在启动 MyService2 时依次调用 onCreate()和 onStartCommand()方法,并在 onStartCommand()方法中成功输出 Intent 对象中的数据。flags 值为 0 说明是正常启动的 Service,startId 值为 1 说明是第一次启动。

再次单击"启动 MyService2"按钮,Logcat 输出如下:

```
11-30 05:16:31.452: I/MyService2(4503): onStartCommand
11-30 05:16:31.452: I/MyService2(4503): intent:hello!,flags:0,startId:2
```

从输出结果可以看出,第二次调用 startService()方法时,并没有调用 onCreate()方法,而是直接调用 onStartCommand()方法。startId 值变为 2,说明是第二次启动这个 Service。

多次单击"启动 MyService2"按钮,Logcat 输出如下:

```
11-30 05:36:09.109: I/MyService2(4503): onStartCommand
11-30 05:36:09.109: I/MyService2(4503): intent:hello!,flags:0,startId:3
11-30 05:36:09.669: I/MyService2(4503): onStartCommand
11-30 05:36:09.669: I/MyService2(4503): intent:hello!,flags:0,startId:4
11-30 05:36:10.079: I/MyService2(4503): onStartCommand
11-30 05:36:10.079: I/MyService2(4503): intent:hello!,flags:0,startId:5
```

可以看到,与第二次启动完全相同,没有执行 onCreate()方法,而且 startId 启动次数一直在增加。

单击"停止 MyService2"按钮,Logcat 输出如下:

```
11-30 05:37:33.191: I/MyService2(4503): onDestroy
```

从输出结果可以看出,调用 stopService()方法后,触发了 Service 的 onDestroy()方法,此时 Service 被系统销毁,Service 生命周期结束。此时如果再次单击"启动 MyService2"按钮时,则会重新开始一个新的生命周期,依次执行 onCreate()→onStartCommand()→onStartCommand()→…→onDestroy()的循环过程。

在 Service 内部也提供了可以结束自己的方法,具体如下。

- **public final void stopSelf()**:用于销毁当前 Service,在销毁之前会执行 onDestroy()方法。
- **public final boolean stopSelfResult(int startId)**:调用该方法时,系统将检查 startId 参数值是否和最后一次启动 Service 时自动生成的请求 ID 相同。如果相同则调用 onDestroy()方法销毁当前 Service 并返回 true,否则不做任何操作并返回 false。
- **public final void stopSelf(int startId)**:该方法是 stopSelfResult(int startId)的早期版本,没有返回值。

下面修改 MyService2 代码,在 onStartCommand()方法中调用 stopSelf()方法停止服务。

【代码 8-7】 MyService2.java

```java
public class MyService2 extends Service {
    @Override
    public int onStartCommand(Intent intent, int flags, int startId) {
        Log.i("MyService2", "onStartCommand");
        final String message = intent.getStringExtra("message");
        Log.i("MyService2", "intent:" + message + ",flags:" + flags
                + ",startId:" + startId);
        stopSelf();
        return super.onStartCommand(intent, flags, startId);
    }
    …//省略其他方法
}
```

单击"启动 MyService2"按钮,Logcat 输出信息如下:

```
11-30 11:30:26.790: I/MyService2(9806): onCreate
11-30 11:30:26.790: I/MyService2(9806): onStartCommand
11-30 11:30:26.790: I/MyService2(9806): intent:hello!,flags:0,startId:1
11-30 11:30:26.810: I/MyService2(9806): onDestroy
```

从输出结果可以看出,调用 stopSelf()方法时系统自动调用 onDestroy()方法,说明 Service 已被销毁。

因为同一个 Service 的 onStartCommand()多次调用都是运行于同一个线程(UI 线程)中的,所以在调用 stopSelf()方法时可能有已经收到但是还未来得及处理的启动请求,而调用 stopSelf()方法后,无论是否有未处理的启动请求,Service 都会被立即销毁。为了更安全地停止 Service,可以通过 stopSelfResult()方法停止 Service。stopSelfResult()方法要求一个 startId 参数,系统会检查 startId 参数值是否是最后一次请求启动此 Service 时所自动生成的请求 ID,如果相同才会停止 Service。

修改 MyService2 代码,将调用 stopSelf()方法改为 stopSelfResult()方法,代码如下所示。

【代码 8-8】 MyService2.java

```java
public class MyService2 extends Service {
    @Override
    public int onStartCommand(Intent intent, int flags, int startId) {
        Log.i("MyService2", "onStartCommand");
        final String message = intent.getStringExtra("message");
        Log.i("MyService2", "intent:" + message + ",flags:" + flags
                + ",startId:" + startId);
        stopSelfResult(startId);
        return super.onStartCommand(intent, flags, startId);}
    …//省略其他方法
}
```

改用 stopSelfResult()方法后,在调用 stopSelfResult(startId)方法时,如果 Service 已经收到了新的启动请求,此时 startId 与新请求的 ID 不同,则不会终止 Service,方法返回 false 表示停止 Service 失败。

注意

> Service 并没有提供类似于 onStop()之类的回调方法,当停止 Service 时如果该 Service 没有被绑定,则会被立即销毁。以 Bind 绑定方式启动 Service 将在后续内容中介绍。

Service 运行于应用程序主线程中,与 Activity 等可见组件运行于同一个线程,因此如果 Service 需要执行耗时较长的操作时,应该新开线程执行,否则会引起应用程序无响应(Application Not Responding,ANR)错误。

下例修改 MyService2 代码,模拟耗时操作处理,代码如下所示。

【代码 8-9】 MyService2.java

```java
public class MyService2 extends Service {
    @Override
    public int onStartCommand(Intent intent, int flags, int startId) {
        Log.i("MyService2", "onStartCommand");
        final String message = intent.getStringExtra("message");
        Log.i("MyService2", "intent:" + message + ",flags:" + flags
                + ",startId:" + startId);
        try {
            Thread.sleep(20000);
        } catch (InterruptedException e) {
            e.printStackTrace();}
        return super.onStartCommand(intent, flags, startId);}
    …//省略其他方法
}
```

上述 MyService2 代码中,在 onStartCommand()方法中通过 Thread.sleep()方法模拟了一个 20s 的耗时操作。

运行应用程序,单击"启动 MyService2"按钮,过一段时间就会弹出如图 8-3 所示提示窗口。

图 8-3 耗时操作使 MyService2 造成 ANR 错误

> 无论 Start 还是 Bind 方式启动的 Service,都是运行于 UI 线程中的,需要避免直接在当前线程进行耗时操作。

8.2.2 Bind 方式启动 Service

通过调用 Context 的 bindService()方法也可以启动 Service。使用 Bind 方式启动的 Service 会和启动它的组件关联在一起,并可以进行通信,组件可以通过 unbindService()方法来解除绑定关系。Bind 方式启动 Service 时的生命周期如图 8-4 所示。

Bind 方式启动 Service 时会自动调用 onBind()方法。如果该 Service 是第一次启动,则会首先调用 onCreate()方法,然后再调用 onBind()方法,否则直接调用 onBind()方法。在组件和 Service 解除绑定时会触发 Service 的 onUnbind()方法,一个 Service 可以被多个组件绑定,当所有的绑定组件都解除绑定时,该 Service 将被销毁,并执行 onDestroy()方法。同样地,如果系统资源不足,Android 也随时有可能销毁这个 Service。一个组件绑定 Service 后,如果这个组件被销毁,系统会自动解除与之对应的 Service 绑定。

图 8-4 Bind 方式启动的 Service 的生命周期

Service 的 onBind()方法详的语法结构如下所示。

【语法】

```
public abstract IBinder onBind(Intent intent)
```

onBind()方法是一个抽象方法,其参数 intent 为绑定这个 Service 时传入的 Intent 对象,其返回值是一个 android.os.IBinder 对象。onBind()方法返回的 IBinder 对象会被传递到所绑定 Service 的组件中,通过 IBinder 对象来实现组件与 Service 之间的交互。因此,还需要编写一个实现 IBinder 接口的类作为 onBind()方法的返回类型,而直接实现 IBinder 接口非常复杂,通常继承 IBinder 接口的实现类 android.os.Binder 即可。

Service 的 onUnbind()方法的语法结构如下所示。

【语法】

```
public boolean onUnbind(Intent intent)
```

onUnbind()方法相对比较简单,其参数 intent 代表需要解除绑定的 Service。onUnbind()方法的返回值可以用于混合使用 Start 和 Bind 方式的 Service 中,下一小节中将详细介绍。

注意

> 为实现进程间通信,Android 提供了 IBinder 接口,其专门用于跨进程的远程方法调用。

针对 Bind 方式启动 Service,Context 中提供了 bindService()和 unbindService()方法,分别用于绑定 Service 和解除与 Service 的绑定。

其中,bindService()方法的语法结构如下所示。

【语法】

```
public boolean bindService(Intent service, ServiceConnection conn, int flags)
```

bindService()方法用于绑定 Service,其返回值代表是否绑定成功,其参数如下。
- 参数 intent 是在绑定 Service 时所传入的 Intent 对象。
- 参数 conn 是一个 ServiceConnection 接口类型的对象,在绑定或解除绑定时,系统会调用 ServiceConnection 接口中对应的回调方法,ServiceConnection 接口包含以下两个方法:

① **void onServiceConnected**(**ComponentName name**,**IBinder service**):当绑定成功时会自动调用 onServiceConnected()方法,其中,参数 name 为绑定的 Service 的 ComponentName,参数 service 为绑定 Service 的 onBind()方法的返回值。

② **void onServiceDisconnected**(**ComponentName name**):当系统资源不足时,Android 可能会销毁 Service,此时会调用此方法。
- 参数 flags 用于决定 Service 的一些行为规则,常用的取值有 0、BIND_AUTO_CREATE、BIND_NOT_FOREGROUND、BIND_WAIVE_PRIORITY、BIND_IMPORTANT、BIND_ABOVE_CLIENT 和 BIND_ADJUST_WITH_ACTIVITY。

① **0**:当 flags 为 0 时,bindService()方法会返回 true。此时如果 Service 已被以 Start 方式启动,则绑定成功;否则不会创建 Service,但在 Service 使用 Start 方式启动时自动绑定。

② **Context.BIND_AUTO_CREATE**:在使用 bindService()绑定时,如果 Service 尚未被创建则创建 Service,即执行 Service 的 onCreate()方法;否则不会执行 onCreate()方法。

③ **Context.BIND_NOT_FOREGROUND**:表示所绑定的 Service 不允许拥有前台优先

级。默认情况下,绑定一个 Service 后系统会提升其优先级,flags 设为 BIND_NOT_FOREGROUND 后,会限制对其优先级的提升。

④ Context.BIND_WAIVE_PRIORITY：在绑定 Service 时不会改变其优先级。

⑤ Context.BIND_IMPORTANT：当所绑定 Service 的组件位于前台时,该 Service 也会提升为前台优先级。

⑥ Context.BIND_ABOVE_CLIENT：与 Context.BIND_IMPORTANT 类似,但当系统资源不足时,Android 会在终止 Service 之前先终止与其绑定的客户端组件。

⑦ Context.BIND_ADJUST_WITH_ACTIVITY：系统将根据 Activity 的优先级调整被绑定的 Service 的优先级,当 Activity 运行在前台时 Service 优先级提升,当 Activity 运行在后台时 Servcie 优先级相对降低。

unbindService()方法用于解除与 Service 的绑定,其语法结构如下所示。

【语法】

```
public void unbindService(ServiceConnection conn)
```

其中,参数 conn 是调用 bindService()方法绑定 Service 时所传入的 ServiceConnection 对象。需要注意的是：如果尚未绑定 Service 或者已解除绑定,则调用 unbindService()方法会抛出异常。

下列代码是以 Bind 方式演示 Service 的用法。首先自定义一个 Service 类 MyService3,代码如下所示。

【代码 8-10】 MyService3.java

```java
public class MyService3 extends Service {
    private MyBinder myBinder = new MyBinder();
    @Override
    public void onCreate() {
        Log.i("MyService3", "onCreate");}
    @Override
    public void onDestroy() {
        Log.i("MyService3", "onDestroy");}
    @Override
    public IBinder onBind(Intent intent) {
        Log.i("MyService3", "onBind");
        final String message = intent.getStringExtra("message");
        Log.i("MyService3", "intent:" + message);
        return myBinder;}
    @Override
    public boolean onUnbind(Intent intent) {
        Log.i("MyService3", "onUnbind");
        return false;}
    public String doSomeOperation(String param) {
        Log.i("MyService3", "doSomeOperation: param = " + param);
        return "return value";}
    public class MyBinder extends Binder {
        public MyService3 getService() {
            return MyService3.this;}}
}
```

在上述代码中,重写了 onBind()和 onUnbind()等方法,在各个生命周期方法中都输出相关 Log 信息。在 MyService3 中还声明了一个内部类 MyBinder,该类继承了 Binder,并提供了 getService()方法用于返回 MyService3 的当前实例。在 MyService3 中定义了一个 MyBinder 类型的属性 myBinder,使用 onBind()方法可以返回该 myBinder 属性。MyService3 中的 doSomeOperation()方法用于模拟业务操作。

在 AndroidManifest.xml 中配置 MyService3,代码如下所示。

【代码 8-11】 AndroidManifest.xml 中配置 MyService3

```xml
<service android:name="com.qst.chapter08.MyService3" />
```

修改 MainActivity 代码,提供按钮的事件处理方法来完成对 MyService3 的绑定、调用、解除绑定操作,代码如下所示。

【代码 8-12】 MainActivity.java

```java
public class MainActivity extends AppCompatActivity {
    ...//省略其他属性声明
    private MyService3 myService3;
    private ServiceConnection myService3Connection = new ServiceConnection() {
        @Override
        public void onServiceDisconnected(ComponentName name) {
            Log.i("MainActivity",
                    "myService3Connection.onServiceDisconnected():name = " + name);
            myService3 = null;}
        @Override
        public void onServiceConnected(ComponentName name, IBinder service) {
            Log.i("MainActivity",
                    "myService3Connection.onServiceConnected():name = " + name);
            myService3 = ((MyService3.MyBinder) service).getService();}};
    @Override
    protected void onCreate(Bundle savedInstanceState) {
        super.onCreate(savedInstanceState);
        setContentView(R.layout.activity_main);
        startButton = (Button) findViewById(R.id.startButton);
        stopButton = (Button) findViewById(R.id.stopButton);
        bindButton = (Button) findViewById(R.id.bindButton);
        operateButton = (Button) findViewById(R.id.operateButton);
        unbindButton = (Button) findViewById(R.id.unbindButton);
        bindButton.setOnClickListener(new OnClickListener() {
            @Override
            public void onClick(View v) {
                Intent intent = new Intent(MainActivity.this, MyService3.class);
                intent.putExtra("message", "hello!");
                bindService(intent, myService3Connection,
                        Context.BIND_AUTO_CREATE);}});
        operateButton.setOnClickListener(new OnClickListener() {
            @Override
            public void onClick(View v) {
                if (myService3 == null)
                    return;
```

```
                String returnValue = myService3.doSomeOperation("test");
                Log.i("MainActivity", "myService3.doSomeOperation:"
                    + returnValue);}});
        unbindButton.setOnClickListener(new OnClickListener() {
            @Override
            public void onClick(View v) {
                unbindService(myService3Connection);}});}
}
```

上述代码中，bindButton、operateButton 和 unbindButton 按钮分别用于完成绑定 MyService3、MyService3 方法的调用、解除与 MyService3 的绑定，还声明了 myService3Connection 和 myService3 两个属性。

在创建 myService3Connection 时，重写了 onServiceConnected() 和 onServiceDisconnected() 方法；在 onServiceConnected() 中将 service 参数强制转化为 MyService3.MyBinder 类型，并调用其 getService() 方法来获取 MySercie3 对象，最后赋值给 myService3 属性；在 onServiceDisconnected() 方法中将 myService3 属性赋值为 null。

在 bindButton 按钮的事件处理方法中，创建了一个 Intent 对象，用于指定组件的类型为 MyService3.class，并传入 message 附加数据，然后调用 bindService() 方法启动 MyService3。调用 bindService() 方法时传入了 Intent 对象和 myService3Connection 对象，并指定 flags 为 Context.BIND_AUTO_CREATE。

在 unbindButton 按钮的事件处理方法中，调用 unbindService() 方法解除与 MyService3 的绑定，其中传入了绑定 MyService3 时所使用的 myService3Connection 属性。

在 operateButton 按钮事件处理方法中，调用已绑定的 myService3 对象的 doSomeOperation() 方法完成业务逻辑处理。

MainActivity 的界面结构非常简单，不再列出其布局 XML 代码。运行程序，结果如图 8-5 所示。

图 8-5 MainActivity 界面

单击"绑定 MyService3"按钮，Logcat 输出如下：

```
12-02 04:02:44.151: I/MyService3(4108): onCreate
12-02 04:02:44.151: I/MyService3(4108): onBind
12-02 04:02:44.151: I/MyService3(4108): intent:hello!
12-02 04:02:44.171: I/MainActivity(4108):
myService3Connection.onServiceConnected():name = ComponentInfo{com.qst.chapter08/com.qst.chapter08.MyService3}
```

通过执行结果可以看到，执行绑定操作 bindService() 后，依次触发了 MyService3 的 onCreate()、onBind() 方法，并执行了 bindService() 时所指定的 ServiceConnection 对象的 onServiceConnected() 方法，各个方法中都正确输出了信息。

再多次单击"绑定 MyService3"按钮，Logcat 没有新的输出，说明在已经绑定 Service 的情况下，再次调用 bindService() 方法不会触发该 Service 的 onBind() 方法。

单击"操作 MyService3"按钮，Logcat 输出如下：

```
12-02 04:07:53.135: I/MyService3(4108): doSomeOperation: param = test
12-02 04:07:53.135: I/MainActivity(4108):
myService3.doSomeOperation:return value
```

通过执行结果可以看到，绑定成功后，即可调用 myService3 对象的业务逻辑方法。

单击"解绑 MyService3"按钮，Logcat 输出如下：

```
12-02 04:09:58.277: I/MyService3(4108): onUnbind
12-02 04:09:58.277: I/MyService3(4108): onDestroy
```

通过执行结果可以看到，当调用 unbindService() 方法解除绑定时，调用了 Service 的 onUnbind() 方法。由于 MyService3 只被当前应用程序绑定过一次，解绑后已没有绑定的客户端，因此还执行了 onDestroy() 方法，说明 Service 已被销毁。

8.2.3 混合方式的 Service

如果一个 Service 被一个或多个客户端以 Start 方式和 Bind 方式都启动过，则其生命周期将变得复杂，须同时满足两种方式的终止条件才会终止，如图 8-6 所示。

图 8-6 混合使用 Start 和 Bind 方式的 Service 生命周期

无论 Service 是先 Start 后 Bind，还是先 Bind 后 Start，onCreate() 方法只会执行一次，该 Service 将会一直运行，其中 onStartCommand() 方法调用的次数与 startService() 相同。

混合方式的 Service,当调用 stopService()或 unbindService()时不一定会被终止,需要同时满足 Start 和 Bind 两种方式的终止条件时 Service 才会终止。当 Service 所绑定的客户端都调用 unbindService()后,然后再调用 stopService()时该 Service 才会停止;同样,只调用 stopService()也不会终止 Service,还需要所有绑定的客户端都调用 unbindService()或者这些绑定客户端都终止之后服务才会自动停止。

下列示例演示混合使用 Start 和 Bind 方式来启动和停止 Service,代码如下所示。

【代码 8-13】 MyService4.java

```java
public class MyService4 extends Service {
    private MyBinder myBinder = new MyBinder();
    @Override
    public void onCreate() {
        Log.i("MyService4", "onCreate");}
    @Override
    public void onDestroy() {
        Log.i("MyService4", "onDestroy");}
    @Override
    public int onStartCommand(Intent intent, int flags, int startId) {
        Log.i("MyService4", "onStartCommand");
        return super.onStartCommand(intent, flags, startId); }
    @Override
    public IBinder onBind(Intent intent) {
        Log.i("MyService4", "onBind");
        return myBinder; }
    @Override
    public void onRebind(Intent intent) {
        Log.i("MyService4", "onRebind");}
    @Override
    public boolean onUnbind(Intent intent) {
        Log.i("MyService4", "onUnbind");
        return false;}
    public class MyBinder extends Binder {
    }
}
```

在上述代码中,重写了 Service 的各个生命周期方法,并使用 Logcat 输出相关信息。在 AndroidManifest.xml 中配置 MyService4,代码如下所示。

【代码 8-14】 AndroidManifest.xml 中配置 MyService4

```xml
<service android:name="com.qst.chapter08.MyService4" />
```

修改 MyActivity 代码,添加对 MyService4 的 start、stop、bind、unbind 的操作方法,代码如下所示。

【代码 8-15】 MyActivity.java

```java
public class MainActivity extends AppCompatActivity {
    ...//省略其他属性声明
    private Button start4Button;
    private Button stop4Button;
    private Button bind4Button;
```

```java
        private Button unbind4Button;
        private ServiceConnection myService4Connection = new ServiceConnection() {
            @Override
            public void onServiceDisconnected(ComponentName name) {
                Log.i("MainActivity",
                        "myService4Connection.onServiceDisconnected()");}
            @Override
            public void onServiceConnected(ComponentName name, IBinder service) {
                Log.i("MainActivity", "myService4Connection.onServiceConnected()");}};
    @Override
    protected void onCreate(Bundle savedInstanceState) {
        ...//省略
        start4Button = (Button) findViewById(R.id.start4Button);
        stop4Button = (Button) findViewById(R.id.stop4Button);
        bind4Button = (Button) findViewById(R.id.bind4Button);
        unbind4Button = (Button) findViewById(R.id.unbind4Button);
        start4Button.setOnClickListener(new OnClickListener() {
            @Override
            public void onClick(View v) {
                Intent intent = new Intent(MainActivity.this, MyService4.class);
                startService(intent); }});
        stop4Button.setOnClickListener(new OnClickListener() {
            @Override
            public void onClick(View v) {
                Intent intent = new Intent(MainActivity.this, MyService4.class);
                stopService(intent); } });
        bind4Button.setOnClickListener(new OnClickListener() {
            @Override
            public void onClick(View v) {
                Intent intent = new Intent(MainActivity.this, MyService4.class);
                bindService(intent, myService4Connection,
                        Context.BIND_AUTO_CREATE); } });
        unbind4Button.setOnClickListener(new OnClickListener() {
            @Override
            public void onClick(View v) {
                unbindService(myService4Connection); } });}
}
```

上述代码中,添加了 4 个按钮,单击时分别针对 MyService4 调用 startService()、stopService()、bindService()、unbindService()方法;而 ServiceConnection 类型的 myService4Connection 属性用于绑定 MyService4 时的连接对象。

运行应用程序,结果如图 8-7 所示。

首先单击"启动 MyService4"按钮,Logcat 输出如下:

```
12-02 10:25:55.979: I/MyService4(4174): onCreate
12-02 10:25:55.979: I/MyService4(4174): onStartCommand
```

然后单击"绑定 MyService4"按钮,Logcat 输出如下:

```
12-02 10:26:32.019: I/MyService4(4174): onBind
12-02 10:26:32.019: I/MainActivity(4174):
myService4Connection.onServiceConnected()
```

图 8-7 混合使用 Start 和 Bind 方式启动 Service

从输出结果可以看出,在已启动 Service 的情况下,绑定 Service 并不会执行其 onCreate()方法。单击"解绑 MyService4"按钮,Logcat 输出如下:

```
12-02 10:40:36.682: I/MyService4(4174): onUnbind
```

可见在已启动 Service 的情况下,解除绑定只会执行 onUnbind()方法,而不会执行 onDestroy()方法。单击"停止 MyService4"按钮,Logcat 输出如下:

```
12-02 10:43:27.625: I/MyService4(4174): onDestroy
```

此时,由于与 Service 绑定的所有客户端都已解除绑定,所以 stopService()执行了 onDestroy()方法。

接下来,依次单击"绑定 MyService4""启动 MyService4""停止 MyService4""解绑 MyService4"按钮,Logcat 输出如下:

```
12-02 10:45:31.927: I/MyService4(4174): onCreate
12-02 10:45:31.927: I/MyService4(4174): onBind
12-02 10:45:31.927: I/MainActivity(4174):
myService4Connection.onServiceConnected()
12-02 10:45:34.257: I/MyService4(4174): onStartCommand
12-02 10:45:37.277: I/MyService4(4174): onUnbind
12-02 10:45:37.277: I/MyService4(4174): onDestroy
```

当存在绑定 Service 的客户端时,startService()方法不会触发 onCreate()方法,stopService()方法也不会触发 onDestroy()方法。对于 Start 和 Bind 混合方式启动的 Service,调用 stopService()方法和解除所有客户端的绑定是 Service 销毁的必要条件。

当客户端和 Service 解除绑定后,如果 Service 仍处于启动状态,客户端再次绑定 Service 时仍会执行 onBind()方法;但是如果 onUnbind()方法返回 true,则再次绑定

Service 时不会执行 onBind()方法,而是执行 onRebind()方法。修改 MyService4 代码,将 onUnbind()方法返回值改为 true,代码如下所示。

【代码 8-16】 **MyService4.java 相关代码**

```
public boolean onUnbind(Intent intent) {
    Log.i("MyService4", "onUnbind");
    return true;
}
```

然后重新运行应用程序,依次单击"启动 MyService4""绑定 MyService4""解绑 MyService4""停止 MyService4""绑定 MyService4"按钮,Logcat 输出如下:

```
12-02 10:57:20.647: I/MyService4(5747): onCreate
12-02 10:57:20.647: I/MyService4(5747): onStartCommand
12-02 10:57:21.777: I/MyService4(5747): onBind
12-02 10:57:21.777: I/MainActivity(5747):
myService4Connection.onServiceConnected()
12-02 10:57:23.597: I/MyService4(5747): onUnbind
12-02 10:57:26.457: I/MainActivity(5747):
myService4Connection.onServiceConnected()
12-02 10:57:26.457: I/MyService4(5747): onRebind
```

可以看到,在再次绑定 Service 时会执行 onRebind()方法。因此,当 onUnbind()方法返回值为 true 时,可以在 onRebind()方法中专门处理重新绑定的情况。

8.2.4 前台 Service

Android 系统对运行的进程按照优先级进行了归类,当高优先级的进程所需内存不足时,Android 会终止优先级低的进程以释放内存,从而保证高优先级进程顺利运行。下面按照优先级从高到低列出了常见的进程类型。

- **前台进程**(**Foreground Process**):前台进程具有最高优先级,通常前台进程的数量很少,前台进程几乎不会被系统终止,只有当内存极低以致无法保证所有的前台进程同时运行时,系统才会选择终止某个前台进程。以下状态的进程属于前台进程:进程中包含处于前台的正与用户交互的 Activity,进程中包含与前台 Activity 绑定的 Service,进程中包含调用了 startForeground()方法的 Service,进程中包含正在执行 onCreate()、onStart()或 onDestroy()方法的 Service,进程中包含正在执行 onReceive()方法的 BroadcastReceiver。
- **可见进程**(**Visible Process**):可见进程是指界面可见但处于暂停状态的进程,除非要为前台进程释放内存,否则系统不会终止可见进程。可见进程包括:进程中包含处于暂停状态的 Activity,即调用了 onPause()方法的 Activity;进程中包含绑定到暂停状态 Activity 的 Service。
- **服务进程**(**Service Process**):服务进程是指通过 startService()方法启动的 Service 所在的进程。
- **后台进程**(**Background Process**):后台进程是指包含处于停止状态的 Activity 的进程,即调用了 onStop()方法的 Activity 所在的进程。由于后台进程通常不会直接影

第8章 Service服务

响用户体验,因此为了保证更高优先级进程的顺利运行,Android 可能随时终止后台进程。在 Activity 中应该根据需要在 onStop()方法中保存当前状态,以备重新启动时能够恢复到用户之前使用时的状态。

Service 启动后,其所在进程默认是服务进程,优先级并不高,如果该 Service 非常重要,则可以通过 Service 的 startForeground()方法将其改为前台进程。调用 startForeground()方法后,Service 运行时会在通知栏显示一个通知(Notification),Service 停止后通知会消失。

startForeground()方法声明格式如下。

【语法】

```
public final void startForeground(int id, Notification notification)
```

参数 id 为通知的 ID;参数 notification 为需要显示的通知。

当 Service 成为前台进程后,如需恢复原有的优先级,可以调用 stopForeground()方法取消其前台状态,从而允许系统在内存不足时更容易终止这个 Service。stopForeground()方法声明格式如下。

【语法】

```
public final void stopForeground(boolean removeNotification)
```

stopForeground()方法只有一个参数,用于降低 Service 的前台优先级时是否移除 startForeground()方法所创建的通知。

下面编写 MyService5,调用 startForeground()方法使其成为前台服务,代码如下所示。

【代码 8-17】 MyService5.java

```java
public class MyService5 extends Service {
    @Override
    public void onCreate() {
        Notification.Builder builder = new Notification.Builder(this);
        builder.setSmallIcon(R.drawable.btn_star_big_on_pressed);
        builder.setLargeIcon(BitmapFactory.decodeResource(getResources(),
                R.drawable.ic_launcher));
        builder.setContentTitle("MyService5");
        builder.setContentText("MyService5 正在运行...");
        Notification notification = builder.build();
        notification.flags = Notification.FLAG_AUTO_CANCEL;
        startForeground(1, notification);}
    @Override
    public IBinder onBind(Intent intent) {
        return null;}
}
```

上述代码中,在 onCreate()方法中构造了一个 Notification 对象,并调用 Service 的 startForeground()方法,将构造的 Notification 对象作为该方法的参数。

在 AndroidManifest.xml 中配置 MyService5,代码如下所示。

【代码 8-18】 AndroidManifest.xml 中配置 MyService5

```xml
<service android:name="com.qst.chapter08.MyService5" />
```

修改 MyActivity 代码，添加两个按钮，分别用于启动和停止 MyService5，代码如下所示。

【代码 8-19】 **MainActivity.java**

```java
public class MainActivity extends AppCompatActivity {
    @Override
    protected void onCreate(Bundle savedInstanceState) {
        ...//省略
        start5Button.setOnClickListener(new OnClickListener() {
            @Override
            public void onClick(View v) {
                Intent intent = new Intent(MainActivity.this, MyService5.class);
                startService(intent);}});
        stop5Button.setOnClickListener(new OnClickListener() {
            @Override
            public void onClick(View v) {
                Intent intent = new Intent(MainActivity.this, MyService5.class);
                stopService(intent);}}); }
}
```

运行应用程序，然后单击"前台启动 MyService5"按钮，通知栏会显示 MyService5 中定义的通知，如图 8-8 所示。

图 8-8　Service 成为前台状态并显示通知

单击"停止 MyService5"按钮时，MyService5 停止，相应的通知也会自动消失。

startForeground()方法可以在任何位置调用，也可以多次调用。因此，如果需要在启动服务时指定通知的内容，可以将创建 Notification 和调用 startForeground()操作移到 onStartCommand()或 onBind()方法内，通过 Intent 将通知内容传递给 Service。按照此种方式修改 MyService5 代码，代码如下所示。

【代码 8-20】 **MyService5.java**

```java
public class MyService5 extends Service {
    @Override
    public int onStartCommand(Intent intent, int flags, int startId) {
        Notification.Builder builder = new Notification.Builder(this);
        builder.setSmallIcon(R.drawable.btn_star_big_on_pressed);
        builder.setLargeIcon(BitmapFactory.decodeResource(getResources(),
```

```
                R.drawable.ic_launcher));
        builder.setContentTitle(intent.getStringExtra("notice title"));
        builder.setContentText(intent.getStringExtra("notice text"));
        Notification notification = builder.build();
        notification.flags = Notification.FLAG_AUTO_CANCEL;
        startForeground(1, notification);
        return super.onStartCommand(intent, flags, startId);}
    @Override
    public IBinder onBind(Intent intent) {
        return null; }
}
```

上述代码中,在 onStartCommand()方法中创建 Notification 通知和调用 startForeground()方法,通过 Intent 参数获取 Service 客户端信息并赋给 Notification 的属性。运行应用程序,如图 8-9 所示。

除了通知的标题和内容外,实际上可以通过 Intent 传递各种配置信息,例如 Notification 的图标,以及是否取消 Service 的前台状态等。

图 8-9　调用 startForeground() 方法时定制通知内容

本小节前面所述内容都是以 Start 方式来启动 Service,实际上 Bind 方式所启动的 Service 同样可以调用 startForeground()和 stopForeground()方法来改变 Service 的前台状态和取消前台状态。但是,客户端使用 Bind 方式启动 Service 时可以直接获取 Service 的实例,从而能够更方便快捷地操作 Service。下面的示例演示以 Bind 方式启动前台 Service,并在通知栏中模拟显示一个进度条,修改 MyService5 代码如下所示。

【代码 8-21】　MyService5.java

```
public class MyService5 extends Service {
    private MyBinder myBinder = new MyBinder();
    private Notification.Builder builder;
    @Override
    public IBinder onBind(Intent intent) {
        builder = new Notification.Builder(this);
        builder.setSmallIcon(R.drawable.btn_star_big_on_pressed);
        builder.setLargeIcon(BitmapFactory.decodeResource(getResources(),
                R.drawable.ic_launcher));
        builder.setContentTitle("MyService5");
        return myBinder;}
    public void setProgress(int progress) {
        builder.setContentText("进度: " + progress + "%");
        builder.setProgress(100, progress, false);
        Notification notification = builder.build();
        startForeground(1, notification);}
    public class MyBinder extends Binder {
        public MyService5 getService() {
            return MyService5.this;}}
}
```

上述代码中声明了 MyBinder 内部类并继承 Binder 类,其中包含 getService()方法用于返回 MyService5 的当前实例。在 MyService5 中,onBind()方法用于返回 myBinder 属性;在 setProgress()方法中通过 builder 的 setProgress()方法来设定进度条式通知栏的进度值,然后调用 startForeground()方法来更新通知。修改 MainActivity 代码,添加绑定 MyService5 的操作,代码如下所示。

【代码 8-22】 MainActivity.java

```java
public class MainActivity extends AppCompatActivity {
    ...//省略
    private Button bind5Button;
    private Button progress5Button;
    private MyService5 myService5;
    private ServiceConnection myService5Connection = new ServiceConnection() {
        @Override
        public void onServiceDisconnected(ComponentName name) {
            myService5 = null;}
        @Override
        public void onServiceConnected(ComponentName name, IBinder service) {
            myService5 = ((MyService5.MyBinder) service).getService();}};
    @Override
    protected void onCreate(Bundle savedInstanceState) {
        ...//省略
        bind5Button.setOnClickListener(new OnClickListener() {
            @Override
            public void onClick(View v) {
                Intent intent = new Intent(MainActivity.this, MyService5.class);
                bindService(intent, myService5Connection,
                    Context.BIND_AUTO_CREATE);}});
        progress5Button.setOnClickListener(new OnClickListener() {
            @Override
            public void onClick(View v) {
                new Thread() {
                    public void run() {
                        for (int i = 0; i <= 100; i++) {
                            final int p = i;
                            runOnUiThread(new Runnable() {
                                public void run() {
                                    myService5.setProgress(p); }});
                            try {
                                sleep(100);
                            } catch (InterruptedException e) {
                                e.printStackTrace();}}}
                }.start();}});}
}
```

上述代码中,在 bind5Button 的单击事件中绑定 MyService5。在 progress5Button 的单击事件中,每隔 100ms 调用一次 myService5 的 setProgress()方法实现进度条的增长效果。

运行应用程序,首先单击"前台绑定 MyService5"按钮,然后单击"开始 MyService5 进度"按钮,可以在状态栏中观察到 MyService5 的进度条在逐渐增加,如图 8-10 所示。

图 8-10　调用 startForeground()方法并在通知中显示进度

8.2.5　Service 中执行耗时任务

Service 运行于 UI 线程中，如果直接在 UI 线程中执行耗时或可能被阻塞的任务，会造成界面无响应甚至 ANR 错误，因此这种耗时任务通常都需要新开线程执行。下列代码演示了如何在 Service 中执行耗时任务。

【代码 8-23】　MainService6.java

```java
public class MyService6 extends Service {
    @Override
    public int onStartCommand(Intent intent, int flags, final int startId) {
        new Thread() {
            public void run() {
                Log.i("MyService6","任务" + startId + "开始运行");
                for (int i = 0; i < 5; i++) {
                    Log.i("MyService6","任务" + startId + "正在运行:" + i);
                    try {
                        Thread.sleep(2000);
                    } catch (InterruptedException e) {
                    } }
                Log.i("MyService6","任务" + startId + "运行结束");}
        }.start();
        return super.onStartCommand(intent, flags, startId);}
    @Override
    public void onDestroy() {
        Log.i("MyService6", "onDestroy");
        super.onDestroy();}
    @Override
    public IBinder onBind(Intent intent) {
        return null;}
}
```

上述代码中，在 onStartCommand()方法中新开线程执行一个模拟的耗时任务，任务中循环 5 次，每次循环持续 2s，并在 Logcat 中输出执行任务状态，提供 startId 参数可以看出所执行的是哪一次 startService()方法。

在 AndroidManifest.xml 中配置 MyService6,代码如下所示。

【代码 8-24】 **AndroidManifest.xml 中配置 MyService6**

```xml
<service android:name="com.qst.chapter08.MyService6" />
```

修改 MainActivity 代码,添加启动、停止 MyService6 的按钮,代码如下所示。

【代码 8-25】 **MainActivity.java**

```java
public class MainActivity extends AppCompatActivity {
    …//省略
    private Button start6Button;
    private Button stop6Button;
    @Override
    protected void onCreate(Bundle savedInstanceState) {
        …//省略
        start6Button.setOnClickListener(new OnClickListener() {
            @Override
            public void onClick(View v) {
                Intent intent = new Intent(MainActivity.this, MyService6.class);
                startService(intent);} });
        stop6Button.setOnClickListener(new OnClickListener() {
            @Override
            public void onClick(View v) {
                Intent intent = new Intent(MainActivity.this, MyService6.class);
                stopService(intent); }});}
}
```

运行应用程序(界面结构非常简单,此处不再演示界面),单击"启动 MyService6"按钮,Logcat 输出如下:

```
12-06 13:35:11.590: I/MyService6(2133):任务 1 开始运行
12-06 13:35:11.590: I/MyService6(2133):任务 1 正在运行:0
12-06 13:35:13.600: I/MyService6(2133):任务 1 正在运行:1
12-06 13:35:15.610: I/MyService6(2133):任务 1 正在运行:2
12-06 13:35:17.620: I/MyService6(2133):任务 1 正在运行:3
12-06 13:35:19.630: I/MyService6(2133):任务 1 正在运行:4
12-06 13:35:21.640: I/MyService6(2133):任务 1 运行结束
```

可以看到,在 MyService6 的 onStartCommand()方法中线程成功运行。实际操作时,会发现单击"启动 MyService6"按钮后,按钮立即变为可用状态,这是由于任务在新线程中运行,所以按钮的单击事件能够立即结束。

需要注意,onStartCommand()方法中启动了新的任务线程,这个线程是一个完全独立的普通线程,与启动它的 Service 没有任何关系,该线程会一直运行直到自己结束,或者由于进程被终止而提前结束,而不会因为 Service 的销毁而停止。例如,在单击"启动 MyService6"按钮后,新线程开始运行,此时单击"停止 MyService6"按钮,Logcat 输出如下:

```
12-06 14:03:59.810: I/MyService6(2327): onStartCommand: startId=1
12-06 14:03:59.810: I/MyService6(2327):任务 1 开始运行
12-06 14:03:59.810: I/MyService6(2327):任务 1 正在运行:0
```

```
12-06 14:04:01.820: I/MyService6(2327):任务 1 正在运行: 1
12-06 14:04:03.080: I/MyService6(2327): onDestroy
12-06 14:04:03.830: I/MyService6(2327):任务 1 正在运行: 2
12-06 14:04:05.840: I/MyService6(2327):任务 1 正在运行: 3
12-06 14:04:07.850: I/MyService6(2327):任务 1 正在运行: 4
12-06 14:04:09.860: I/MyService6(2327):任务 1 运行结束
```

可以看到，虽然 MyService6 的 onDestroy() 方法已被调用，但是线程仍在运行。

再次运行应用程序，连续多次单击"启动 MyService6"按钮，Logcat 输出如下：

```
12-06 14:08:05.640: I/MyService6(2327): onStartCommand: startId = 1
12-06 14:08:05.640: I/MyService6(2327):任务 1 开始运行
12-06 14:08:05.640: I/MyService6(2327):任务 1 正在运行: 0
12-06 14:08:07.650: I/MyService6(2327):任务 1 正在运行: 1
12-06 14:08:09.000: I/MyService6(2327): onStartCommand: startId = 2
12-06 14:08:09.000: I/MyService6(2327):任务 2 开始运行
12-06 14:08:09.000: I/MyService6(2327):任务 2 正在运行: 0
12-06 14:08:09.660: I/MyService6(2327):任务 1 正在运行: 2
12-06 14:08:11.010: I/MyService6(2327):任务 2 正在运行: 1
12-06 14:08:11.670: I/MyService6(2327):任务 1 正在运行: 3
12-06 14:08:13.020: I/MyService6(2327):任务 2 正在运行: 2
12-06 14:08:13.640: I/MyService6(2327): onStartCommand: startId = 3
12-06 14:08:13.640: I/MyService6(2327):任务 3 开始运行
12-06 14:08:13.640: I/MyService6(2327):任务 3 正在运行: 0
12-06 14:08:13.680: I/MyService6(2327):任务 1 正在运行: 4
12-06 14:08:15.030: I/MyService6(2327):任务 2 正在运行: 3
12-06 14:08:15.650: I/MyService6(2327):任务 3 正在运行: 1
12-06 14:08:15.690: I/MyService6(2327):任务 1 运行结束
12-06 14:08:17.040: I/MyService6(2327):任务 2 正在运行: 4
12-06 14:08:17.660: I/MyService6(2327):任务 3 正在运行: 2
12-06 14:08:19.050: I/MyService6(2327):任务 2 运行结束
12-06 14:08:19.670: I/MyService6(2327):任务 3 正在运行: 3
12-06 14:08:21.680: I/MyService6(2327):任务 3 正在运行: 4
12-06 14:08:23.690: I/MyService6(2327):任务 3 运行结束
```

从运行结果可以看到，每次调用 startService() 后，都会执行 Service 的 onStartCommand() 方法，进而启动新线程。需要注意，三次调用开启的新线程是并发运行的，它们的执行顺序是由系统调度的。

注意

为完成耗时任务，在 onStartCommand() 方法中启动了新线程，如果使用 Bind 方式启动的 Service，则可以在 onBind() 方法中启动新线程。在新线程中执行耗时任务时，与 Service 以 Start 还是 Bind 方式启动是没有关系的。

针对在 Service 中执行耗时任务，Android 还专门提供了一种特殊的 Service：IntentService。抽象类 android.app.IntentService 是 Service 的子类，其内部会自动开始一个新线程来执行任务，并在任务执行完毕后停止 Service。当有多个任务时，IntentService 会将任务加到一个队列中，按照次序依次执行，直到所有任务执行完毕后停止 Service。

使用 IntentService 非常简单，只须继承 IntentService 并重写 onHandleIntent()方法即可，onHandleIntent()方法的语法格式如下所示。

【语法】

```
protected abstract void onHandleIntent(Intent intent)
```

其中，参数 intent 是 Service 客户端以 Start 方式启动 Service 时 startService()方法所传入的 intent 对象。

下例演示了 IntentService 的用法，编写 MyService7 并继承 IntentService，代码如下所示。

【代码 8-26】 MyService7.java

```java
public class MyService7 extends IntentService {
    public MyService7() {
        super("IntentService 测试");}
    @Override
    public int onStartCommand(Intent intent, int flags, int startId) {
        Log.i("MyService7", "onStartCommand: startId = " + startId);
        intent.putExtra("startId", startId);
        return super.onStartCommand(intent, flags, startId);}
    @Override
    public void onDestroy() {
        Log.i("MyService7", "onDestroy");
        super.onDestroy();}
    @Override
    protected void onHandleIntent(Intent intent) {
        int startId = intent.getIntExtra("startId", 0);
        Log.i("MyService7", "任务" + startId + "开始运行");
        for (int i = 0; i < 5; i++) {
            Log.i("MyService7", "任务" + startId + "正在运行：" + i);
            try {
                Thread.sleep(2000);
            } catch (InterruptedException e) {
            }}
        Log.i("MyService7", "任务" + startId + "运行结束");}
}
```

上述代码中，MyService7 继承了 IntentService，并重写 onHandleIntent()方法实现模拟了一个 10s 的耗时操作。为了输出任务编号，在 onStartCommand()方法中将 startId 存入 intent，在 onHandleIntent()方法中从 intent 中获取了 startId 作为任务的编号。

在 AndroidManifest.xml 中配置 MyService7，代码如下所示。

【代码 8-27】 AndroidManifest.xml 中配置 MyService7

```
<service android:name="com.qst.chapter08.MyService7" />
```

修改 MainActivity 代码，添加启动 MyService7 的按钮，代码如下所示。

【代码 8-28】 MainActivity.java

```java
public class MainActivity extendsAppCompat Activity {
    ...//省略
```

```
    private Button start7Button;
    @Override
    protected void onCreate(Bundle savedInstanceState) {
        ...//省略
        start7Button.setOnClickListener(new OnClickListener() {
            @Override
            public void onClick(View v) {
                Intent intent = new Intent(MainActivity.this, MyService7.class);
                startService(intent);}});}
}
```

运行应用程序(界面结构非常简单,此处不再演示界面),单击"启动 MyService7"按钮,Logcat 输出如下:

```
12-06 17:20:25.540: I/MyService7(2327): onStartCommand: startId = 1
12-06 17:20:25.540: I/MyService7(2327):任务 1 开始运行
12-06 17:20:25.540: I/MyService7(2327):任务 1 正在运行: 0
12-06 17:20:27.550: I/MyService7(2327):任务 1 正在运行: 1
12-06 17:20:29.560: I/MyService7(2327):任务 1 正在运行: 2
12-06 17:20:31.570: I/MyService7(2327):任务 1 正在运行: 3
12-06 17:20:33.580: I/MyService7(2327):任务 1 正在运行: 4
12-06 17:20:35.590: I/MyService7(2327):任务 1 运行结束
12-06 17:20:35.590: I/MyService7(2327): onDestroy
```

可以看到,任务正常执行,并且在执行完毕后调用 Service 的 onDestroy()方法,说明 IntentService 会在任务执行完毕后自动销毁。

多次单击"启动 MyService7"按钮,Logcat 输出如下:

```
12-06 17:22:44.430: I/MyService7(2327): onStartCommand: startId = 1
12-06 17:22:44.430: I/MyService7(2327):任务 1 开始运行
12-06 17:22:44.430: I/MyService7(2327):任务 1 正在运行: 0
12-06 17:22:45.220: I/MyService7(2327): onStartCommand: startId = 2
12-06 17:22:46.440: I/MyService7(2327):任务 1 正在运行: 1
12-06 17:22:48.450: I/MyService7(2327):任务 1 正在运行: 2
12-06 17:22:50.460: I/MyService7(2327):任务 1 正在运行: 3
12-06 17:22:51.460: I/MyService7(2327): onStartCommand: startId = 3
12-06 17:22:52.470: I/MyService7(2327):任务 1 正在运行: 4
12-06 17:22:54.480: I/MyService7(2327):任务 1 运行结束
12-06 17:22:54.480: I/MyService7(2327):任务 2 开始运行
12-06 17:22:54.480: I/MyService7(2327):任务 2 正在运行: 0
12-06 17:22:56.490: I/MyService7(2327):任务 2 正在运行: 1
12-06 17:22:58.500: I/MyService7(2327):任务 2 正在运行: 2
12-06 17:23:00.510: I/MyService7(2327):任务 2 正在运行: 3
12-06 17:23:02.520: I/MyService7(2327):任务 2 正在运行: 4
12-06 17:23:04.530: I/MyService7(2327):任务 2 运行结束
12-06 17:23:04.530: I/MyService7(2327):任务 3 开始运行
12-06 17:23:04.530: I/MyService7(2327):任务 3 正在运行: 0
12-06 17:23:06.540: I/MyService7(2327):任务 3 正在运行: 1
12-06 17:23:08.550: I/MyService7(2327):任务 3 正在运行: 2
12-06 17:23:10.560: I/MyService7(2327):任务 3 正在运行: 3
12-06 17:23:12.570: I/MyService7(2327):任务 3 正在运行: 4
```

```
12-06 17:23:14.580: I/MyService7(2327):任务 3 运行结束
12-06 17:23:14.580: I/MyService7(2327): onDestroy
```

可以看到,每次调用 startService()后,都会立即执行 Service 的 onStartCommand()方法,但是对应的任务并没有马上执行,而是按照调用的次序依次执行,在所有任务都执行完毕后调用 Service 的 onDestroy()方法。

当需要执行耗时的后台任务时,使用 IntentService 是一种合适的选择,开发者可以避免启动和管理新线程,使用方便,代码简洁。但是 IntentService 也有其局限性,由于将任务按照调用次序排队依次执行,因此损失了并发性。如果多个任务并没有执行次序的要求,或者多个任务明确地需要并发执行,此时手动启动多个并发的新线程将是更好的选择。

8.2.6 远程 Service

本章前面所介绍的都是本地 Service,即与客户端运行于同一个进程中的 Service。实际上,有时还需要一种 Service,通常用于提供一些通用的系统级服务,需要运行于独立的进程中,并为其他进程提供服务,为此 Android 提供了远程 Service,即允许被另一个进程中的组件访问的 Service。

为使远程 Service 能被其他进程访问,需要一种进程间通信的机制。进程是操作系统的概念,因此跨进程通信需要将传递的对象分解成操作系统可以理解的基本单元,并且有序地通过进程边界。通过代码实现进程间通信数据的解析和传输需要编写冗长的模板式代码,为此,Android 提供了 AIDL 工具来完成这项工作。

注意

> 进程间通信是一个"历史悠久"的概念,有很多种进程通信的技术,例如 CORBA、COM、Java RMI、EJB、WebService 等,对进程间通信的完整介绍超出了本书的范畴,感兴趣的读者可以参阅更具针对性的资料。

Android 接口定义语言(Android Interface Definition Language,AIDL)是 Android 提供的一种专门用于描述进程间通信接口的语言,使用 AIDL 可以简化在进程间交换数据的代码,使客户端可以像本地 Service 那样直接绑定远程 Service。

下列示例演示如何通过 AIDL 实现远程 Service。首先编写 AIDL 的接口文件,aidl 文件的语法与 Java Interface 的语法几乎相同,在源代码目录的 com.qst.chapter08 包中创建 MyService8.aidl 文件,代码如下所示。

【代码 8-29】 MyService8.aidl

```
package com.qst.chapter08;
interface MyService8 {
    int sum(int a, int b);}
```

上述代码中,首先使用 package 来声明 aidl 文件所在的包,然后使用 interface 来定义了名为 MyService8 的 AIDL 接口,其中提供了一个 sum()方法,接收两个整数,返回一个整数。需要注意,aidl 文件是客户端访问远程 Service 的接口描述文件,因此需要将该文件复制到客户端所在的项目中。

下一步需要编写一个类,同时实现 MyService8 接口和 android.os.IBinder 接口,该类负责完成进程间通信的数据传输。实际上,在使用 Android Studio 开发环境时,aidl 文件编写完成后,选择 Android Studio 上方菜单栏"Build→Rebuild Project",编译后的项目结构如图 8-11 所示。

编译时 Android Studio 会在 build 目录下自动生成一个与 aidl 文件同名的 Java 接口文件 MyService8.java,在 Project 项目结构中查看项目的目录结构,如图 8-12 所示。

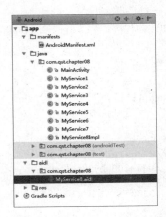
图 8-11 Android Studio 自动生成 AIDL 实现类

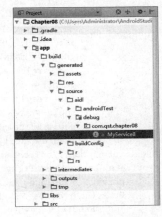
图 8-12 接口文件 MyService8.java

使用 Android Studio 生成的 MyService8.java 的代码如下所示。

【代码 8-30】 Android Studio 生成的 MyService8.java

```java
package com.qst.chapter08;
public interface MyService8 extends android.os.IInterface {
    /** Local-side IPC implementation stub class. */
    public static abstract class Stub extends android.os.Binder implements
            com.qst.chapter08.MyService8 {
        private static final java.lang.String DESCRIPTOR
                = "com.qst.chapter08.MyService8";
        /** Construct the stub at attach it to the interface. */
        public Stub() {
            this.attachInterface(this, DESCRIPTOR);}
        /**
         * Cast an IBinder object into an com.qst.chapter08.MyService8
         * interface, generating a proxy if needed.
         */
        public static com.qst.chapter08.MyService8 asInterface(
                android.os.IBinder obj) {
            if ((obj == null)) {
                return null; }
            android.os.IInterface iin = obj.queryLocalInterface(DESCRIPTOR);
            if (((iin != null)
                    && (iin instanceof com.qst.chapter08.MyService8))) {
                return ((com.qst.chapter08.MyService8) iin);}
            return new com.qst.chapter08.MyService8.Stub.Proxy(obj);}
        @Override
```

```java
            public android.os.IBinder asBinder() {
                return this;}
            @Override
            public boolean onTransact(int code, android.os.Parcel data,
                    android.os.Parcel reply, int flags)
                    throws android.os.RemoteException {
                switch (code) {
                    case INTERFACE_TRANSACTION: {
                        reply.writeString(DESCRIPTOR);
                        return true;}
                    case TRANSACTION_sum: {
                        data.enforceInterface(DESCRIPTOR);
                        int _arg0;
                        _arg0 = data.readInt();
                        int _arg1;
                        _arg1 = data.readInt();
                        int _result = this.sum(_arg0, _arg1);
                        reply.writeNoException();
                        reply.writeInt(_result);
                        return true;}}
                return super.onTransact(code, data, reply, flags);}
            private static class Proxy implements com.qst.chapter08.MyService8 {
                private android.os.IBinder mRemote;
                Proxy(android.os.IBinder remote) {
                    mRemote = remote;}
                @Override
                public android.os.IBinder asBinder() {
                    return mRemote;}
                public java.lang.String getInterfaceDescriptor() {
                    return DESCRIPTOR;}
                @Override
                public int sum(int a, int b) throws android.os.RemoteException {
                    android.os.Parcel _data = android.os.Parcel.obtain();
                    android.os.Parcel _reply = android.os.Parcel.obtain();
                    int _result;
                    try {
                        _data.writeInterfaceToken(DESCRIPTOR);
                        _data.writeInt(a);
                        _data.writeInt(b);
                        mRemote.transact(Stub.TRANSACTION_sum, _data, _reply, 0);
                        _reply.readException();
                        _result = _reply.readInt();
                    } finally {
                        _reply.recycle();
                        _data.recycle();}
                    return _result;}}
            static final int TRANSACTION_sum
                    = (android.os.IBinder.FIRST_CALL_TRANSACTION + 0); }
    public int sum(int a, int b) throws android.os.RemoteException;
}
```

Android Studio 自动生成的 MyService8.java 主要完成以下功能：

（1）MyService8 接口继承了 android.os.IInterface 接口。

（2）MyService8 接口中声明了内部抽象类 Stub,其继承 android.os.Binder 并实现了

MyService8 接口。

（3）Stub 中还有一个内部类 Proxy，用于负责代理远程客户端的调用请求，Stub 实现了 IInterface 接口的 asInterface()方法并返回 Proxy 的实例，在 Proxy 类中对 aidl 文件中所声明的每个方法都声明了一个整数编号。

（4）在 Stub 类中重写 Binder 的 onTransact()方法时调用抽象方法 sum()。因此，实际的 Service 实现类中，onBind()方法需要返回 Stub 类的子类的实例。

完成 aidl 文件并由 Android Studio 重新编译成功后，还需要编写 Service 来实现服务功能。编写 MyService8Impl 类，代码如下。

【代码 8-31】 MyService8Impl.java

```java
public class MyService8Impl extends Service {
    private final MyService8.Stub binder = new MyService8.Stub() {
        @Override
        public int sum(int a, int b) throws RemoteException {
            Log.i("MyService8Impl", "sum(" + a + ", " + b + ")");
            return a + b; }};
    @Override
    public void onCreate() {
        Log.i("MyService8Impl", "onCreate()");}
    @Override
    public IBinder onBind(Intent arg0) {
        Log.i("MyService8Impl", "onBind()");
        return binder; }
    @Override
    public boolean onUnbind(Intent intent) {
        Log.i("MyService8Impl", "onUnbind()");
        return false;}
    @Override
    public void onDestroy() {
        Log.i("MyService8Impl", "onDestroy()");};
}
```

上述 MyService8Impl 代码中，声明了 MyService8.Stub 类型的 binder 属性，其中重写了 MyService8.Stub 类的业务处理方法 sum()，然后在 MyService8Impl 的 onBind()方法中返回该 binder 对象。

最后，还需要在 AndroidManifest.xml 中配置 MyService8Impl。需要注意，MyService8Impl 是提供给远程客户端使用的 Service，而远程客户端是无法直接获取 MyService8Impl 的类型的，即无法通过 new Intent(context, MyService8Impl.class)的方式绑定 MyService8Impl，而只能通过隐式 Intent 访问。因此，MyService8Impl 必须配置<intent-filter>，配置代码如下。

【代码 8-32】 AndroidManifest.xml

```xml
<service android:name = "com.qst.chapter08.MyService8Impl">
    <intent-filter>
        <action android:name = "com.qst.service.MyService8" />
    </intent-filter>
</service>
```

在上述配置中，为 MyService8Impl 的<intent-filter>元素指定了一个名为 com.qst.service.MyService8 的<action>子元素。

至此，远程 Service 编写完毕，运行应用程序，远程 Service 即发布成功。

下面编写客户端来连接这个远程 Service，为了模拟在另一个线程中访问此远程 Service，客户端需要在新的项目中进行编写。

新建项目 chapter08_RemoteServiceClient，并将 Project 目录下 main 文件夹中的整个 aidl 文件复制到新项目中的 main 文件中，然后重新编译并由 Android Studio 会自动生成 MyService8.java 代码，如图 8-13 所示。

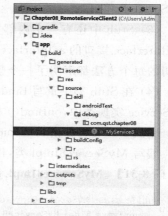

图 8-13　复制 aidl 到客户端项目中

在 MainActivity 中添加绑定远程 Service 的操作，代码如下所示。

【代码 8-33】 MainActivity.java

```java
public class MainActivity extends AppCompatActivity {
    private Button bindMyService8Button;
    private Button callMyService8Button;
    private TextView remoteCallResultTextView;
    private MyService8 myService8;
    private ServiceConnection myService8Connection = new ServiceConnection() {
        @Override
        public void onServiceDisconnected(ComponentName name) {
            Log.i("MainActivity", "onServiceDisconnected");
            myService8 = null;}
        @Override
        public void onServiceConnected(ComponentName name, IBinder service) {
            Log.i("MainActivity", "onServiceConnected");
            myService8 = MyService8.Stub.asInterface(service); }};
    @Override
    protected void onCreate(Bundle savedInstanceState) {
        super.onCreate(savedInstanceState);
        setContentView(R.layout.activity_main);
        bindMyService8Button = (Button) findViewById(R.id.bindMyService8Button);
        callMyService8Button = (Button) findViewById(R.id.callMyService8Button);
        remoteCallResultTextView = (TextView) findViewById(
                R.id.remoteCallResultTextView);
        bindMyService8Button.setOnClickListener(new OnClickListener() {
            @Override
            public void onClick(View v) {
                Intent intent = new Intent("com.qst.chapter08.MyService8");
                intent.setPackage("com.qst.chapter08");
                bindService(intent, myService8Connection,
                    Context.BIND_AUTO_CREATE);}});
        callMyService8Button.setOnClickListener(new OnClickListener() {
            @Override
```

```
            public void onClick(View v) {
                try {
                    int result = myService8.sum(123, 456);
                    Log.i("MainActivity", "远程调用结果为:" + result);
                    remoteCallResultTextView.setText("远程调用结果为:" + result);
                } catch (RemoteException e) {
                    e.printStackTrace();
                    Toast.makeText(MainActivity.this, "远程调用错误.",
                        Toast.LENGTH_SHORT).show();}}});}
}
```

上述 MainActivity 代码中，声明了 MyService8 类型的属性 myService8；声明了 ServiceConnection 类型属性 myService8Connection，在其 onServiceConnected()方法中通过 MyService8.Stub.asInterface()方法来获取远程 Service 的本地代理对象，并赋值给 myService8 属性；在 bindMyService8Button 的单击事件中，通过指定 action 和 package 的方式构造了 Intent 对象，并调用 bindService()方法绑定了远程 Service；在 callMyService8Button 的单击事件中，通过调用 myService8 的 sum()方法实现了远程 Service 的 sum()方法的调用。

至此，访问远程 Service 的客户端编写完毕。运行 chapter08_RemoteServiceClient 项目，单击"绑定远程 MySercice08"按钮时，Logcat 输出如下：

```
12-07 17:43:34.360: I/MainActivity(2500): onServiceConnected
```

此时，在 chapter08 项目的 LogCat 窗口输出以下内容，说明远程 Service 绑定成功。

```
12-07 17:43:34.350: I/MyService8Impl(2449): onCreate()
12-07 17:43:34.350: I/MyService8Impl(2449): onBind()
```

单击"调用远程 MySercice08"按钮，在 chapter08_RemoteServiceClient 项目的 Logcat 窗口中输出如下信息。

```
12-07 17:48:53.800: I/MainActivity(2555):远程调用结果为: 579
```

此时，在 chapter08 项目的 LogCat 窗口中输出如下信息，说明远程 Service 调用成功。

```
12-07 17:48:53.800: I/MyService8Impl(2449): sum(123, 456)
```

8.3 系统自带 Service

Android 提供了许多系统级别的 Service，通过这些服务，应用程序可以方便地调用系统功能。系统服务都是通过 Context.getSystemService(String serviceName)方法获取的，其中参数 serviceName 表示需要传入的服务名称，而系统服务的名称都在 Context 类中定义了常量。通常 getSystemService()方法会返回一个特定服务的管理器对象，使用此对象可完成服务调用功能。常用的系统服务如表 8-2 所示。

表 8-2　Android 常用系统服务

服务对象	Context 中对应的服务名称常量	功　　能
AccessibilityManager	ACCESSIBILITY_SERVICE	通过已注册的事件监听器将 UI 事件反馈给用户
AccountManager	ACCOUNT_SERVICE	账户服务
ActivityManager	ACTIVITY_SERVICE	管理 Activity、Service 等各种组件
AlarmManager	ALARM_SERVICE	闹钟服务
AppOpsManager	APP_OPS_SERVICE	在设备操作时跟踪应用
AudioManager	AUDIO_SERVICE	音频服务
BluetoothAdapter	BLUETOOTH_SERVICE	蓝牙服务
ClipboardManager	CLIPBOARD_SERVICE	剪切板服务
ConnectivityManager	CONNECTIVITY_SERVICE	网络连接服务
ConsumerIrManager	CONSUMER_IR_SERVICE	红外信号服务
DevicePolicyManager	DEVICE_POLICY_SERVICE	设备监听服务
DisplayManager	DISPLAY_SERVICE	显示设备管理
DownloadManager	DOWNLOAD_SERVICE	针对 HTTP 的下载服务
DropBoxManager	DROPBOX_SERVICE	获取 DropBoxManager 实例以记录诊断日志
InputMethodManager	INPUT_METHOD_SERVICE	输入法的管理服务程序
InputManager	INPUT_SERVICE	输入设备管理
NotificationManager	KEYGUARD_SERVICE	键盘锁服务
LayoutInflater	LAYOUT_INFLATER_SERVICE	根据 XML 生成布局的服务
LocationManager	LOCATION_SERVICE	GPS 定位服务等
NfcManager	NFC_SERVICE	NFC 服务
NotificationManager	NOTIFICATION_SERVICE	通知服务
PowerManager	POWER_SERVICE	电源服务
PrintManager	PRINT_SERVICE	打印服务
SearchManager	SEARCH_SERVICE	搜索服务
SensorManager	SENSOR_SERVICE	传感器服务
StorageManager	STORAGE_SERVICE	系统存储服务
TelephonyManager	TELEPHONY_SERVICE	电话服务
TextServicesManager	TEXT_SERVICES_MANAGER_SERVICE	文字服务，如拼写检查等
UiModeManager	UI_MODE_SERVICE	界面模式服务，如夜间模式、驾车模式等
UsbManager	USB_SERVICE	USB 管理服务
UserManager	USER_SERVICE	用户管理服务
Vibrator	VIBRATOR_SERVICE	震动器服务
WallpaperService	WALLPAPER_SERVICE	壁纸服务
WifiP2pManager	WIFI_P2P_SERVICE	WiFi-P2P 连接服务
WifiManager	WIFI_SERVICE	WiFi 服务
WindowManager	WINDOW_SERVICE	系统窗口服务

Android 提供的系统服务非常多，基本涵盖了移动设备可能涉及的方方面面。本章只选取常用的 NotificationManager 和 DownloadManager 服务进行介绍。

8.3.1 NotificationManager

NotificationManager 用于在界面顶部的通知栏显示消息，以一种标准的方式向用户显示提示信息，下列示例演示了 NotificationManager 的用法。

修改 MainActivity 代码，添加 NotificationManager 按钮，单击时通过 NotificationManager 显示通知信息，代码如下：

【代码 8-34】 MainActivity.java

```java
private void notification() {
    Intent intent = new Intent(android.provider.Settings.ACTION_SETTINGS);
    PendingIntent pendingIntent = PendingIntent.getActivity(this, 1,
                intent, PendingIntent.FLAG_UPDATE_CURRENT);
    Notification notification = new Notification.Builder(this)
            .setSmallIcon(R.drawable.ic_launcher)
            .setLargeIcon(
                    BitmapFactory.decodeResource(getResources(),
                            R.drawable.ic_launcher))
            .setContentTitle("通知标题")
            .setContentText("通知内容.")
            .setContentIntent(pendingIntent)
            .setNumber(1)
            .build();
    notification.flags |= Notification.FLAG_AUTO_CANCEL;
    NotificationManager notificationManager
        = (NotificationManager)getSystemService(Context.NOTIFICATION_SERVICE);
    notificationManager.notify(1, notification);
}
```

上述代码中，notification()方法用于在单击"NotificationManager"按钮时执行，其中，声明一个 Intent 对象 intent，其 Action 为 Settings.ACTION_SETTINGS，代表系统的设置 Activity；声明一个 PendingIntent 对象 pendingIntent，在构造时将 intent 对象传入；通过 Notification.Builder 构造了一个 Notification 对象 notification，其中定义了通知的标题、内容、图标等，并指定了单击通知时需要执行 pendingIntent；调用 getSystemService()方法来获取 NotificationManager 对象，并调用该对象的 notify()方法来显示通知。

运行修改后的 MainActivity，结果如图 8-14 所示。

图 8-14 NotificationManager

8.3.2 DownloadManager

DownloadManager 提供了一种标准简洁的 HTTP 下载解决方案,借助 DownloadManager 系统服务,开发者只须编写少量代码即可完成在后台下载文件。下列示例演示了 DownloadManager 的用法。

修改 MainActivity 代码,添加 DownloadManager 按钮,单击时通过 DownloadManager 下载指定地址的文件,代码如下。

【代码 8-35】 MainActivity.java

```java
private void download() {
    Uri uri = Uri.parse(
            "http://dl.ops.baidu.com/baidusearch_AndroidPhone_757p.apk");
    DownloadManager.Request request = new DownloadManager.Request(uri);
    request.setTitle("下载示例");                              //标题
    request.setDescription("下载说明");                        //说明
    request.setDestinationInExternalFilesDir(this,
            Environment.DIRECTORY_DOWNLOADS, "temp.apk");   //下载文件保存路径
    //下载完毕后通知不消失
    request.setNotificationVisibility(
        DownloadManager.Request.VISIBILITY_VISIBLE_NOTIFY_COMPLETED);
    final DownloadManager downloadManager
        = (DownloadManager) getSystemService(Context.DOWNLOAD_SERVICE);
    final long downloadId = downloadManager.enqueue(request);
    IntentFilter filter = new IntentFilter();
    filter.addAction(DownloadManager.ACTION_DOWNLOAD_COMPLETE);
    filter.addAction(DownloadManager.ACTION_NOTIFICATION_CLICKED);
    final BroadcastReceiver receiver = new BroadcastReceiver() {
        @Override
        public void onReceive(Context context, Intent intent) {
            String action = intent.getAction();
            long id = intent.getLongExtra(
                    DownloadManager.EXTRA_DOWNLOAD_ID, -1);
            if (DownloadManager.ACTION_DOWNLOAD_COMPLETE.equals(action)) {
                if (id == downloadId) {
                    Uri uri = downloadManager
                            .getUriForDownloadedFile(downloadId);
                    Log.i("MainActivity", "下载完毕: " + uri.toString());}
            } else if (DownloadManager.ACTION_NOTIFICATION_CLICKED
                    .equals(action)) {
                Log.i("MainActivity", "取消下载");
                downloadManager.remove(id);}}};
    registerReceiver(receiver, filter);
}
```

上述代码中,download()方法用于在单击 DownloadManager 按钮时执行,其中:声明了一个 Uri 对象 uri,用于封装待下载文件(百度 App)的地址;声明了一个 DownloadManager.Request 对象 request,在构造时传入文件的 uri 地址,并指定了标题、说明、保存地址等几个属性;使用 getSystemService(Context.DOWNLOAD_SERVICE)方法获取 downloadManager 对象;调用了 downloadManager 的 enqueue()方法,将下载请求

request 加入下载队列,并准备下载指定的文件;为监听下载的状态,还声明了一个 BroadcastReceiver,用于接收 ACTION_DOWNLOAD_COMPLETE 和 ACTION_NOTIFICATION_CLICKED 广播,分别是下载完毕和下载过程中单击通知时的广播通知;在接收到广播时,使用 Logcat 输出相应信息。

运行修改后的 MainActivity,结果如图 8-15 所示。

图 8-15　DownloadManager

本章总结

- 按照运行的进程不同,可以将 Service 分为本地 Service 和远程 Service;按照运行的形式分为前台 Service 和后台 Service;按照使用 Service 的方式可以分为启动方式 Service、绑定方式 Service、和混合式 Service。
- Service 组件需要通过 Context 对象启动,有两种启动方式,Start 方式和 Bind 绑定方式,分别对应于 Context 的 startService() 和 bindService() 方法。
- 无论是 Start 还是 Bind 方式启动 Service,都会经历 onCreate() 和 onDestroy() 方法。如果是 Start 方式启动,在启动时会调用 onStartCommand() 方法;如果是 Bind 方式启动,在启动时会调用 onBind() 方法,取消绑定时会调用 onUnbind() 方法,重新绑定时会调用 onRebind() 方法。
- Start 方式启动的 Service 必须自己管理生命周期,并会一直运行下去,除非 Service 调用自身的 stopSelf() 方法,或其他组件对该 Service 调用 stopService() 方法。
- Bind 方式启动的 Service 会和启动它的组件关联在一起并可以进行通信。在启动时自动调用 onBind() 方法,如果该 Service 是第一次启动,则在调用 onBind() 方法前还会调用 onCreate() 方法。组件和 Service 解除绑定时会触发 Service 的 onUnbind() 方法,一个 Service 可以被多个组件绑定,当所有的绑定组件都解除绑定时,该

Service 将被销毁，并执行 onDestroy()方法。同样地，如果系统资源不足，Android 也随时有可能销毁这个 Service。一个组件绑定 Service 后，如果这个组件被销毁，系统会自动解除其和对应 Service 的绑定。

- Service 启动后，其所在进程默认是服务进程，优先级并不高，如果是非常重要的 Service，可以通过调用 Service 的 startForeground()方法将其改为前台进程。
- Service 运行于 UI 线程中，如果直接在当前线程中执行耗时或可能被阻塞的任务，会造成界面无响应甚至 ANR 错误，因此，这种耗时任务通常都需要新开线程执行。
- IntentService 是 Service 的子类，其内部会自动开始一个新线程执行任务，并在任务执行完毕后停止 Service。当有多个任务时，IntentService 会将任务加到一个队列中，按照次序依次执行，直到所有任务执行完毕后停止 Service。
- 远程 Service 是指运行于独立的进程中，并为其他进程提供服务的 Service。调用远程 Service 时需要使用 AIDL。
- Android 提供了许多系统级的 Service，利用这些服务，应用程序可以方便地调用系统功能，通过 Context.getSystemService(String serviceName)方法可以获取这些服务对象。

Q&A

1. 问题：简述 Service 的作用和分类。

回答：在 Android 中，Service 组件表示一种服务，专门用于执行一些持续性的、耗时的并且无需用户界面交互的操作。按照运行的进程不同，可以将 Service 分为本地 Service 和远程 Service；按照运行的形式分为前台 Service 和后台 Service；按照使用 Service 的方式可以分为启动方式 Service、绑定方式 Service 和混合式 Service。

2. 问题：简述 Service 的生命周期。

回答：Start 方式启动的 Service 必须自己管理生命周期，其会一直运行下去，除非 Service 调用自身的 stopSelf()方法，或其他组件对这个 Service 调用 stopService()方法。Bind 方式启动的 Service 会和启动该 Service 的组件关联在一起并可以进行通信。在启动时自动调用 onBind()方法，如果该 Service 是第一次启动，则在调用 onBind()方法前还会调用 onCreate()方法。组件和 Service 解除绑定时会触发 Service 的 onUnbind()方法，一个 Service 可以被多个组件绑定，当所有的绑定组件都解除绑定时，该 Service 将被销毁，并执行 onDestroy()方法。同样地，如果系统资源不足，Android 也随时有可能销毁这个 Service。一个组件绑定 Service 后，如果这个组件被销毁，系统会自动解除其和对应 Service 的绑定。

章节练习

习题

1. 下列关于 Service 的描述错误的是_____。

　A. Service 是由系统管理的组件，具有复杂的生命周期

B. Service 具有比 Activity 更高的优先级

C. Service 无法显示界面,最多只能显示一个通知

D. Service 适合执行一些需要持续运行并无需界面的操作

2. 下列关于 Service 生命周期的说法中正确的是_____。

   A. Start 方式启动的 Service,如果不调用 stopSelf() 或 stopService() 方法,则这个 Service 会一直运行,不会终止

   B. Bind 方式启动的 Service,只要还存在未跟这个 Service 解除绑定的组件,则这个 Service 会一直运行,不会终止

   C. 无论何种 Service,当系统资源不足时,都有可能被强制终止

   D. 如果 Service 被系统强制终止,则只能通过 Start 或 Bind 方式才能使它再次运行

3. 关于 Start 方式启动 Service 的说法错误的是_____。

   A. Start 方式启动的 Service 与启动它的组件没有关联,组件的生命周期与这个 Service 的生命周期无关

   B. Start 方式启动的 Service 生命周期包括 onCreate()、onStartCommand()、onDestroy() 三个方法

   C. Start 方式启动的 Service 被系统强制终止后,系统是否自动重新启动这个 Service 取决于 onStartCommand() 方法的返回值

   D. 对于 Start 方式启动的 Service,我们无法判断它是应用程序中启动的还是系统将其强制终止后又由系统重新启动的

4. 关于 Bind 方式启动 Service 的说法正确的是_____。

   A. Bind 方式启动的 Service 与启动它的组件没有关联,组件的生命周期与这个 Service 的生命周期无关

   B. Bind 方式启动的 Service 生命周期包括 onCreate()、onStart()、onBind()、onUnbind() 三个方法

   C. Bind 方式启动的 Service,如果与其绑定的所有组件都已解除绑定,则此 Service 将会销毁

   D. 对于 Bind 方式启动的 Service,当资源不足时也会被系统强制终止,但是此时绑定它的组件得不到任何通知

5. 下列关于 Service 的说法正确的是_____。

   A. 由于 Service 没有界面,因此可以执行长时间的任务而不会影响用户的界面操作

   B. 通过调用 startForeground() 方法可以将 Service 变为前台 Service,对于前台 Service 就可以像 Activity 那样设计复杂的界面了

   C. IntentService 是 Service 的子类,是一种专用于执行耗时任务的 Service

   D. 无论以何种方式启动 Service,它都只能被同一进程中的其他组件访问

6. 简述混合使用 Start 和 Bind 方式的 Service 的生命周期。

7. 请列举出 5 个以上特别适合使用 Service 的应用场景。

上机

1. 训练目标：Service 应用。

培养能力	熟练使用 Service 实现功能		
掌握程度	★★★★★	难度	中
代码行数	200	实施方式	重复编码
结束条件	熟练使用 Service 实现功能，理解 Service 生命周期		

参考训练内容
(1) 编写名为 TestService 的 Service，并在 onCreate()、onStartCommand()、onBind()、onUnbind()、onDestroy()方法中使用 Log 输出日志；
(2) 编写 Activity，放置 4 个按钮分别用于使用 Start 方式的启动和停止 TestService、使用 Bind 方式绑定和解除绑定 TestServive；
(3) 多次运行该应用程序，分别以不同的次序单击 4 个按钮，观察 Logcat 的输出信息，加深对 Service 生命周期的理解。

2. 训练目标：系统 Service 应用。

培养能力	熟练使用 DownloadManager 完成 APK 下载功能		
掌握程度	★★★★★	难度	中
代码行数	100	实施方式	重复编码
结束条件	熟练使用 DownloadManager		

参考训练内容
(1) 编写 Activity，其中放置一个按钮，单击时下载某个网络上的 APK 文件；
(2) 下载功能通过 DownloadManager 实现，并在通知栏显示下载进度。

第 9 章 网络编程

本章任务是完成"GIFT-EMS 礼记"App 与服务器端的交互：

- 【任务 9-1】 编写 HttpUtils 类封装 HTTP 对服务器的请求调用。
- 【任务 9-2】 修改 BaseActivity，完成与服务器交互数据的 Handler 模板。
- 【任务 9-3】 修改登录 Activity，改为从服务器验证登录。
- 【任务 9-4】 引入 Android-Universal-Image-Loader 库，用于显示网络图片。
- 【任务 9-5】 修改礼物类型列表 Activity，改为从服务器查询数据。

知 识 点	Listen(听)	Know(懂)	Do(做)	Revise(复习)	Master(精通)
网络编程简介	★	★			
基于 TCP 协议的网络通信	★	★	★	★	★
HttpURLConnection	★	★	★	★	★
HttpClient 组件	★	★	★		
WebView 组件	★	★	★		

9.1 网络编程简介

如今人类的生活已经离不开网络,而无线网络的产生也为人类的生活提供了便利条件,例如可以通过无线网络上网、视频通话、信息检索等。因此,网络支持对于手机应用的重要性不言而喻。

Android 完全支持 JDK 本身所提供的 TCP、UDP 网络通信的 API,也支持 URL、URLConnection 等网络通信 API。Java 网络编程经验完全适用于 Android 网络编程。

Android 中常用的网络编程有如下几种方式:① 针对 TCP/IP 协议的 Socket 和 ServerSocket;② 针对 HTTP 协议的网络编程,如 HttpURLConnection 和 HttpClient;③ 直接使用 WebKit 访问网络。

> 由于篇幅有限,本书不涉及 UDP 协议编程的相关内容,需要掌握 UDP 协议的读者可以自己查阅相关参考资料。

9.2 基于 TCP 协议的网络通信

在计算机网络中实现通信必须遵守一些约定,即通信协议。通信协议是用来管理数据通信的一组规则,用于规范传输速率、传输代码、代码结构、传输控制步骤、出错控制等。如同人与人之间沟通交流需要遵循一定的语言约定一样,两台计算机之间相互通信也需要共同遵守通信协议,这样才能进行信息交换。

通信协议规定了通信的内容、方式和通信时间,其核心要素由三部分组成。

- **语义**:用于决定双方对话的类型,即规定通信双方要发出何种控制信息、完成何种动作以及做出何种应答;
- **语法**:用于决定双方对话的格式,即规定数据与控制信息的结构和格式;
- **时序**:用于决定通信双方的实现顺序,即确定通信状态的变化和过程,如通信双方的应答关系。

常见的通信协议包括 TCP/IP 协议、IPX/SPX 协议、NetBEUI 协议、RS-232-C 协议、V.35 等。其中,传输控制协议/互联网络协议(Transmission Control Protocol/Internet Protocol,TCP/IP)是最基本的通信协议,也是网络中最常用的协议之一。如果访问 Internet,则必须在网络协议中添加 TCP/IP 协议。IPX/SPX 则一般用于局域网中。

TCP/IP 协议规范了网络上所有通信设备之间的数据往来格式以及传送方式。TCP/IP 是一组协议,包括 TCP、IP、UDP、ICMP、RIP、TELNETFTP、SMTP、ARP、TFTP 等许多协议,通常这些协议一起称为 TCP/IP 协议族。TCP/IP 协议最早出现在 UNIX 操作系统中,现在几乎所有的操作系统都支持 TCP/IP 协议,因此 TCP/IP 协议也是 Internet 中最常用的基础协议之一。TCP/IP 协议提供一种数据打包和寻址的标准方法,可以在 Internet 中无差错地传送数据。对于普通用户,无须了解网络协议的整个结构,仅了解 IP 的地址格式,即可与世界各地进行网络通信。

TCP/IP 通信协议是一种可靠的、双向的、持续的、点对点的网络协议。使用 TCP/IP 协议进行通信时,会在通信的两端各建立一个 Socket(套接字),从而在通信的两端之间形成网络虚拟链路,其通信原理如图 9-1 所示。

图 9-1　TCP/IP 协议通信原理

Java 对基于 TCP 的网络通信提供了良好的封装,使用 Socket 对象代表两端的通信端口。Socket 对象屏蔽了网络的底层细节,例如媒体类型、信息包的大小、网络地址、信息的重发等。Socket 允许应用程序将网络连接当成一个 I/O 流,既可以向 I/O 流中写数据,也可以从 I/O 流中读取数据。通过 Socket 对象可以建立 Java 的 I/O 系统到其他 Internet 上任何机器(包括本机)的程序的连接。

java.net 包中包含了网络编程所需的类型,其中基于 TCP 协议的网络编程主要使用以下两种 Socket:①ServerSocket 是服务器套接字,用于监听并接收来自客户端的 Socket 连接;②Socket 是客户端套接字,用于实现两台计算机之间的通信。

9.2.1　Socket

使用 Socket 套接字可以方便地在网络上传递数据,从而实现两台计算机之间的通信。通常客户端使用 Socket 的构造方法来连接指定的服务器,常用的 Socket 构造方法有以下两种。

- **Socket(String host, int port)**:创建连接到指定远程主机、远程端口的 Socket 对象,该构造方法没有指定本地地址和本地端口,默认使用本地主机 IP 地址和系统动态分配的端口。此外,参数 host 也可以是 InetAddress 类型。
- **Socket(String host, int port, InetAddress localAddr, int localPort)**:创建连接到指定远程主机、远程端口的 Socket,并指定本地 IP 地址和本地端口,适用于本地主机有多个 IP 地址的情况。此外,参数 host 也可以是 InetAddress 类型。

注意

> 上述两个 Socket 构造方法都声明抛出 IOException 异常,因此在创建 Socket 对象时必须捕获或抛出异常。最好选择注册端口(范围是 1024~49 151 的数),通常应用程序使用这个范围内的端口,以防止发生冲突。

【示例】 创建 Socket 对象

```
try{
    Socket s = new Socket("192.168.1.128" , 28888);
    ...            //Socket 通信
}catch (IOException e) {
    e.printStackTrace();
}
```

除了构造方法，Socket 类常用的其他方法如表 9-1 所示。

表 9-1　Socket 类常用方法

方　　法	功 能 描 述
public InetAddress getInetAddress()	返回连接到远程主机的地址，如果连接失败，则返回以前连接的主机
public int getPort()	返回 Socket 连接到远程主机的端口号
public int getLocalPort()	返回本地连接终端的端口号
public InputStream getInputStream()	返回一个输入流，从 Socket 中读取数据
public OutputStream getOutputStream()	返回一个输出流，往 Socket 中写数据
public synchronized void close()	关闭当前 Socket 连接

9.2.2　ServerSocket

ServerSocket 是服务器套接字，运行在服务器端，在指定的端口上主动监听来自客户端的 Socket 连接。当客户端发送 Socket 请求并与服务器指定的端口建立连接时，服务器将验证并接收客户端的 Socket，从而建立客户端与服务器之间的网络虚拟链路，一旦两端的实体之间建立了虚拟链路，两者之间就可以相互传送数据了。

ServerSocket 类常用的构造方法如下。

- **ServerSocket(int port)**：根据指定的端口创建一个 ServerSocket 对象。
- **ServerSocket(int port, int backlog)**：创建一个 ServerSocket 对象，并指定端口和连接队列长度，参数 backlog 用于指定连接队列的长度。
- **ServerSocket(int port, int backlog, InetAddress localAddr)**：创建一个 ServerSocket 对象，指定端口、连接队列长度和 IP 地址，在机器存在多个 IP 地址时才允许使用 localAddr 参数将 ServerSocket 绑定到特定端口。

注意

ServerSocket 类的构造方法都声明抛出 IOException 异常，因此在创建 ServerSocket 对象时必须捕获或抛出异常。另外，在选择端口号时，最好选择注册端口（范围是 1024～49 151 的数），通常应用程序使用这个范围内的端口，以防止发生冲突。

【示例】 创建 ServerSocket 对象

```
try {
    ServerSocket server = new ServerSocket(28888);
```

```
    } catch (IOException e) {
        e.printStackTrace();
    }
```

ServerSocket 类常用的方法如表 9-2 所示。

表 9-2 ServerSocket 类常用方法

方　　法	功　能　说　明
public Socket accept()	接收客户端 Socket 连接请求,并返回一个与客户端 Socket 对应的 Socket 实例,该方法是一个阻塞方法,如果没有接收到客户端发送的 Socket,则一直处于等待状态,线程也会被阻塞
public InetAddress getInetAddress()	返回当前 ServerSocket 实例的地址信息
public int getLocalPort()	返回当前 ServerSocket 实例的服务端口
public void close()	关闭当前 ServerSocket 实例

通常使用 ServerSocket 进行网络通信的具体步骤如下:
(1) 根据指定端口实例化一个 ServerSocket 对象。
(2) 调用 ServerSocket 对象的 accept()方法接收客户端发送的 Socket 对象。
(3) 调用 Socket 对象的 getInputStream()/getOutputStream()方法建立与客户端进行交互的 I/O 流。
(4) 服务器与客户端根据一定的协议进行交互,直到关闭连接。
(5) 关闭服务器端的 Socket。
(6) 回到第(2)步,继续监听下一次客户端发送的 Socket 请求连接。

下述代码演示创建服务器端 ServerSocket 的过程。

【代码 9-1】 Server.java

```
public class Server {
    private int ServerPort = 9898;                       //定义端口
    private ServerSocket serversocket = null;            //声明服务器套接字
    private OutputStream outputStream = null;            //声明输出流
    private InputStream inputStream = null;              //声明输入流
    private PrintWriter printWriter = null;              //声明打印流,用于将数据发送给对方
    private Socket socket = null;                        //声明套接字,注意同服务器套接字不同
    private BufferedReader reader = null;                //声明缓冲流,用于读取接收的数据
    /* Server 类的构造函数 */
    public Server() {
        try {
            //根据指定的端口号,创建套接字
            serversocket = new ServerSocket(ServerPort);
            System.out.println("服务启动...");
            socket = serversocket.accept();              //用 accept 方法等待客户端的连接
            System.out.println("客户已连接...\n");
        } catch (Exception ex) {
            ex.printStackTrace();                        //打印异常信息
        }
        try {
            outputStream = socket.getOutputStream();//获取套接字输出流
            inputStream = socket.getInputStream();   //获取套接字输入流
            //根据 outputStream 创建 PrintWriter 对象
```

```java
            printWriter = new PrintWriter(outputStream, true);
            //根据 inputStream 创建 BufferedReader 对象
            reader = new BufferedReader(new InputStreamReader(inputStream));
            //根据 System.in 创建 BufferedReader 对象
            BufferedReader in = new BufferedReader(new InputStreamReader(
                    System.in));
            while (true) {
                String message = reader.readLine();        //读客户端的传输信息
                //将接收的信息打印出来
                System.out.println("来自客户端的信息: " + message);
                //若消息为 Bye 或者 bye,则结束通信
                if (message.equals("Bye") || message.equals("bye"))
                    break;
                message = in.readLine();                   //接收键盘输入
                printWriter.println(message);              //将输入的信息向客户端输出
            }
            outputStream.close();                          //关闭输出流
            inputStream.close();                           //关闭输入流
            socket.close();                                //关闭套接字
            serversocket.close();                          //关闭服务器套接字
            System.out.println("客户端关闭连接");
        } catch (Exception e) {
            e.printStackTrace();                           //打印异常信息
        } finally {
        }}
    /* 程序入口,程序从 main 函数开始执行 */
    public static void main(String[] args) {
        new Server();}
}
```

上述代码作为服务器端,用于响应客户端的连接,注意此程序是一个通过 main()方法启动的标准 Java 应用程序,而不是 Android 应用,因此需要运行在 Windows(或其他)系统的 JRE 中,而不是 Android 系统中。

下述代码使用 Socket 实现客户端网络通信。

【代码 9-2】 **ClientActivity.java**

```java
public class ClientActivity extends AppCompatActivity implements Runnable {
    //声明文本视图 chatmessage,用于显示聊天记录
    private TextView chatmessage = null;
    //声明编辑框 sendmessage,用于用户输入短信内容
    private EditText sendmessage = null;
    //声明按钮 send_button,用于发送短信
    private Button send_button = null;
    private static final String HOST = "192.168.2.152";   //服务器的 IP 地址
    private static final int PORT = 9898;                 //服务器端口号
    private Socket socket = null;                         //声明套接字类,传输数据
    private BufferedReader bufferedReader = null;
    private PrintWriter printWriter = null;
    private String string = "";
    public void onCreate(Bundle savedInstanceState) {
        super.onCreate(savedInstanceState);
```

```java
        setContentView(R.layout.main);
        chatmessage = (TextView) findViewById(R.id.chatmessage);
        sendmessage = (EditText) findViewById(R.id.sendmessage);
        send_button = (Button) findViewById(R.id.sendbutton);
        try {
            //指定 IP 和端口号创建套接字
            socket = new Socket(HOST, PORT);
            //使用套接字的输入流构造 BufferedReader 对象
            bufferedReader = new BufferedReader(new InputStreamReader(socket
                    .getInputStream()));
            //使用套接字的输出流构造 PrintWriter 对象
            printWriter = new PrintWriter(new BufferedWriter(
                    new OutputStreamWriter(socket.getOutputStream())), true);
        } catch (Exception e) {
            e.printStackTrace();                    //打印异常
            CreateDialog(e.getMessage());           //调用 CreateDialog()方法生成对话框
        }
        /* 注册 send_button 的鼠标单击监听器.当单击按钮时,发送指定的信息 */
        send_button.setOnClickListener(new View.OnClickListener() {
            public void onClick(View view) {
                //获取输入框的内容
                String message = sendmessage.getText().toString();
                //判断 socket 是否连接
                if (socket.isConnected()) {
                    if (!socket.isOutputShutdown()) {
                        printWriter.println(message);           //将输入框的内容发送到服务器
                        //设置 chatmessage 的内容
                        chatmessage.setText(chatmessage.getText().toString()
                                + "\n" + "发送: " + message);
                        //清空 sendmessage 的内容,以便下次输入
                        sendmessage.setText("");}}}});
        new Thread(this).start();                   //启动线程
    }
    /* CreateDialog 产生对话框 */
    public void CreateDialog(String msmessage) {
        android.app.AlertDialog.Builder builder = new AlertDialog.Builder(this);
        //首先获取 AlertDialog 的 Builder 类,该 Builder 对象用于构造对话框
        builder.setTitle("异常");                    //指定对话框的标题
        builder.setMessage(msmessage);              //设置显示的信息
        builder.setPositiveButton("是", new DialogInterface.OnClickListener(){
            public void onClick(DialogInterface dialog, int which) {
            }});
        //设置 PositiveButton 的名称以及监听器
        builder.setNegativeButton("否", new DialogInterface.OnClickListener() {
            public void onClick(DialogInterface dialog, int which) {
            }});
        //设置 NegativeButton 的名称以及监听器
        builder.show();                             //显示对话框
    }
    /*线程运行的代码*/
    public void run() {
        try {
```

```
                while (true) {
                    //若套接字同服务器的链接存在且输入流也存在,则发送消息
                    if (socket.isConnected()) {
                        if (!socket.isInputShutdown()) {
                            if ((string = bufferedReader.readLine()) != null) {
                                Log.i("TAG", "++" + string);
                                string += " ";
                                messegner.sendMessage(messegner.obtainMessage());
                            } else {
                            }}}}
            } catch (Exception ex) {
                ex.printStackTrace(); //显示异常信息
                Log.w("TAG", "-- " + ex.toString()); //将信息输出到日志里,级别为 Warn
            }}
    /* 创建 Handler 对象 messegner */
    public Handler messegner = new Handler() {
        public void handleMessage(Message msg) {
            super.handleMessage(msg);
            Log.i("TAG", "-- " + msg); //将消息打印到日志里,级别为 Inf
            chatmessage.setText(chatmessage.getText().toString() + "\n"
                + "来自服务端的消息: " + string);}};
}
```

上述代码作为客户端运行在 Android 系统中,客户端通过套接字绑定服务器端的 IP 地址和端口号。注意,这里的 IP 地址是服务器的 IP 地址,即使服务器端和 Android 的模拟器在同一机器上运行,也不能使用回环地址(127.0.0.1)作为服务器的 IP 地址,否则,程序会出现拒绝连接的错误。

对于服务器端的消息,使用 Handler 消息处理机制进行处理,这种处理机制是异步的。对于发送和接收信息,Handler 有不同的处理方式,向消息队列发送消息时则会立即返回,而从消息队列中接收消息时则会阻塞。其中读取消息时会执行 Handler 中的 handleMessage(Message msg)方法,因此,在创建 Handler 时应该重写该方法,在该方法中写上读取到消息后的操作,使用 Handler 的 obtainMessage()来获得消息对象。

要让客户端能够访问服务器,必须在 AndroidManifest.xml 配置文件中增加如下权限。

【示例】 授权应用程序能够访问网络

```
<uses-permission android:name="android.permission.INTERNET"></uses-permission>
```

先启动 Server 服务器,再运行客户端 ClientActivity 程序,在客户端界面的文本框中输入信息并单击"发送"按钮,信息会发送给服务器;在服务器端通过键盘输入信息,并对客户端进行响应,来完成服务器与客户端之间的通信。客户端的显示结果如图 9-2 所示。

服务器端的输出结果如下:

图 9-2 客户端显示结果

```
服务启动…
客户已连接…
来自客户端的信息: Hi, I am Tom
```

```
Hi,Tom,welcome to use QST system
来自客户端的信息：Nice to meet you
Nice to meet you,too.
```

当服务器端向客户端发送消息时，两台设备必须在同一 IP 环境下才能进行消息传输，否则服务器端无法将消息发送到客户端。

9.3 使用 HttpURLConnection

9.3.1 URL 和 URLConnection

统一资源定位器（Uniform Resource Locator，URL）用于表示互联网上资源的唯一地址。Java 中的 java.net.URL 类封装了针对 URL 的操作，URL 类常用方法如表 9-3 所示。

表 9-3　URL 类常用方法

方　　法	功　能　描　述
public URL(String spec)	构造方法，根据指定的字符串创建一个 URL 对象
public URL(String protocol, String host, int port, String file)	构造方法，根据指定的协议、主机名、端口号和文件资源创建一个 URL 对象
public URL(String protocol, String host, String file)	构造方法，根据指定的协议、主机名和文件资源创建 URL 对象
public String getProtocol()	返回协议名
public String getHost()	返回主机名
public int getPort()	返回端口号，如果没有设置端口，则返回 -1
public String getFile()	返回文件名
public String getRef()	返回 URL 的锚
public String getQuery()	返回 URL 的查询信息
public String getPath()	返回 URL 的路径
public URLConnection openConnection()	返回一个 URLConnection 对象
public final InputStream openStream()	返回一个用于读取该 URL 资源的 InputStream 流

其中，openConnection()方法返回一个 URLConnection 对象，该对象表示应用程序和 URL 之间的通信连接。URLConnection 是一个抽象类，其常用方法如表 9-4 所示。

表 9-4　URLConnection 常用方法

方　　法	功　能　描　述
public int getContentLength()	获得文件的长度
public String getContentType()	获得文件的类型
public long getDate()	获得文件创建的时间
public long getLastModified()	获得文件最后修改的时间
public InputStream getInputStream()	获得输入流，以便读取文件的数据

续表

方　法	功　能　描　述
public OutputStream getOutputStream()	获得输出流,以便输出数据
public void setRequestProperty(String key,String value)	设置请求属性值

下述代码演示 URLConnection 的应用。

【代码 9-3】 GetPostUtil.java

```java
public class GetPostUtil{
    /* 向指定 URL 发送 GET 方法的请求
     * @param url 发送请求的 URL
     * @param params 请求参数,请求参数应该是 name1=value1&name2=value2 的形式
     * @return URL 代表远程资源的响应 */
    public static String sendGet(String url, String params) {
        String result = "";
        BufferedReader in = null;
        try{
            String urlName = url + "?" + params;
            URL realUrl = new URL(urlName);
            URLConnection conn = realUrl.openConnection();     //打开和 URL 之间的连接
            //设置通用的请求属性
            conn.setRequestProperty("accept", "*/*");
            conn.setRequestProperty("connection", "Keep-Alive");
            conn.setRequestProperty("user-agent",
                    "Mozilla/4.0 (compatible; MSIE 6.0; Windows NT 5.1; SV1)");
            //建立实际的连接
            conn.connect();   //①
            //获取所有响应头字段
            Map<String, List<String>> map = conn.getHeaderFields();
            //遍历所有的响应头字段
            for (String key : map.keySet()){
                System.out.println(key + "--->" + map.get(key));
            }
            //定义 BufferedReader 输入流来读取 URL 的响应
            in = new BufferedReader(
                    new InputStreamReader(conn.getInputStream()));
            String line;
            while ((line = in.readLine()) != null) {
                result += "\n" + line; }}
        catch (Exception e) {
            System.out.println("发送 GET 请求出现异常!" + e);
            e.printStackTrace();}
        //使用 finally 块来关闭输入流
        finally{
            try{
                if (in != null) {
                    in.close(); }}
            catch (IOException ex) {
                ex.printStackTrace();}}
        return result;}
    /* 向指定 URL 发送 POST 方法的请求
     * @param url 发送请求的 URL
```

* @param params 请求参数,请求参数应该是 name1 = value1&name2 = value2 的形式。
* @return URL 所代表远程资源的响应 */
```java
public static String sendPost(String url, String params) {
    PrintWriter out = null;
    BufferedReader in = null;
    String result = "";
    try{
        URL realUrl = new URL(url);
        URLConnection conn = realUrl.openConnection();        //打开和 URL 之间的连接
        //设置通用的请求属性
        conn.setRequestProperty("accept", "*/*");
        conn.setRequestProperty("connection", "Keep-Alive");
        conn.setRequestProperty("user-agent",
                "Mozilla/4.0 (compatible; MSIE 6.0; Windows NT 5.1; SV1)");
        //发送 POST 请求必须设置如下两行
        conn.setDoOutput(true);
        conn.setDoInput(true);
        //获取 URLConnection 对象对应的输出流
        out = new PrintWriter(conn.getOutputStream());
        //发送请求参数
        out.print(params);                    //②
        //flush 输出流的缓冲
        out.flush();
        //定义 BufferedReader 输入流来读取 URL 的响应
        in = new BufferedReader(
                new InputStreamReader(conn.getInputStream()));
        String line;
        while ((line = in.readLine()) != null) {
            result += "\n" + line;}}
    catch (Exception e) {
        System.out.println("发送 POST 请求出现异常!" + e);
        e.printStackTrace();}
    //使用 finally 块来关闭输出流、输入流
    finally{
        try{
            if (out != null) {
                out.close();}
            if (in != null) {
                in.close(); }}
        catch (IOException ex) {
            ex.printStackTrace();}}
    return result;}
}
```

上述 GetPostUtil 类是一个提供了发送 GET 请求、POST 请求的工具类。接下来在 Activity 中使用 GetPostUtil 工具类实现向服务器发送请求。

【代码 9-4】 URLConnectionActivity.java

```java
public class URLConnectionActivity extends AppCompatActivity{
    Button get, post;
    TextView show;
    //代表服务器响应的字符串
```

```java
    String response;
    Handler handler = new Handler(){
        @Override
        public void handleMessage(Message msg)
        {
            if(msg.what == 0x123)
            {
                //设置 show 组件显示服务器响应
                show.setText(response);}}};
    @Override
    public void onCreate(Bundle savedInstanceState) {
        super.onCreate(savedInstanceState);
        setContentView(R.layout.main);
        get = (Button) findViewById(R.id.get);
        post = (Button) findViewById(R.id.post);
        show = (TextView)findViewById(R.id.show);
        get.setOnClickListener(new OnClickListener(){
            @Override
            public void onClick(View v) {
                new Thread(){
                    @Override
                    public void run(){
                        response = GetPostUtil.sendGet(
                                "http://192.168.52.13:8080/index.jsp",
                                null);
                        //发送消息通知 UI 线程更新 UI 组件
                        handler.sendEmptyMessage(0x123);}
                }.start();
        }});
        post.setOnClickListener(new OnClickListener(){
            @Override
            public void onClick(View v) {
                new Thread(){
                    @Override
                    public void run(){
                        response = GetPostUtil.sendPost(
                                "http://192.168.52.13:8080/judgeUser.jsp",
                                "userName = abc&passWord = 123");}
                }.start();
                //发送消息通知 UI 线程更新 UI 组件
                handler.sendEmptyMessage(0x123);}});
}
```

运行上述代码，结果如图 9-3 所示。

图 9-3　URLConnection 的使用

9.3.2 HttpURLConnection

HTTP 是最常见的应用层网络协议之一，Internet 上的大部分资源都是基于 HTTP 的。Java 提供了 java.net.HttpURLConnection 类专门用于处理 HTTP 的请求和响应。HttpURLConnection 继承自 URLConnection 类，每个 HttpURLConnection 实例都可生成单个请求，以透明的共享方式连接到 HTTP 服务器。HttpURLConnection 常用的方法如表 9-5 所示。

表 9-5 HttpURLConnection 常用方法

方　　法	功　能　描　述
InputStream getInputStream()	返回从此处打开的连接读取的输入流
OutputStream getOutputStream()	返回写入到此连接的输出流
String getRequestMethod()	获取请求方法
int getResponseCode()	获取状态码，如 HTTP_OK、HTTP_UNAUTHORIZED
void setRequestMethod(String method)	设置 URL 请求的方法
void setDoInput(boolean doinput)	设置输入流，如果使用 URL 连接进行输入，则将 DoInput 标志设置为 true（默认值）；如果不打算使用，则设置为 false
void setDoOutput(boolean dooutput)	设置输出流，如果使用 URL 连接进行输出，则将 DoOutput 标志设置为 true；如果不打算使用，则设置为 false（默认值）
void setUseCaches(boolean usecaches)	设置连接是否使用任何可用的缓存
void disconnect()	关闭连接

HttpURLConnection 是一个抽象类，无法直接实例化，通常使用 URL 的 openConnection() 方法获得 HttpURLConnection 实例。下述代码示例获取一个 HttpURLConnection 连接。

【示例】 获取 HttpURLConnection 对象

```
URL url = new URL("http://www.google.com/");            //创建 URL
HttpURLConnection urlConn = (HttpURLConnection)url.openConnection();
                                                        //获取 HttpURLConnection 连接
```

在进行连接操作之前，可以对 HttpURLConnection 的连接属性进行设置。

【示例】 设置 HttpURLConnection 属性

```
//设置输出、输入流
urlConn.setDoOutput(true);
urlConn.setDoInput(true);
urlConn.setRequestMethod("POST");      //设置方式为 POST
urlConn.setUseCaches(false);           //请求不能使用缓存
```

连接完成之后可以关闭连接，代码如下所示。

【示例】 关闭 HttpURLConnection 连接

```
urlConn.disconnect();
```

下述代码演示 HttpURLConnection 的应用。

【代码 9-5】 http_layout.xml

```
<RelativeLayout xmlns:android = "http://schemas.android.com/apk/res/android" …>
```

```xml
<FrameLayout
    android:layout_width = "match_parent"
    android:layout_height = "match_parent" >
    <TextView
        android:id = "@+id/textview_show"
        android:layout_width = "wrap_content"
        android:layout_height = "wrap_content"
        android:text = "hello_world" />
    <ImageView
        android:id = "@+id/imagview_show"
        android:layout_width = "wrap_content"
        android:layout_height = "wrap_content"
        android:layout_gravity = "center" />
    <Button
        android:id = "@+id/btn_download_img"
        android:layout_width = "wrap_content"
        android:layout_height = "wrap_content"
        android:layout_alignParentBottom = "true"
        android:layout_toRightOf = "@+id/btn_visit_web"
        android:text = "下载图片"
        android:layout_gravity = "right|bottom" />
</FrameLayout>
<Button
    android:id = "@+id/btn_visit_web"
    android:layout_width = "wrap_content"
    android:layout_height = "wrap_content"
    android:layout_alignParentBottom = "true"
    android:layout_alignParentLeft = "true"
    android:text = "访问百度" />
</RelativeLayout>
```

【代码9-6】 HttpURLConnectionActivity.java

```java
public class HttpURLConnecttionActivity extends AppCompatActivity{
    Button visitWebBtn = null;
    Button downImgBtn = null;
    TextView showTextView = null;
    ImageView showImageView = null;
    String resultStr = "";
    ProgressBar progressBar = null;
    ViewGroup viewGroup = null;
    @Override
    protected void onCreate(Bundle savedInstanceState) {
        super.onCreate(savedInstanceState);
        setContentView(R.layout.http_layout);
        initUI();
        visitWebBtn.setOnClickListener(new View.OnClickListener() {
            @Override
            public void onClick(View v) {
                //TODO Auto-generated method stub
                showImageView.setVisibility(View.GONE);
                showTextView.setVisibility(View.VISIBLE);
                Thread visitBaiduThread = new Thread(new VisitWebRunnable());
```

```java
            visitBaiduThread.start();
            try {
                visitBaiduThread.join();
                if(!resultStr.equals("")){
                    showTextView.setText(resultStr); }
            } catch (InterruptedException e) {
                //TODO Auto-generated catch block
                e.printStackTrace();}}});
    downImgBtn.setOnClickListener(new View.OnClickListener() {
        @Override
        public void onClick(View v) {
            //TODO Auto-generated method stub
            showImageView.setVisibility(View.VISIBLE);
            showTextView.setVisibility(View.GONE);
            String imgUrl = "http://pic.5442.com/2013/0204/08/01.jpg";
            new DownImgAsyncTask().execute(imgUrl); }});}
public void initUI(){
    showTextView = (TextView)findViewById(R.id.textview_show);
    showImageView = (ImageView)findViewById(R.id.imagview_show);
    downImgBtn = (Button)findViewById(R.id.btn_download_img);
    visitWebBtn = (Button)findViewById(R.id.btn_visit_web);}
/* 获取指定 URL 的响应字符串
 * @param urlString
 * @return */
private String getURLResponse(String urlString){
    HttpURLConnection conn = null;              //连接对象
    InputStream is = null;
    String resultData = "";
    try {
        URL url = new URL(urlString);           //URL 对象
        conn = (HttpURLConnection)url.openConnection();        //使用 URL 打开一个链接
        conn.setDoInput(true);                  //允许输入流,即允许下载
        conn.setDoOutput(true);                 //允许输出流,即允许上传
        conn.setUseCaches(false);               //不使用缓冲
        conn.setRequestMethod("GET");           //使用 get 请求
        is = conn.getInputStream();             //获取输入流,此时才真正建立链接
        InputStreamReader isr = new InputStreamReader(is);
        BufferedReader bufferReader = new BufferedReader(isr);
        String inputLine = "";
        while((inputLine = bufferReader.readLine()) != null){
            resultData += inputLine + "\n";}
    } catch (MalformedURLException e) {
        //TODO Auto-generated catch block
        e.printStackTrace();
    }catch (IOException e) {
        //TODO Auto-generated catch block
        e.printStackTrace();
    }finally{
        if(is != null){
            try {
                is.close();
            } catch (IOException e) {
                e.printStackTrace();}}
```

```java
        if(conn != null){
            conn.disconnect(); }}
        return resultData;}
    /* 从指定 URL 获取图片
     * @param url
     * @return */
    private Bitmap getImageBitmap(String url){
        URL imgUrl = null;
        Bitmap bitmap = null;
        try {
            imgUrl = new URL(url);
            HttpURLConnection conn
                        = (HttpURLConnection)imgUrl.openConnection();
            conn.setDoInput(true);
            conn.connect();
            InputStream is = conn.getInputStream();
            bitmap = BitmapFactory.decodeStream(is);
            is.close();
        } catch (MalformedURLException e) {
            //TODO Auto-generated catch block
            e.printStackTrace();
        }catch(IOException e){
            e.printStackTrace();}
        return bitmap;}
    class VisitWebRunnable implements Runnable{
        @Override
        public void run() {
            String data = getURLResponse("http://www.baidu.com/");
            resultStr = data;}}
    class DownImgAsyncTask extends AsyncTask<String, Void, Bitmap> {
        @Override
        protected void onPreExecute() {
            //TODO Auto-generated method stub
            super.onPreExecute();
            showImageView.setImageBitmap(null);
            showProgressBar();                     //显示进度条提示框
        }
        @Override
        protected Bitmap doInBackground(String... params) {
            //TODO Auto-generated method stub
            Bitmap b = getImageBitmap(params[0]);
            return b;}
        @Override
        protected void onPostExecute(Bitmap result) {
            //TODO Auto-generated method stub
            super.onPostExecute(result);
            if(result!= null){
                dismissProgressBar();
                showImageView.setImageBitmap(result);}}}
    /* 在母布局中间显示进度条 */
    private void showProgressBar(){
        progressBar = new ProgressBar(this, null,
                android.R.attr.progressBarStyleLarge);
```

```
                RelativeLayout.LayoutParams params
                    = new RelativeLayout
                        .LayoutParams(ViewGroup.LayoutParams.WRAP_CONTENT,
                            ViewGroup.LayoutParams.WRAP_CONTENT);
                params.addRule(RelativeLayout.CENTER_IN_PARENT, RelativeLayout.TRUE);
                progressBar.setVisibility(View.VISIBLE);
                Context context = getApplicationContext();
                viewGroup = (ViewGroup)findViewById(R.id.parent_view);
                viewGroup.addView(progressBar, params);}
        /*隐藏进度条*/
        private void dismissProgressBar(){
            if(progressBar != null){
                progressBar.setVisibility(View.GONE);
                viewGroup.removeView(progressBar);
                progressBar = null;}}
}
```

运行上述代码,结果如图 9-4 所示。

图 9-4　使用 HttpURLConnection 进行网络访问

9.4　使用 HttpClient

　　Apache 提供了 HTTP 客户端组件 HttpClient,该组件对 java.net 中的类进行封装和抽象,更适合在 Android 上开发网络应用,使得 HTTP 编程更加方便、高效。HttpClient 本身不是一个浏览器,而是一个客户端的 HTTP 传输库,其目的是为了让 HttpClient 接收和发送 HTTP 消息。HttpClient 最重要的作用是执行 HTTP 方法,当执行一个 HTTP 方法时可能涉及一个或几个 HTTP 请求或响应的交互。根据请求方法的不同选择 HttpGet 或 HttpPost 对象来封装请求数据,而 HttpClient 负责将数据请求转送到目标服务器,并返回一个相应的响应对象。因此,HttpClient API 的主要部分是定义了上述功能的 HttpClient 接口。HttpClient 接口是最基本的 HTTP 请求执行规约。HttpClient 在请求执行的过程中没有任何限制或特定的具体细节,无须关心连接管理细节、状态管理细节,也无须关心认证和重定向处理的具体实现。

通常使用 HttpClient 的子类 DefaultHttpClient 进行操作，DefaultHttpClient 是 HttpClient 的默认实现类，用来负责处理 HTTP 协议的某一方面功能，如重定向或认证处理、关于保持连接和保活时间的决策等。用户可以选择性地使用特定的应用来替换默认的功能。

下述示例代码演示使用 HttpClient 请求执行的过程。

【示例】　HttpClient 请求执行

```
HttpClient httpclient = new DefaultHttpClient();
                                    //使用 DefaultHttpClient 生成一个 HttpClient 对象
String uri = "http://test/";//定义一个 URL 地址
HttpGethttpget = new HttpGet(uri);           //定义一个以 Get 方式提交的 HttpGet 请求对象
//执行 HttpClient 对象的 execute()方法，即将请求对象提交给服务器，并返回一个响应对象
HttpResponsehttpesponse = httpclient.execute(httpget);
//获取响应信息
HttpEntityhttpentity = httpresponse.getEntity();
...
```

下述代码演示 HttpClient 的应用。

【代码 9-7】　HttpClientActivity.java

```java
public class HttpClientActivity extends AppCompatActivity{
    private Button requestButton;
    private HttpResponse httpResponse;
    private HttpEntity entity;
    @Override
    protected void onCreate(Bundle savedInstanceState) {
        super.onCreate(savedInstanceState);
        setContentView(R.layout.client_layout);
        requestButton = (Button) findViewById(R.id.requestButton);
        requestButton.setOnClickListener(new View.OnClickListener() {
            public void onClick(View v) {
                new Thread(new Downtest()).start();}});}
    class Downtest implements Runnable{
        public void run() {
            HttpGet get = new HttpGet("http://www.baidu.com");  //生成一个请求对象，请求
            HttpClient hClient = new DefaultHttpClient();        //生成一个 Http 客户端对象
            //使用 Http 客户端发送请求对象
            InputStream inputStream = null;
            try {
                httpResponse = hClient.execute(get);              //httpResponse 返回的响应
                //返回的响应数据就放在里边
                entity = httpResponse.getEntity();
                inputStream = entity.getContent();
                BufferedReader reader
                    = new BufferedReader(new InputStreamReader(inputStream));
                String result = "";
                String line = "";
                while((line = reader.readLine())!= null){
                    result = result+line;}
                Log.d("HttpClient",result);
            } catch (Exception e) {
```

```
                    //TODO Auto-generated catch block
                    e.printStackTrace();
            }finally{
                try{
                    inputStream.close();
                }catch(Exception e){
                    e.printStackTrace();}}} }
}
```

运行上述代码,单击"发送请求"按钮,Logcat 输出信息如下:

```
10-24 03:09:24.880 3669-3790/com.qst.chapter09 D/HttpClient: <!DOCTYPE html><!—STATUS
OK--><html>…
```

9.5 使用 WebView 视图浏览网页

WebView 是专门用来浏览网页的视图组件。从表面看,WebView 组件与普通的 ImageView 相似,但实际上 WebView 本身就是一个浏览器实现。Android 5.0 增强的 WebView 基于 Chromium M37,直接支持 WebRTC、WebAudio 和 WebGL。

WebView 作为应用程序的 UI 接口,为用户提供了一系列的网页浏览、用户交互接口,通过这些接口显示和处理请求的网络资源。WebView 具有以下几个优点:

- 功能强大,支持 CSS、JavaScript 和 HTML,并很好地融入布局,使页面更加美观;
- 能够对浏览器控件进行详细的设置,例如字体、背景颜色、滚动条样式;
- 能够捕捉到所有浏览器的操作,例如单击、打开或关闭 URL。

WebView 提供了一些浏览器方法,如表 9-6 所示。

表 9-6 WebView 常用的方法

方　　法	功 能 描 述
loadUrl(String url)	打开一个指定的 Web 资源页面
loadData(String data, String mimeType, String encoding)	显示 HTML 格式的网页内容
getSettings()	获取 WebView 的设置对象
addJavascriptInterface()	将一个对象添加到 JavaScript 的全局对象 Window 中,这样可以通过 Window.XXX 进行调用,与 JavaScript 进行交互
clearCache()	清除缓存
destroy()	销毁 WebView

下述代码基于 WebView 开发一个简单的浏览器,该程序的界面布局代码如下所示。

【代码 9-8】 webviewbrowser_activity.xml

```
<LinearLayout xmlns:android = "http://schemas.android.com/apk/res/android" …>
    <LinearLayout
        android:orientation = "horizontal"
        android:layout_height = "wrap_content"
        android:layout_width = "match_parent">
```

```xml
        <EditText
            android:layout_width = "match_parent"
            android:layout_height = "wrap_content"
            android:id = "@ + id/url"
            android:layout_weight = "1"
            android:text = "www.baidu.com" />
        <Button
            android:text = "连接地址"
            android:layout_width = "99dp"
            android:layout_height = "wrap_content"
            android:id = "@ + id/connect" />
    </LinearLayout>
    <!-- 显示页面的 WebView 组件 -->
    <WebView
        android:layout_width = "match_parent"
        android:layout_height = "match_parent"
        android:id = "@ + id/show"
        android:layout_marginTop = "10dp">
    </WebView>
</LinearLayout>
```

在 AndroidManifest.xml 配置文件中添加能够访问网络的权限。

【代码 9-9】 授权应用程序访问网络的权限

```xml
<uses-permission android:name = "android.permission.INTERNET"></uses-permission>
```

创建一个名为 WebViewActivity 的 Activity，其代码如下。

【代码 9-10】 WebViewActivity.java

```java
public class WebViewActivity extends AppCompatActivity{
    private EditText url;
    private WebView show;
    private Button connect;
    @Override
    public void onCreate(Bundle savedInstanceState) {
        super.onCreate(savedInstanceState);
        setContentView(R.layout.main);
        //获取页面中文本框、WebView组件
        url = (EditText) findViewById(R.id.url);
        show = (WebView) findViewById(R.id.show);
        connect = (Button) findViewById(R.id.connect);
        WebSettings webSettings = show.getSettings();
        webSettings.setJavaScriptEnabled(true);
        webSettings.setAllowFileAccess(true);           //设置可以访问文件
        webSettings.setBuiltInZoomControls(true);       //设置支持缩放
        //设置 WebViewClient
        show.setWebViewClient(new WebViewClient(){
            public boolean shouldOverrideUrlLoading(WebView view, String url) {
                view.loadUrl(url);
                return true;}
            @Override
            public void onPageFinished(WebView view, String url) {
```

```
                super.onPageFinished(view, url);}
            @Override
            public void onPageStarted(WebView view,String url,Bitmap favicon) {
                super.onPageStarted(view, url, favicon);}});
        connect.setOnClickListener(new View.OnClickListener() {
            @Override
            public void onClick(View v) {
                String urlStr = url.getText().toString();
                show.loadUrl("http://" + urlStr); }});}
}
```

上述代码中,使用 WebView 组件的 loadUrl()方法直接打开 Web 网页,如图 9-5 所示。

图 9-5 使用 WebView 控件浏览网络

WebView 组件还提供了 loadData()方法用于加载 HTML 片段,该方法的语法格式如下所示。

【语法】

```
void loadData(String data,String mimeType,String encoding)
```

其中:参数 data 是 html 内容;参数 mimeType 是 MIME 类型,如 text/html 指明文本类型是 HTML 格式;参数 encoding 是编码字符集。

【示例】 使用 WebView 加载 HTML 页面

```
String html = "";
html += "<html>";
html += "<body>";
html += "<a href = http://www.google.com>Google Home</a>";
html += "</body>";
html += "</html>";
webView.loadData(html, "text/html", "utf-8");
```

上述代码中,通过 loadData()方法将 HTML 片段信息显示在 WebView 组件中。

9.6　Volley 框架

Volley 框架是 Google 在 2013 年发布的基于 Android 平台的网络通信库，能使网络通信更快、更简单、更健壮。

在没有引入 Volley 框架之前，从网上下载图片的步骤可能是如下流程：在 ListAdapter 的 getView()中开始图像的读取；通过使用 HttpURLConnection 从服务器获取图片资源；设置相应 ImageView 的属性。

在没有使用 Volley 框架时，屏幕旋转会导致再次从网络取得数据，为了避免这种不必要的网络访问，可能需要编写各种情况的 cache 处理。例如用户滚动 ListView 过快，可能导致在网络请求返回时已滚过某个位置，相应的数据根本没必要显示在 List 中，但是这些没有必要的数据会浪费系统的各种资源。

Volley 将 HttpClient 和 Universal-Image-Loader 的优点集于一身，既可以像 HttpClient 一样方便地进行 HTTP 通信，也可以像 Universal-Image-Loader 一样轻松加载网络上的图片。除了简单易用之外，Volley 在性能方面也进行了大幅度的调整，其设计目标就是非常适合进行数据量不大、通信频繁的网络操作。

Volley 提供以下几个功能：JSON、图像等资源的异步下载；网络请求的排序及优先级处理；缓存和多级别取消请求；与 Activity 生命周期联动，Activity 生命周期结束时取消所有的网络请求。

注意

> Volley 在性能方面适合数据量不大、通信频繁的网络操作，而对于大数据量的网络操作，例如文件下载等，Volley 的性能表现则非常糟糕。

使用 Volley 框架前，需要将 Volley.jar 文件导入到项目中，具体步骤如下：

（1）在 Project 项目结构目录下，选择 app→src→main→libs 将 Volley.jar 文件复制到 libs 文件夹中；

（2）右击 Volle.jar 文件，在弹出的菜单中单击 Add As Library；

（3）选择 module，单击 OK 按钮。

使用 Volley 框架实现网络数据请求，主要有以下三个步骤：

（1）创建 RequestQueue 对象，定义网络请求队列；

（2）创建 xxxRequest 对象（xxx 代表 String、JSON 和 Image 等），定义网络数据请求的详细过程；

（3）将 xxxRequest 对象添加到 RequestQueue 中，开始执行网络请求。

下述示例代码演示 Volley 框架的使用过程。

【代码 9-11】　MyApplication.java

```
public class MyApplication extends Application {
    //建立请求队列
```

```
    public static RequestQueue queue;
    @Override
    public void onCreate() {
        super.onCreate();
        queue = Volley.newRequestQueue(getApplicationContext());}
    public static RequestQueue getHttpQueue() {
        return queue;}
}
```

上述代码中,使用 newRequestQueue()方法创建一个 RequestQueue 对象,该对象是整个 App 内使用的全局性对象,用作网络请求队列,因此本书建议将创建过程写入 Application 类中。

创建完毕以后,还需要修改 AndroidManifest.xml 文件,修改该程序的 Application 对象为 MyApplication,并添加 INTERNET 权限。

【代码9-12】 AndroidManifest.xml

```xml
<?xml version = "1.0" encoding = "utf-8"?>
<manifest xmlns:android = "http://schemas.android.com/apk/res/android"
    package = "com.qst.chapter09">
    <uses-permission android:name = "android.permission.INTERNET"/>
    <application
        android:name = ".MyApplication"
        android:allowBackup = "true"
        android:icon = "@mipmap/ic_launcher"
        android:label = "@string/app_name"
        android:supportsRtl = "true"
        android:theme = "@style/AppTheme">
        <activity android:name = ".VolleyActivity">
            <intent-filter>
                <action android:name = "android.intent.action.MAIN" />
                <category android:name = "android.intent.category.LAUNCHER" />
            </intent-filter>
        </activity>
    </application>
</manifest>
```

创建 WebViewActivity,实现 Get 和 Post 两种方式的数据请求,代码如下所示。

【代码9-13】 VolleyActivity.java

```java
public class VolleyActivity extends AppCompatActivity {
    private Button get_btn;
    private Button post_btn;
    @Override
    protected void onCreate(Bundle savedInstanceState) {
        super.onCreate(savedInstanceState);
        setContentView(R.layout.volley_layout);
        get_btn = (Button) findViewById(R.id.get_btn);
        post_btn = (Button) findViewById(R.id.post_btn);
        get_btn.setOnClickListener(new View.OnClickListener() {
            @Override
```

```java
            public void onClick(View v) {
                VolleyGet();}});                              //GET 请求
        post_btn.setOnClickListener(new View.OnClickListener() {
            @Override
            public void onClick(View v) {
                VolleyPost();}});}                            //POST 请求
    private void VolleyPost() {                               //定义 POST 请求的方法
        String url = "http://www.itshixun.com/index";         //请求地址
        //创建 StringRequest,定义字符串请求的请求方式为 POST
        StringRequest request = new StringRequest(Request.Method.POST, url,
                new Response.Listener<String>() {
                    //请求成功后执行的函数
                    @Override
                    public void onResponse(String response) {
                        Log.d("VolleyPost","Post" + response);}   //打印出 POST 请求返回的字符串
                }, new Response.ErrorListener() {
                    //请求失败时执行的函数
                    @Override
                    public void onErrorResponse(VolleyError volleyError) {
                }}){
            //定义请求数据
            @Override
            protected Map<String, String> getParams() throws AuthFailureError {
                Map<String, String> hashMap = new HashMap<String, String>();
                hashMap.put("phone", "400-658-0166");
                return hashMap; }};
        request.setTag("postRequest");                        //设置该请求的标签
        MyApplication.getHttpQueue().add(request);}           //将请求添加到队列中
    private void VolleyGet() {                                //定义 GET 请求的方法
        String url = "http://www.itshixun.com/index";         //定义请求地址
        StringRequest request = new StringRequest(Request.Method.GET, url,
                new Response.Listener<String>(){
                    @Override
                    public void onResponse(String response) {
                        Log.d("VolleyGet","Get" + response);}  //打印出 GET 请求返回的字符串
                }, new Response.ErrorListener() {
                    @Override
                    public void onErrorResponse(VolleyError volleyError) {
                }});
        request.setTag("getRequest");                         //设置该请求的标签
        MyApplication.getHttpQueue().add(request);}           //将请求添加到队列中
}
```

Volley 提供了 JsonObjectRequest、JsonArrayRequest 和 StringRequest 等形式的 Request：JsonObjectRequest 形式用于返回 JSONObject 对象；JsonArrayRequest 形式用于返回 JsonArray 对象；StringRequest 形式用于返回 String 对象。

布局文件较为简单，就不再列出详细代码。运行程序后，单击"发送 Post 请求"按钮，Logcat 输出信息如图 9-6 所示。

第9章 网络编程

图 9-6　使用 Volley 发送 Post 请求

单击"发送 Get 请求"按钮，Logcat 输出信息如图 9-7 所示。

图 9-7　使用 Volley 发送 Get 请求

本章总结

小结

- Java 中的网络编程经验完全适用于 Android 应用的网络编程。
- Android 完全支持 JDK 本身的 TCP、UDP 网络通信的 API，也支持 URL、URLConnection 等网络通信 API。
- 通信协议是用来管理数据通信的一组规则，用于规范传输速率、传输代码、代码结构、传输控制步骤、出错控制等。
- 通信协议规定了通信的内容、方式和通信时间，其核心要素由语义、语法和时序三部分组成。
- 传输控制协议/互联网络协议（Transmission Control Protocol/Internet Protocol，TCP/IP）是最基本的通信协议，也是网络中最常用的协议。
- TCP/IP 通信协议是一种可靠的、双向的、持续的、点对点的网络协议。
- ServerSocket 是服务器套接字，用于监听并接收来自客户端的 Socket 连接。
- Socket 是客户端套接字，用于实现两台计算机之间的通信。
- 统一资源定位器（Uniform Resource Locator，URL）表示互联网上某一资源的地址。
- URL 的 openConnection()方法返回一个 URLConnection 对象。
- HttpURLConnection 继承 URLConnection，每个 HttpURLConnection 实例都可用于生成单个请求，可以透明地共享连接到 HTTP 服务器的基础网络。
- DefaultHttpClient 是 HttpClient 的默认实现类，用来负责处理 HTTP 协议的某一方面功能。

- WebView是专门用来浏览网页的视图组件,为用户提供了一系列的网页浏览、用户交互接口,通过这些接口显示和处理请求的网络资源。

Q&A

1. 问题:简述Android中常用的网络编程方式。

回答:针对TCP/IP协议的Socket和ServerSocket;针对HTTP协议的网络编程,如HttpURLConnection和HttpClient;直接使用WebKit访问网络。

2. 问题:简述WebView组件的优点。

回答:功能强大,支持CSS、JavaScript和HTML,并很好地融入布局,使页面更加美观;能够对浏览器控件进行详细的设置,例如字体、背景颜色、滚动条样式;能够捕捉到所有浏览器的操作,例如,单击、打开或关闭URL。

章节练习

习题

1. 下面_____不属于Android中常用的网络编程方式。
   A. Socket和ServerSocket　　　　　　B. HttpClient
   C. HttpURLConnection　　　　　　　D. Firefox浏览器

2. 下面关于Socket和ServerSocket的说法不正确的是_____。
   A. Socket是客户端套接字,用于实现两台计算机之间的通信
   B. ServerSocket是服务器套接字,用于监听并接收来自客户端的Socket连接
   C. 服务器端无须使用Socket
   D. 客户端无须使用ServerSocket

3. 下面关于HttpURLConnection的说法不正确的是_____。
   A. HttpURLConnection继承URLConnection
   B. HttpURLConnection是一个抽象类,无法直接实例化
   C. 通过URL的openConnection()方法获得一个HttpURLConnection连接
   D. 通过URLConnection的openConnection()方法获得一个HttpURLConnection连接

上机

1. 训练目标:网络编程。

培养能力	掌握基于TCP协议的网络通信		
掌握程度	★★★★★	难度	难
代码行数	600	实施方式	重复编码
结束条件	实现基于TCP协议的网络通信功能		
参考训练内容			
使用Socket和ServerSocket实现聊天室			

2. 训练目标：使用 HttpURLConnection。

培养能力	熟练使用 HttpURLConnection		
掌握程度	★★★★	难度	中
代码行数	200	实施方式	重复编码
结束条件	使用 HttpURLConnection 实现网络通信功能		

参考训练内容
使用 HttpURLConnection 访问指定网络数据

附录 A　Android版本新特性

A.1　Android 5.0 新特性

Android 5.0 是 Google 公司于 2014 年 10 月 15 日发布的 Android 操作系统，根据 Android 系统以往的惯例，每一代新系统会根据其字母代号对应一个甜品的名称，Aadroid 5.0 所对应的英文名为 Lollipop(棒棒糖)，代号 L。

Android 5.0 在设计风格、设备支持、电池续航和安全性等方面进行了加强，具体如下：

(1) 全新 Material Design 设计风格。Android 5.0 系统采用了一种新的 Material Design 设计风格。在界面设计时，加入了五彩缤纷的颜色和流畅的动画效果，呈现出一种清新的风格，如图 A-1 所示。采用这种设计的目的在于统一 Android 设备的外观和使用体验。

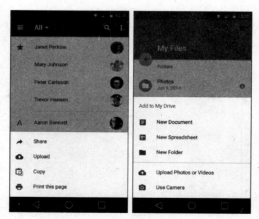

图 A-1　Material Design 设计风格

(2) 支持多种设备。现在无论是智能手机、平板电脑、笔记本电脑、智能电视、智能手表还是其他家用电子产品，Android 5.0 系统几乎能在所有的设备上运行。

(3) 全新的通知中心设计。Google 在 Android 5.0 中加入了全新风格的通知系统。改进后的通知系统会优先显示对用户来说比较重要的信息，将不太紧急的内容隐藏起来。用户只需要向下滑动就可以查看全部的通知内容，并且可以在锁屏状态直接查看通知消息。

(4) 支持 64 位 ART 虚拟机。Android 5.0 放弃了之前一直使用的 Dalvik 虚拟机，改用 ART 模式,实现了真正的跨平台编译。ART 虚拟机编译器在内存占用率和应用程序加载时间方面的性能得到大幅提升。

（5）Project Volta 电池续航改进计划。对开发者而言,Project Volta 计划增加的新工具可以让开发者能够更容易地找出所开发的应用程序对电量产生比较大的影响因素,同时确保在执行某项任务时将手机电量的影响降至最低。对用户而言,Android 5.0 增加了 Battery Saver 模式,该模式将在低电量时自动降低屏幕亮度,限制自动更换背景等耗电功能。

（6）全新的"最近应用程序"界面。全新的"最近应用程序"界面借鉴了 Chrome 浏览器的理念,采用独立的标签展示方式,如图 A-2 所示。更重要的是,Google 已经向开发者开放了 API,开发者可以利用这个改进为特定的应用增加全新的功能。

（7）整体安全性提高。Android 5.0 开启了系统数据加密功能,并且通过 SElinux 执行应用程序,系统变得更加安全。

（8）不同数据独立保存。用户在 Android 5.0 系统下可以通过一台设备就可以搞定所有的工作和娱乐活动,该特性首先将各种数据独立保存,并且让所有生成的新数据都有依据。

（9）Google Search 性能提高。Google 将新系统的搜索功能重点放在了"重新发现"上,这意味着 Google Search 将会更好地识别用户正在做什么。另外,当用户在进行应用搜索时,可以直接展示相似或部分提示,进入选定的应用程序,而无须将内容全部输入。

图 A-2 最近应用程序

（10）新的 API 支持。Android 5.0 还新增加了 API 支持、蓝牙 4.1、USB Audio 外接音响及多人分享等功能。

A.2 Android 6.0 新特性

Android 6.0 是 Google 公司于 2015 年 5 月 28 日发布的 Android 操作系统,英文名为 Marshmallow(棉花糖),其代号为 M。Android 6.0 最大的一个亮点是为用户提供了两套相互独立的解决方案,为每位用户的每一个应用都提供两套数据存储方案,一套专门用来存储用户的工作资料,另一套专门用来存储用户的个人信息,而且这两套存储系统完全相互独立。

Android 6.0 提供了软件权限管理、网页体验提升、安卓支付和指纹支持等方面的新特性,具体如下:

（1）App Permissions(软件权限管理)。在 Android 6.0 中,用户可以自定义应用许可提示,对应用的权限进行管理,例如能否使用位置、相机、麦克风和通讯录等,这些都可以开放给开发者和用户,如图 A-3 所示。

（2）Chrome Custom Tabs(网页体验提升)。Android 6.0 对于 Chrome 网页浏览的体验进行了提升,对登录网站、存储密码、自动补全资料和多线程浏览网页的安全性进行了一系列的优化,如图 A-4 所示。

图 A-3 应用程序权限管理界面

图 A-4 Chrome Custom Tabs

（3）App Links(App 关联)。在 Android 6.0 系统中加强了软件间的关联，例如，在用户手机邮箱中收到一封邮件，内文中有一个 Twitter 链接，用户单击该链接可以直接跳转到 Twitter 应用，而不再是网页。

（4）Android Pay(安卓支付)。Android 支付统一标准，新的 Android 6.0 系统中集成了 Android Pay，其特性在于简洁、安全、可选性。Android Pay 是一个开放性平台，用户可以选择 Google 的服务或者使用银行的 App 来使用它，Android Pay 支持 4.4 版本以后的系统设备，在发布会上 Google 宣布 Android Pay 已经与美国三大运营商、700 多家商店达成合作。可以使用指纹来实现支付功能，在基于 Android 6.0 的 Nexus 产品中提供了指纹识别功能。

（5）Fingerprint Support(指纹支持)。Android 6.0 增加了对指纹的识别 API，Google 开始在 Android 6.0 中提供官方的指纹识别支持，力求统一方案，但目前大部分 Android 产品的指纹识别都是使用非 Google 认证的技术和接口。

（6）Power Charge(电源管理)。Android 6.0 系统的电源管理模块将更为智能，例如，当平板电脑长时间未使用时，Android 6.0 系统将自动关闭一些 App。同时 Android 设备将支持 USB Type-C 接口，新的电源管理将更好地支持 Type-C 接口。

A.3 Android 7.0 新特性

Android 7.0 是 Google 在 2016 年 5 月 18 日正式推出的 Android 智能手机操作系统，英文名为 Nougat(牛轧糖)，代号为 N。新一代的 Android 操作系统将支持无缝升级，能够通过 Vulkan API 在中低硬件设备上实现流畅游戏体验以及更多的 Emoji 表情。最重要的是，Android 7.0 运行环境有了明显改善，软件运行速度提升幅度达到 600%，应用安装提速75%。

Android 7.0提供了分屏多任务、下拉开关页、捆绑通知和支持VR等新特性,具体如下:

(1) 分屏多任务。Android 7.0支持在同一屏幕同时进行两个应用的操作,进入后台多任务管理页面,按住其中一个卡片,然后向上拖动至顶部即可开启分屏多任务,支持上下分栏和左右分栏,允许拖动中间的分割线调整两个App所占的比例,如图A-5所示。

(2) 全新下拉快捷开关页。在Android 7.0中,通过下拉打开顶部通知栏即可显示5个用户常用的快捷开关,并支持单击开关以及长按进入对应设置。如果继续下拉通知栏即可显示全部快捷开关,如图A-6所示。此外,在快捷开关页右下角也会显示一个"修改"按钮,单击之后即可自定义添加或者删除快捷开关,也可以通过拖动进行排序。

图A-5 分屏多任务

图A-6 下拉快捷开关页

(3) 通知栏消息快捷回复。Android 7.0加入了全新的通知消息类API,支持第三方应用通知的快捷操作和回复,例如,来电会以横幅方式在屏幕顶部出现,提供接听和挂断两个按钮;而信息以及社交类应用通知,还可以直接打开键盘,在输入栏中快捷回复,如图A-7所示。

图A-7 通知栏消息快捷回复

(4) 捆绑通知。与按照时间顺序显示通知不同,Android 7.0支持将来自同一应用程序的通知捆绑在一起,例如消息应用的通知,如图A-8所示。对于那些通知的重度用户来讲,这些改变无疑是非常实用的。

(5) 新增流量节省模式。Android 7.0中引入了"流量节省"模式,在接近用户计费周期末,或是流量包本身较小的情况下,减少应用消耗的数据流量。在启用这一模式时,系统将拦截后台的数据使用,并在可能的情况下减少前台运行应用使用的数据量,如图A-9所示。

(6) 新增VR支持。Android 7.0将会是Google公司充分执行其VR计划的操作系统,它内置Google全新VR平台Daydream,如图A-10所示。Daydream是一个虚拟现实平台,由Daydream头盔、手柄和智能手机构成,支持Daydream的智能手机须满足一定的硬件要求。

图 A-8 消息通知捆绑

图 A-9 节省流量模式

图 A-10 DayDream 平台

附录 B 常用的Android Studio选项设置

B.1 Android Studio 基本配置

在 Android Studio 中，在顶部菜单栏中选择 File→Settings 打开设置选项，或者通过快捷键 Ctrl＋Alt＋S 打开设置界面，如图 B-1 所示。

图 B-1 Studio 设置界面

1．主题配置

在设置界面选择 Appearance & Behavior→Appearance 项，在右侧面板中的 UI Options 选项设置，单击"Theme："右边的下拉框，如图 B-2 所示，选择合适的主题，然后单击 OK 按钮完成修改。

2．字体配置

在设置界面选择 Editor→Color&Fonts→Font 项，由于内建的 Default 与 Darcula 两个组合不允许修改，因此需要另存一个新的组合然后再进行修改。首先单击 Save As 按钮，填写存储字体方案的新名称，然后选择合适的字体进行修改，如图 B-3 所示，最后单击 OK 按钮完成修改。

图 B-2　设置主题

图 B-3　字体配置

3. 显示代码行数

在 Android Studio 中，代码行数是默认不显示的，在程序开发过程中，Logcat 提示在某文件某行出现错误时，定位到错误点不方便。在设置界面中选择 Editor→General→Appearance 项，在右侧面板中勾选 Show line numbers 即可，如图 B-4 所示，单击 OK 按钮完成设置。

4. 设置自动化的 Import 功能

在编写程序时，经常需要 import 函数库的程序包名称。在 Android Studio 中，可以通过设置自动导入程序包。首先在设置界面中选择 Editor→General→Auto Import 项，在右侧面板中的 Java 选项中将 Insert imports on paste 选项中的 Ask 改为 All，勾选 Optimize

附录 B　常用的Android Studio选项设置

图 B-4　显示代码行数

imports on the fly（优化 import 语句）和 Add unambiguous imports on the fly（自动加入 import 语句而不询问）两个选项，如图 B-5 所示。单击 OK 按钮完成设置。

图 B-5　设置自动化的 Import 功能

B.2　Android Studio 快捷键

在 Android 应用开发过程中，使用快捷键可以快速书写代码，以提高开发效率。在 Android Studio 开发环境中常用的快捷键如表 B-1 所示。

表 B-1　Android Studio 常用快捷键

快　捷　键	功　能　描　述
Alt＋Up/Down	从一个类方法跳转到临近的一个类方法
Ctrl＋Alt＋Right	将光标向后移到上一次编辑位置
Ctrl＋Alt＋Left	将光标向前移到上一次编辑位置
Ctrl＋Shift＋Enter	代码自动补全
Alt＋Enter	快速修复存在问题的代码
Ctrl＋N	查找项目中的类
Ctrl＋Shift＋N	查找项目中的文件
Shift＋Shift	查找项目中的文件、类和动作

续表

快　捷　键	功　能　描　述
F2	快速定位到出错的地方
Shift+Ctrl+F12	隐藏相关的 Project 面板等窗口
Ctrl+E 或 Ctrl+Shift+E	显示最近浏览或编辑过的文件
Ctrl+P	显示方法的参数信息
Shift+F6	重命名字段和方法名称
Ctrl+X	删除光标所在的行内容
Ctrl+D	复制粘贴光标所在的行内容
Ctrl+Alt+L	格式化代码

B.3　Android Studio 导入 Eclipse ADT 项目

Android Studio 支持 ADT 的项目导入 Android Studio 中进行开发和调试，下面将介绍如何将 Eclipse ADT 项目导入 Android Studio 中。

B.3.1　步骤

（1）打开 Android Studio，在顶部菜单栏中选择 File→Import Project 菜单选项，如图 B-6 所示。

（2）找到 Eclipse ADT 项目（压缩包需要先进行解压），如图 B-7 所示。

图 B-6　导入项目菜单

图 B-7　选择导入的 Eclipse ADT 项目

（3）选中项目后单击 OK 按钮，弹出如图 B-8 所示的对话框，选择该项目要保存的路径。

（4）单击 Next 按钮，弹出 Import Project from ADT 对话框，如图 B-9 所示。在对话框中的所有选项都会默认被选中，如果没有被选中，则需要手动勾选，单击 Finish 按钮完成项目的导入操作。

附录 B　常用的Android Studio选项设置

图 B-8　选择项目保存路径

图 B-9　完成导入操作

B.3.2　常见问题

如果 ADT 项目使用 GBK 编码（或其他非 UTF-8 编码），那么导入 Studio 后程序代码中的文字信息可能会变成乱码，如图 B-10 所示。

图 B-10　中文乱码问题

此时选中该类文件，在 Android 顶部菜单栏选择 File→File Encoding 菜单选项，然后选择程序原来的编码（例如 GBK 或 x-windows-950），弹出如图 B-11 所示的对话框。

图 B-11　重新编译文件编码

单击 Reload 按钮，可以发现该类文件的中文已经恢复正常，如图 B-12 所示。

重复上述操作，选择 UTF-8 编码，在第二步时不再单击 Reload 按钮，单击 Convert 按钮，如图 B-13 所示。如此可避免将来在执行程序时因 GBK 编码被误当成 UTF-8 编码而出现的乱码情况。

· 431 ·

```
MainActivity.java ×
 1    package com.qst.appendixB;
 2
 3    import ...
 7
 8    public class MainActivity extends Activity {
 9
10        @Override
11        protected void onCreate(Bundle savedInstanceState) {
12            super.onCreate(savedInstanceState);
13            setContentView(R.layout.activity_main);
14
15            String[] string1={"床前明月光","疑是地上霜","举头望明月","低头思故乡"};
16            String[] string2={"春眠不觉晓","处处闻啼鸟","夜来风雨声","花落知多少"};
17        }
18
19    }
20
```

图 B-12　重新编译完成

图 B-13　将 GBK 编码更改为 UTF-8 编码

附录 C Android 应用程序签名

Android 系统要求每一个 Android 应用程序必须要经过数字签名才能够安装到系统中,即 Android 系统不会安装没有数字证书的应用,包括模拟器上安装的应用。Android 通过数字签名来标识应用程序的作者以及在应用程序之间建立信任关系,不是用来决定最终用户可以安装哪些应用程序。数字签名由应用程序的作者来完成,并不需要权威的数字证书签名机构认证,通常只是让应用程序实现自我认证。

在 Android 应用开发过程中,为了方便开发人员开发与调试,ADT 会自动通过 debug 密钥为应用程序签名。debug 密钥是一个名为 debug.keystore 的文件,该文件存放在"C:\Users\用户名\.android"文件夹中。对于 apk 签名,可以通过 ADT 提供的图形化工具和 DOS 命令两种方式完成。

C.1 DOS 命令完成 apk 签名

通过 DOS 命令的方式来完成 apk 签名,需要用到 3 个命令工具。

(1) **keytool**:该工具位于 jdk 安装路径的 bin 目录下,用于生成数字证书,即密钥(扩展名为.keystore 的文件)。

(2) **jarsigner**:该工具位于 jdk 安装路径的 bin 目录下,使用数字证书给 apk 文件签名。

(3) **zipalign**:该工具位于 android-sdk-windows/tools/目录下,对签名后的 apk 进行优化,提高与 Android 系统交互的效率。

通常同一用户所开发的所有应用程序都是使用相同的签名,即使用同一个数字证书。如果是第一次做 Android 应用程序签名,则上述 3 个工具都将用到;但如果已经有了数字证书,那么再给其他 apk 签名时,只需要通过 jarsigner 和 zipalign 两个工具就可以完成。

在对 apk 进行签名操作前,首先要生成未经签名的 apk 文件,然后通过下述的操作为 apk 进行签名。

1. 使用 keytool 工具生成数字证书

使用 keytool 工具生成数字证书,命令格式如下所示。

```
keytool -genkey -v -keystore qst.keystore -alias qst.keystore -keyalg RSA -validity 1000
```

对上述命令的参数说明如下:

keytool 是工具名称,-genkey 表示所执行的是生成数字证书操作,-v 表示将生成证书的详细信息打印出来,显示在 DOS 窗口中;keystore qst.keystore 表示生成的数字证书的文件名为 qst.keystore;alias qst.keystore 表示数字证书的别名为 qst.keystore,当然别名可以不和文件名一样;keyalg RSA 表示生成密钥文件所采用的算法为 RSA;validity 1000 表示该数字证书的有效期为 1000 天,超过 1000 天之后该证书将失效。

2. 使用 jarsigner 工具为 Android 应用程序签名

使用 jarsigner 工具为 Android 应用程序签名,命令格式如下所示。

```
jarsigner - verbose - keystore qst.keystore - signedjar notepad_signed.apk notepad.apk qst.keystore
```

对上述命令的参数说明如下:

jarsigner 是工具名称,-verbose 表示将签名过程中的详细信息打印出来,显示在 DOS 窗口中;keystore qst.keystore 表示签名所使用的数字证书及位置;此次没有指明路径,表示在当前目录下;signedjar notepad_signed.apk notepad.apk 表示给 notepad.apk 文件签名,签名后的文件名称为 notepad_signed.apk;qst.keystore 表示证书的别名,即生成数字证书时-alias 参数后面的名称。

通过上述操作即可在 DOS 下完成 Android 应用程序签名。

C.2 在 Android Studio 中完成 apk 签名

除了使用 DOS 命令完成 apk 签名外,还可以使用 Android Studio 中自带的 apk 签名工具对应用程序进行签名。使用 Android Studio 打包 apk 的步骤如下所示。

在 Android Studio 顶部菜单栏单击 Build 菜单,选择 Generate Signed APK 选项,如图 C-1 所示。

弹出的 Generate Signed APK 窗口如图 C-2 所示。

图 C-1 创建签名文件

图 C-2 选择/创建密钥库

在 Generate Signed APK 窗口中,单击 Create new 按钮,弹出 New Key Store 窗口,如图 C-3 所示。

在 New Key Store 窗口的 Key Store path 项中单击 按钮,弹出 Choose keystore file 窗口,如图 C-4 所示。

图 C-3 创建新的密钥库

图 C-4 选择密钥库存放位置

选择秘钥库的存储位置,单击 OK 按钮,返回 Choose keystore file 窗口。接下来填写密钥库的密码、密钥名称、密码以及相关信息,如图 C-5 所示。

单击 OK 按钮,返回 Generate Signed APK 窗口,在窗口中选择所创建的密钥,如图 C-6 所示。单击 Next 按钮,选择 APK Destination Folder 目录后,单击 Finish 按钮,完成 apk 的签名,如图 C-7 所示。

图 C-5 填写其他相关信息

图 C-6 选择密钥

在指定的目录中可以找到所创建的密钥库和签名 apk 文件,如图 C-8 所示。

图 C-7 选择 apk 的保存路径

图 C-8 密钥库及签名 apk 文件

图书资源支持

感谢您一直以来对清华版图书的支持和爱护。为了配合本书的使用,本书提供配套的素材,有需求的用户请到清华大学出版社主页(http://www.tup.com.cn)上查询和下载,也可以拨打电话或发送电子邮件咨询。

如果您在使用本书的过程中遇到了什么问题,或者有相关图书出版计划,也请您发邮件告诉我们,以便我们更好地为您服务。

我们的联系方式:

地　　址: 北京海淀区双清路学研大厦A座707

邮　　编: 100084

电　　话: 010-62770175-4604

资源下载: http://www.tup.com.cn

电子邮件: weijj@tup.tsinghua.edu.cn

QQ: 883604(请写明您的单位和姓名)

用微信扫一扫右边的二维码,即可关注清华大学出版社公众号"书圈"。

扫一扫
资源下载、样书申请
新书推荐、技术交流